FUNDAMENTALS OF SOLIDIFICATION

FUNDAMENTALS OF SOLIDIFICATION

THIRD EDITION

W. KURZ

D.J. FISHER

TRANS TECH PUBLICATIONS 1992
SWITZERLAND - GERMANY - UK - USA

First edition 1984
Second edition 1986
Third revised edition 1989
Reprinted 1992

W. KURZ

Professor
Department of Materials
Swiss Federal Institute of Technology (EPFL)
Lausanne - Switzerland

D.J. FISHER

Associate editor
Trans Tech Publications Ltd
Aedermannsdorf - Switzerland

ISBN 0-87849-522-3

Trans Tech Publications Ltd.
P.O. Box 10
CH-4711 Aedermannsdorf, Switzerland

Printed in the Netherlands

FOREWORD

Solidification phenomena play an important role in many of the processes used in fields ranging from production engineering to solid-state physics. For instance, a metal is usually continuously cast or ingot cast before forming it into bars or sheets. The bars then often serve as input to a sand-, permanent mould-, or precision-casting operation while the sheet is often fabricated into useful items by welding; another solidification process. At the other extreme, silicon is usually first prepared in the form of an impure reduction product and then, for electronic applications, has to be purified by zone-refining (a solidification process) and pulled, as a single crystal, from its melt.

This broad range of interest in solidification, from the large tonnages of continuously cast products, through the intermediate weight output of superalloy precision castings, to the relatively small quantities of high-purity crystals, means that a book such as the present one must cater for the requirements of a very wide range of readers. To begin with, there is the graduate or final-year undergraduate who may eventually find himself dealing with any problem in the above range, and must therefore be thoroughly conversant with the basic principles and mathematical theory of the subject. Then there is the post-graduate researcher who may need to produce metallic specimens having a well-defined microstructure and, in order to do this, must be able to bring to bear all of the current understanding of solidification mechanisms. Finally, there is the foundryman who would like to exert close control over a cast product, but must contend with so many variables and unknown quantities that his work takes on the aspects of an art. It is hoped that this book will be a value to all three groups and, at least, provide the student with an introduction to modern solidification theory; the researcher with the fundamentals of the more quantitative models for predicting solidification microstructures; and the foundryman with a framework into which he can fit his diverse empirical observations.

The ground covered by the present introduction is essentially the same as that covered by textbooks such as Winegard's "Introduction to the Solidification of Metals", and Chalmers' "Principles of Solidification", both of which were published in 1964, and Flemings' "Solidification Processing", published in 1974. Many of the 'loose ends' of solidification theory, whose interrelationships were unclear ten or twenty years ago, have now been drawn together and many of the qualitative arguments which were a feature of the latter books can now be largely replaced by more quantitative models. It is currently possible to present major parts of the theory as a coherent whole, and to fit solidification microstructures into a logical framework. That is not to say that the present-day solidification literature is immediately accessible to the newcomer. Much of the most useful information is buried within a mass of mathematical formulae and scattered among many journals. Therefore, the aim here has been to collect together the key results obtained by the present and other authors, and to derive simpler solutions whenever possible. The sources of the models used can be found in the references at the end of each chapter but, for easier reading, are often not referred to in the text.

In order to obtain the maximum benefit from this book, the reader should note that it is based upon a hierarchical scheme within which the subject matter can be studied at 3 levels. Firstly, an initial feeling for the subject and for the breadth of coverage, can be obtained simply by reading the extensive figure captions. Secondly, the main text describes the principles in more detail, but usually without deriving the necessary equations. Thirdly, the appendices contain detailed derivations and some essential mathematical background. It is stressed that only those readers who are specialising in the subject would usually need to study the appendices in detail.

Within the main text, an essentially self-contained guide to the subject is presented. That is, the reader is introduced to the mechanisms of crystal nucleation and growth which occur at the atomic scale (chapter 2) before being shown how the form of an initially planar solid/liquid interface evolves (chapter 3). Subsequently, the most important single-phase (chapter 4) and multi-phase (chapter 5) solid/liquid interface morphologies are presented. The effect, which solidification has upon the redistribution of solute is then discussed (chapter 6). Finally, the behaviour which is to be expected at high solidification rates is introduced in this **third edition (chapter 7)**. This is a topic of rapidly increasing importance. It is now possible, on the basis of work carried out during the last few years, to discuss the effect of both high and low growth rates upon the microstructure. Consequently, for this edition, it has been necessary to extend the coverage of several of the appendices and to add a new one on the thermodynamics of rapid solidification (appendix 6).

One subject which is not covered, due to the presently very limited understanding of the field, is convection in the melt and its interaction with solidification microstructures.

Each chapter includes a bibliography of key references for further study, as well as exercises which are designed to test the reader's understanding of the contents of the preceding chapter. For certain exercises, it is advisable firstly to work through the corresponding appendices.

The authors hope that, after reading this book, the newcomer will feel confident when delving further into solidification-related subjects, and that the experienced foundryman will also find some thought-provoking points.

W. Kurz, D.J. Fisher Lausanne, August 1989

ACKNOWLEDGEMENTS

The authors express their especial thanks to Dr. M. Rappaz, who contributed generously to this textbook. Furthermore, they wish to thank Prof. G. Abbaschian, Dr. T.W. Clyne, Prof. J. Dantzig, Dr. H. Esaka, P. Gilgien, Dr. M. Gremaud, Prof. H. Jones, Dr. J. Lipton, Dr. M. Lorenz, Dr. P. Magnin, Prof. A. Mortensson, Prof. J.H. Perepezko, Prof. D.R. Poirier, D. Previero, Prof. P.R. Sahm, Dr. T. Sato, J. Satsuta, Dr. P. Thévoz, Prof. R. Trivedi, Dr. M. Wolf and Dr. M. Zimmermann for their comments and contributions concerning the manuscript.

The invaluable aid of Dr. and Mrs. J.-P. Moinat in typesetting the equations and editing the text, and of Mrs. E. Schlosser in preparing the diagrams is also gratefully acknowledged.

CONTENTS

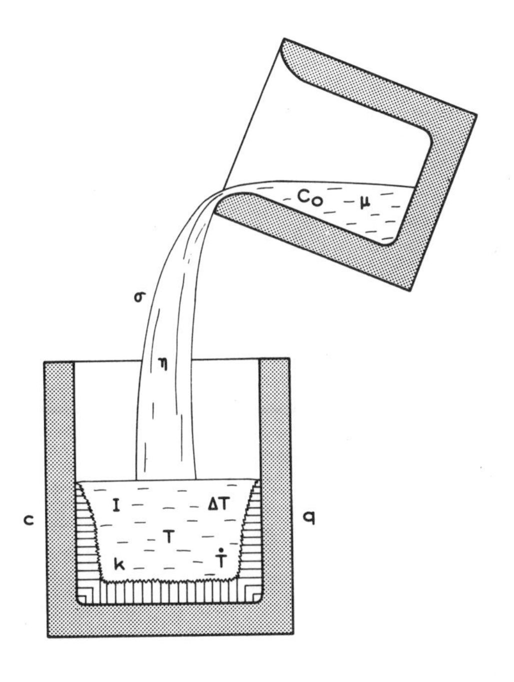

CHAPTER ONE

INTRODUCTION

1.1 The Importance of Solidification

Solidification is a phase transformation which is familiar to everyone, even if the only acquaintance with it involves the making of ice cubes. It is relatively little appreciated that the manufacture of almost every man-made object involves solidification at some stage.

The scope of this book is restricted mainly to a presentation of the theory of solidification as it applies to the most widely-used group of materials, i.e. metallic alloys. Here, solidification is generally accompanied by the formation of crystals; an event which is much rarer during the solidification of ceramic glasses or polymers.

Solidification is of such importance simply because one of its major practical applications, namely casting, is a very economic method of forming a component if the

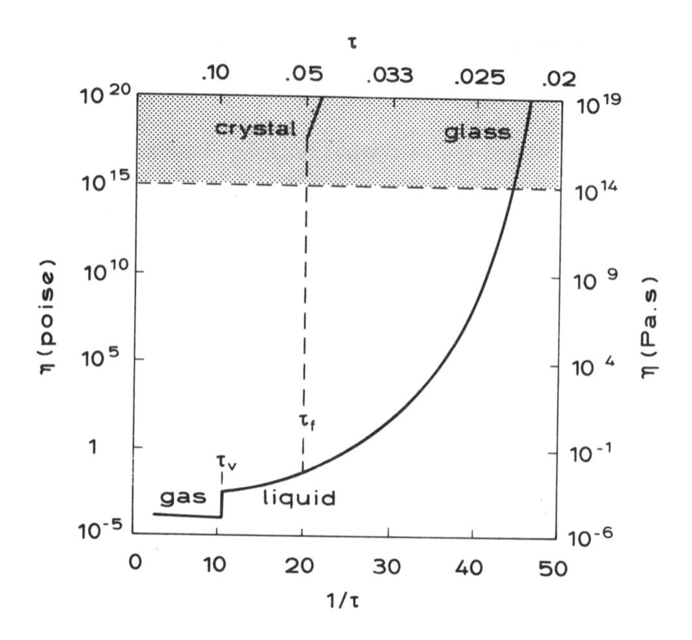

Figure 1.1 Dynamic Viscosity as a Function of Temperature

The fundamental advantage of solidification, as a forming operation, is that it permits metal to be shaped with a minimum of effort since the liquid metal offers very little resistance to shear stresses. When the material solidifies due to a decrease in the temperature, τ, its viscosity increases continuously (glass formation) or discontinuously (crystallisation), by over 20 orders of magnitude, to yield a strong solid, the viscosity of which is defined arbitrarily to be greater than 10^{14}Pas. (The reduced temperature, τ, is used here since it leads to a single curve which is applicable to many substances. The suffix, f or v, indicates the melting point or boiling point, respectively.) [D.Turnbull, Transactions of the Metallurgical Society of AIME **221** (1961) 422].

melting point of the metal is not too high. Nowadays, cast metal products can be economically produced from alloys having melting points as high as 1660°C (Ti).

In the case of metals, melting is accompanied by an enormous decrease in viscosity, of some twenty orders of magnitude, as illustrated in figure 1.1. Thus, instead of expending energy against the typically high flow stress of a solid metal during forging or similar processes, it is only necessary to contend with the essentially zero shear stress of a liquid. If the properties of castings were easier to control, then solidification would be an even more important process. In this respect, solidification theory plays a vital role since it forms the basis for influencing the microstructure and hence improving the quality of cast products.

The effect of solidification is most evident when casting is the final operation since the resultant properties can depend markedly upon the position in the casting

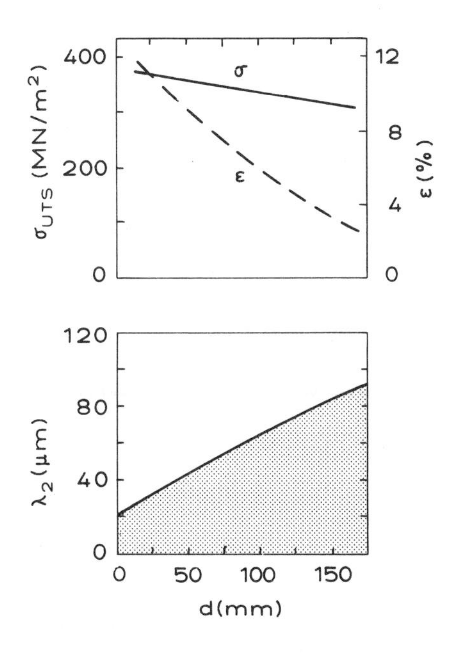

Figure 1.2 Alloy Properties as a Function of Position in a Casting

The use of casting as a production route unfortunately poses its own problems. One of these is the local variation of the microstructure, leading to compositional variations, and this is illustrated for example by the dendrite arm spacing,λ_2, measured as a function of the distance, d, from the surface of the casting. This can lead to a resultant variation in properties such as the ultimate tensile strength and the elongation. Like the weakest link in a chain, the inferior regions of a casting may impair the integrity of the whole. Thus, it is important to understand the factors which influence the microstructure. Finer microstructures generally have superior mechanical properties and finer structures, in turn, generally result from higher solidification rates. Such rates are found at small distances from the surface of the mould, in thin sections, or at laser-remelted surfaces. [M.C. Flemings, Solidification Processing, McGraw-Hill, New York, 1974].

(Fig. 1.2). Its influence is also seen in a finished product, even after heavy working, since a solidification structure and its associated defects are difficult to eliminate once they are created. Solidification defects tend to persist throughout subsequent operations (Fig. 1.3). Good control of the solidification process at the outset is therefore of utmost importance.

Some important processes which involve solidification are:

Casting:	continuous-
	ingot-
	form-
	precision-
	die-

Welding:	arc- resistance- plasma- electron beam- laser- friction- (including the micro-mechanisms of wear)
Soldering/Brazing	
Rapid Solidification Processing:	melt-spinning planar-flow casting atomisation bulk undercooling surface remelting
Directional Solidification:	Bridgman liquid metal cooling Czochralski electroslag remelting

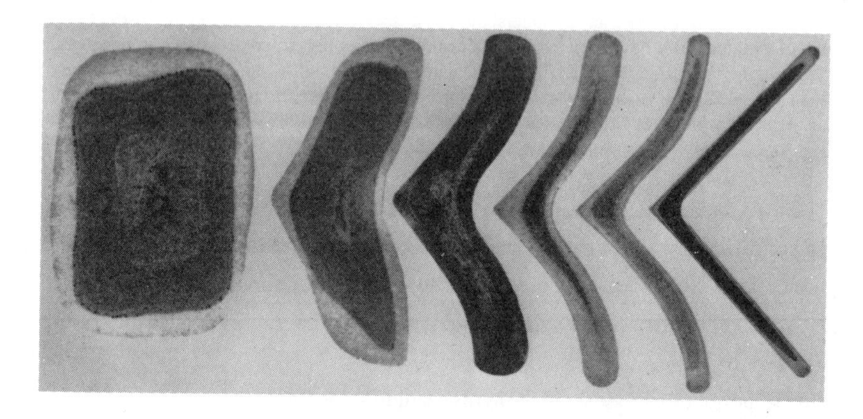

Figure 1.3 Effect of Deformation upon a Cast Microstructure

Often, casting is not the final forming operation. However, subsequent deformation is not a very efficient method of modifying the as-cast microstructure since any initial heterogeneity exhibits a strong tendency to persist. Thus, during the rolling of this L-shaped profile which contains a heavily segregated central region, the latter defect survives the many stages between the cast billet and the final product. This example emphasises the fact that an effective control of product quality must be exercised during solidification. [A.J. Pokorny, De Ferri Metallographia, Vol. III, Luxembourg, 1966, p. 287].

In addition, the crystallisation of certain pure substances is of great importance. For example, the preparation of semiconductor-grade silicon crystals is an essential step in modern solid-state physics and technology. The production of integrated circuits; the basis of any new electronic device (radio, watch, computer, etc), requires the preparation of large single crystals of very high perfection, containing a controlled amount of a uniformly distributed dopant. At the moment, such a crystal can only be produced by growth from the melt. Indeed, the requirements of semiconductor physics have enormously influenced solidification theory and practice. Therefore, during the past thirty years solidification has evolved from being a purely technological, empirical field, to become a science.

Historically, simple cast objects (in copper) first appeared before about 4000BC and were, no doubt, a natural by-product of the potter's skill in handling the clay used in furnace-and mould-making. The production of the renowned and highly sophisticated bronze castings of China began in about 1600BC. However, it is probable that the technique had originally been imported from elsewhere. For instance, the lost-wax process was developed in Mesopotamia as long ago as 3000BC. Iron-casting in China began in about 500BC, but in Europe cast iron did not appear until the 16th century, and achieved acceptance as a constructional material, in England in the 18th century, only under the impetus of the industrial revolution.

Much of the delay in exploiting cast materials probably originated from the complete lack of understanding of the nature of solidification phenomena and of the microstructures produced. In particular, the facets of fracture surfaces were invariably taken to indicate the nature of the 'crystals' of which a casting was composed. In the absence of an adequate picture of the solidification process, casting was bound to remain a black art rather than a science, and vestiges of this attitude still remain today.

1.2 Heat Extraction

The various solidification processes mentioned above involve extraction of heat from the melt in a more or less controlled manner. Heat extraction changes the energy of the phases (solid and liquid) in two ways:

1. There is a decrease in the enthalpy of the liquid or solid, due to cooling, which is given by: $\Delta H = \int c dT$

2. There is a decrease in enthalpy, due to the transformation from liquid to solid, which is equal to the latent heat of fusion, ΔH_f.

Heat extraction is achieved by applying a suitable means of cooling to the melt in order to create an external heat flux, q_e. The resultant cooling rate, dT/dt, can be deduced from a simple heat balance if the metal is isothermal (low cooling rate) and the specific heats of the liquid and the solid are the same. Using the latent heat per unit volume, $\Delta h_f = \Delta H_f/v_m$ (defined to be positive for solidification), and also the specific heat per unit volume, c, in order to conform with the dimensions of the other factors, then:

$$q_e\left(\frac{A'}{v}\right) = -c\left(\frac{dT}{dt}\right) + \Delta h_f\left(\frac{df_s}{dt}\right)$$

so that:

$$\dot{T} = \frac{dT}{dt} = -q_e\left(\frac{A'}{vc}\right) + \left(\frac{df_s}{dt}\right)\left(\frac{\Delta h_f}{c}\right) \qquad [1.1]$$

The first term on the right-hand-side (RHS) of equation 1.1 reflects the effect of casting geometry (ratio of surface area of the casting, A', to its volume, v) upon the extraction of sensible heat, while the second term takes account of the continuing evolution of latent heat of fusion during solidification. It can be seen from this equation that, during solidification, heating will occur if the second term on the RHS of equation 1.1 becomes greater than the first one. This phenomenon is known as recalescence. For an alloy, where solidification occurs over a range of temperatures, the variation of the fraction of solid as a function of time must be calculated from the relationship:

$$\frac{df_s}{dt} = \left(\frac{dT}{dt}\right)\left(\frac{df_s}{dT}\right)$$

since f_s is a function of temperature. In this case:

$$\dot{T} = \frac{-q_e\left(\frac{A'}{vc}\right)}{1 - \left(\frac{\Delta h_f}{c}\right)\left(\frac{df_s}{dT}\right)} \qquad [1.2]$$

It is seen that solidification decreases the cooling rate since df_s/dT is negative.

Figure 1.4 illustrates two fundamentally different solidification processes. In figure 1.4a, the heat is extracted in an almost steady manner by moving the crucible at a fixed rate, V', through the temperature profile imposed by the furnace. Such a process is usually used for single crystal growth or directional solidification. It permits the growth rate of the solid, V (which is not necessarily equal to the rate of crucible movement - see exercise 1.9), and the temperature gradient, G, to be separately controlled. If V' is not too high, both the heat flux and the solidification are unidirectional. The cooling rate at a given location and time is given by:

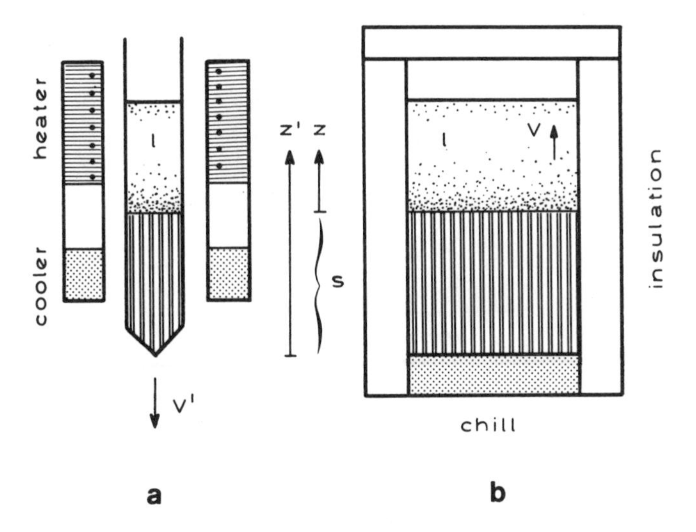

Figure 1.4 Basic Methods of Controlled Solidification

Without heat extraction there is no solidification. The liquid must be cooled to the solidification temperature and then the latent heat of solidification appearing at the growing solid/liquid interface must be extracted. There are several methods of heat extraction. In directional (Bridgman-type) solidification (a): the crucible is drawn downwards through a constant temperature gradient, G, at a uniform rate, V', and therefore the microstructure is highly uniform throughout the specimen. The method is restricted to small specimen diameters and is expensive because it is slow and, paradoxically, heat must be supplied during solidification in order to maintain the imposed positive temperature gradient. For these reasons, it is employed only for research purposes and for the growth of single crystals. In directional casting (b), the benefits of directionality, such as a better control of the properties and an absence of detrimental macrosegregation, are retained but the microstructure is no longer uniform along the specimen because the growth rate, V, and the temperature gradient decrease as the distance from the chill increases. The process is cheaper than that of (a) and is used for the directional solidification of gas-turbine blades, for example. Combined with proper alloy development, these processes result in a higher efficiency and a longer life for the gas-turbines of aircraft.

$$\dot{T}_{s+\varepsilon} = \left(\frac{\partial T}{\partial t}\right)_{s+\varepsilon} = \left(\frac{\partial T}{\partial z'}\frac{\partial z'}{\partial t}\right)_{s+\varepsilon} = G \cdot V \,\Big|_{s+\varepsilon} \qquad [1.3]$$

where the time-dependent position of the solid/liquid interface is $s = z' - z$, and z' is the coordinate with respect to the system (the crucible), while z is the coordinate with respect to the moving s/l interface (Fig. 1.4; see also figure A2.2), and ε is a small quantity with respect to s. Here, V is the rate of movement of the s/l interface and G is the thermal gradient in the liquid when $z' = s + \varepsilon$ (also called G_l) or the thermal gradient in the solid when $z' = s - \varepsilon$ (called G_s). Due to differences in the conductivity of solid and liquid and due to the evolution of latent heat at a moving interface, $G_l \neq G_s$.

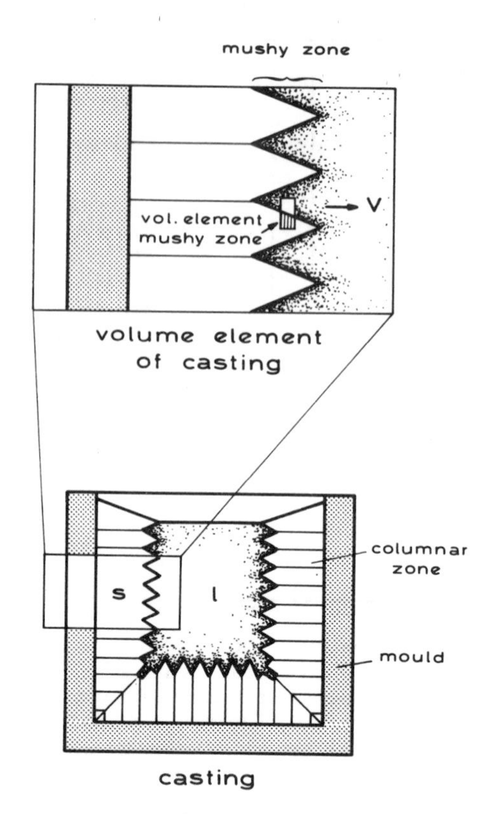

Figure 1.5 Solidification in Conventional Castings and Ingots

In the vast majority of castings, no directionality is imposed upon the overall structure, but the local situation can be seen to be equivalent to that existing in directional casting (Fig. 1.4b). This is true of the way in which the solid advances inwards from the mould wall to form a columnar zone. During the growth of the columnar zone, three regions can be distinguished. These are the liquid, the liquid plus solid (so-called mushy zone), and the solid region. *The mushy zone is the region where all of the microstructural characteristics are determined*, e.g. the shape, size, and distribution of concentration variations, precipitates, and pores. An infinitesimally narrow volume element which is fixed in the mushy zone and is perpendicular to the overall growth direction permits a description of the microscopic solidification process and therefore of the scale and composition of the microstructure.

For reasons of simplicity, the temperature gradient in the liquid will be mostly used in this book and written as $G \equiv G_l$. However, one has to bear in mind that under certain conditions (as explained in chapter 3) the physically meaningful temperature gradient is the mean gradient, $\bar{G} = (G_s\kappa_s+G_l\kappa_l)/(\kappa_s+\kappa_l)$.

Another directional casting process is illustrated by figure 1.4b. Here, heat is extracted via a chill and, as in figure 1.4a, growth occurs in a direction which is parallel, and opposite, to the heat-flux direction. In this situation, the heat-flux decreases with time as do the coupled parameters, G and V. Thus $\dot{T}$ also varies. Heat flow in the mould/metal system leads to an expression for the position, s, of the solid/liquid interface, which is of the type (appendix 1):

$$s = Kt^{1/2} \qquad [1.4]$$

This equation is exact only if the melt is not superheated, if the solid/liquid interface is planar, and if the surface temperature of the casting at the chill drops immediately, at $t = 0$, to a constant value. Figure 1.4b can be regarded as being a volume element in a conventional casting (Fig. 1.5). The difference between the figures is the presence of a dendritic morphology at the solid/liquid interface shown schematically in figure 1.5.

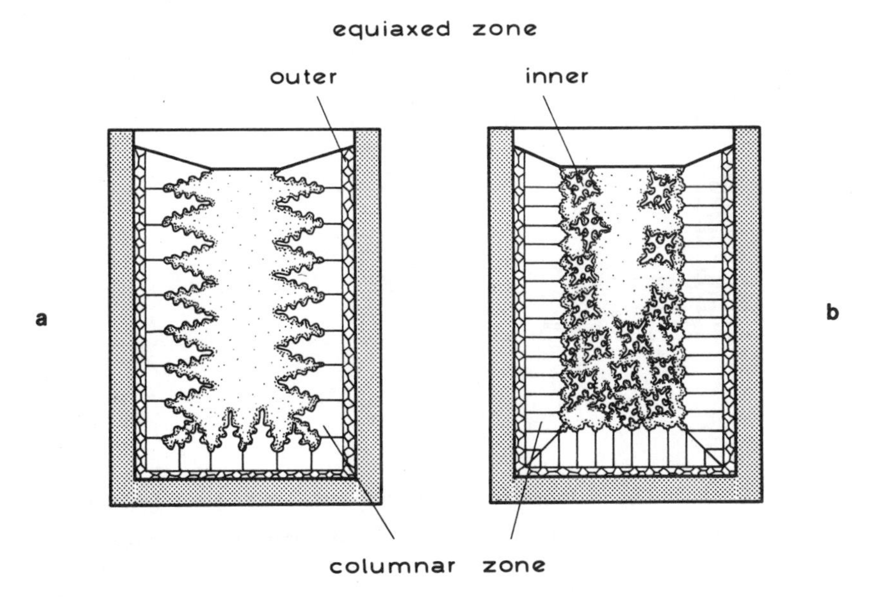

Figure 1.6 Structural Zone Formation in Castings

Firstly, solid nuclei appear in the liquid at, or close to, the mould wall. For a short time, they increase in size and form the outer equiaxed zone. Then, those crystals (dendrites) of the outer equiaxed zone which can grow parallel and opposite to the heat flow direction will advance most rapidly. Other orientations tend to be overgrown, due to mutual competition, leading to the formation of a columnar zone (a). Beyond a certain stage in the development of the columnar dendrites, branches which become detached from the latter can grow independently. These tend to take up an equiaxed shape because their latent heat is extracted radially through the undercooled melt. The solidified region containing them is called the inner equiaxed zone (b). The transition from columnar to equiaxed growth is highly dependent upon the degree of convection in the liquid. In continuous casting machines, electromagnetic stirring is often used to promote this transition and lead to superior soundness at the ingot centre.

This morphology depends (chapter 4) upon the alloy composition, and upon G and V. If it is assumed, for simplicity, that the dendrites can be represented by plates#, then solidification *on a microscopic scale* again takes place directionally (perpendicular to the primary growth axis of the dendrites, as shown in the upper insert of figure 1.5). This representation permits a simple estimation of interdendritic microsegregation (chapter 6).

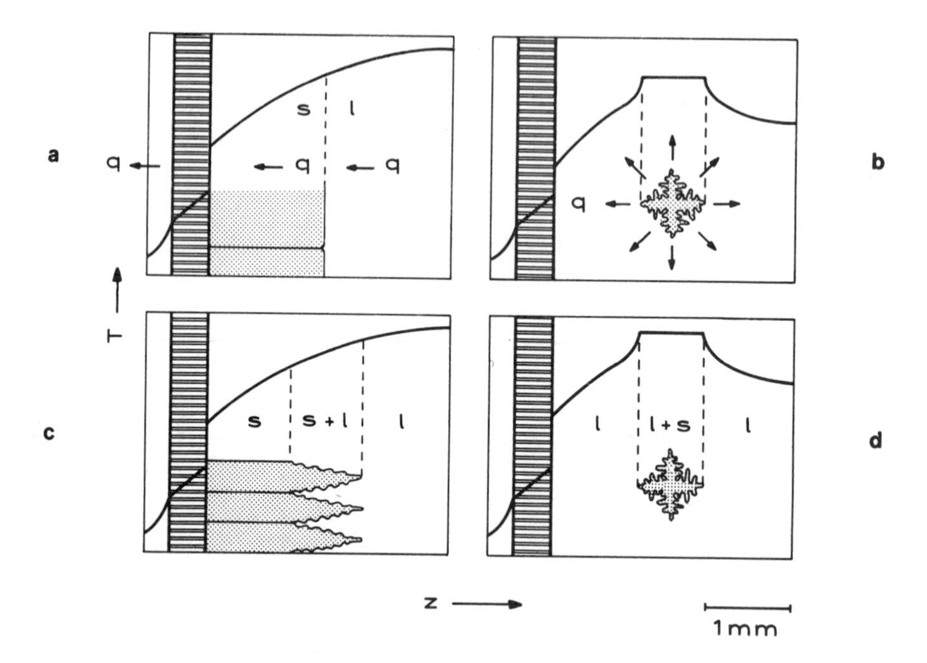

Figure 1.7 Solid/Liquid Interface Morphology and Temperature Distribution

In the case of a pure metal (a,b) which is solidifying inwards from the mould wall, the columnar grains (a) possess an essentially planar interface, and grow in a direction which is antiparallel to that of the heat flow. Within the equiaxed region of pure cast metal (b), the crystals are dendritic and grow radially in the same direction as the heat flow. When alloying elements or impurities are present, the morphology of the columnar crystals (c) is generally dendritic. The equiaxed morphology in alloys (d) is almost indistinguishable from that in pure metals, although a difference may exist in the relative scale of the dendrites. This is because the growth in pure metals is heat-flow-controlled, while the growth in alloys is mainly solute-diffusion-controlled. Note that in columnar growth the hottest part of the system is the melt, while in equiaxed solidification the crystals are the hottest part. It follows that the melt must always be cooled to below the melting point (i.e. undercooled) before equiaxed crystals can grow.

Plate-like primary crystal morphologies are often observed during solid-state precipitation. When metallic primary crystals grow into a melt, they are always rod-like rather than plate-like in form and possess many branches, leading to the characteristic dendrite form.

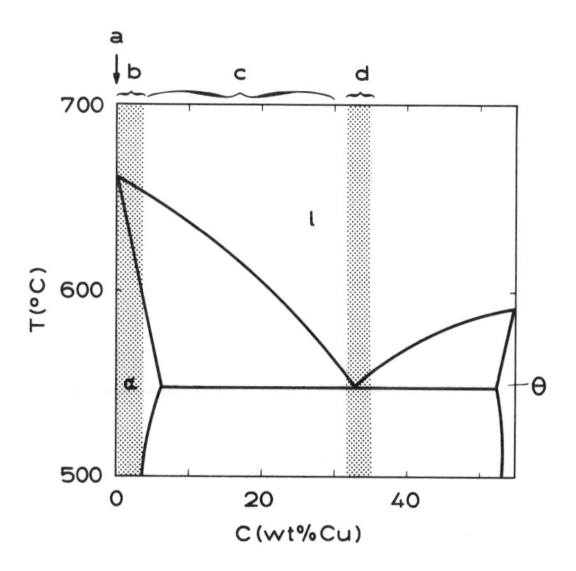

Figure 1.8 Principal Alloy Types

It is important to understand how the various microstructures are influenced by the alloy composition and by the solidification conditions. Fortunately, this can usually be reduced to the study of two basic morphological forms: dendritic and eutectic. Thus, one can distinguish: a) pure substances, which solidify in a planar or dendritic manner, b) solid-solution dendrites (with or without interdendritic precipitates), c) dendrites plus interdendritic eutectic, and d) eutectic. The latter group includes the familiar 'cast iron' and 'plumber's solder' type of alloy. In general, the design of casting alloys is governed by the twin aims of obtaining the required properties and good castability (i.e. easy mould filling, low shrinkage, small hot tearing tendency, etc.). Castability is greatest for pure metals and alloys of eutectic composition. The diagram represents the Al-Cu system between Al and the intermetallic (theta) phase, Al_2Cu.

1.3 Solidification Microstructures

In an ingot or casting, three zones of solidification behaviour can generally be distinguished (Fig. 1.6). At the mould/metal interface, the cooling rate is at its highest due to the initially low relative temperature of the mould. Consequently, many small grains having random orientations are nucleated at the mould surface and an 'outer equiaxed' zone is formed. These grains rapidly become dendritic, and develop arms which grow along preferred crystallographic directions (<001> in the case of cubic crystals). Competitive growth between the randomly oriented outer equiaxed grains causes those which have a preferred growth direction (parallel and opposite to the direction of heat flow) to eliminate the others. This is because their higher growth rate allows them to dominate the solid/liquid interface morphology, thus leading to the formation of the characteristic columnar zone. It is often observed that another equiaxed zone forms in the centre of the casting, mainly as a result of the growth of detached dendrite arms within the remaining, slightly-undercooled liquid.

Figure 1.7 shows the temperature fields in the various cast structures which one might encounter. These are planar interface (columnar grains - a) or thermal dendrites (equiaxed grains - b) in pure materials, and solutal (constitutional) dendrites in alloys (c,d). It can be seen that *columnar* grains must always grow out from the mould (which is the heat sink) in a direction which is *opposite* to that of the heat flow, while equiaxed grains grow in a supercooled melt which acts as their heat sink. Thus, the growth direction and the heat flow direction are the *same* in equiaxed growth.

The form of a solidification microstructure depends not only upon the cooling conditions, but also upon the alloy composition (Fig. 1.8). There are essentially two basic growth morphologies which can exist during alloy solidification. These are the dendritic and eutectic morphologies (peritectic alloys grow in a dendritic manner). Generally, a mixture of both morphologies will be present. It is reassuring, in the face of the apparent microstructural complexity, to remember that it is only necessary to understand these two growth forms in order to interpret the solidification microstructure of almost any alloy.

Figure 1.9 illustrates the various stages of equiaxed solidification — from nucleus to grain — for the two major growth morphologies: dendrites and eutectic. Each grain has one nucleus at its origin. (In the literature of cast iron, an eutectic grain is often called an eutectic cell. This definition will not be adopted because the term, cell, is here reserved for another morphology.)

The transformation of liquid into solid involves the creation of curved solid/liquid interfaces (leading to capillarity effects) and the microscopic flow of heat (and also solute in the case of alloys).

1.4 Capillarity Effects

With any solid/liquid interface of area, A, is associated an excess (interface) energy which is required for its creation. Therefore, heterogeneous systems or parts of systems which possess a high A/v ratio will be in a state of higher energy and therefore unstable with respect to a system of lower A/v ratio. The relative stability can be expressed by the equilibrium temperature between both phases (melting point). As shown in appendix 3, the change in melting point due to this curvature effect, often called the curvature or Gibbs-Thomson undercooling, is given by:

$$\Delta T_r = K \Gamma \qquad [1.5]$$

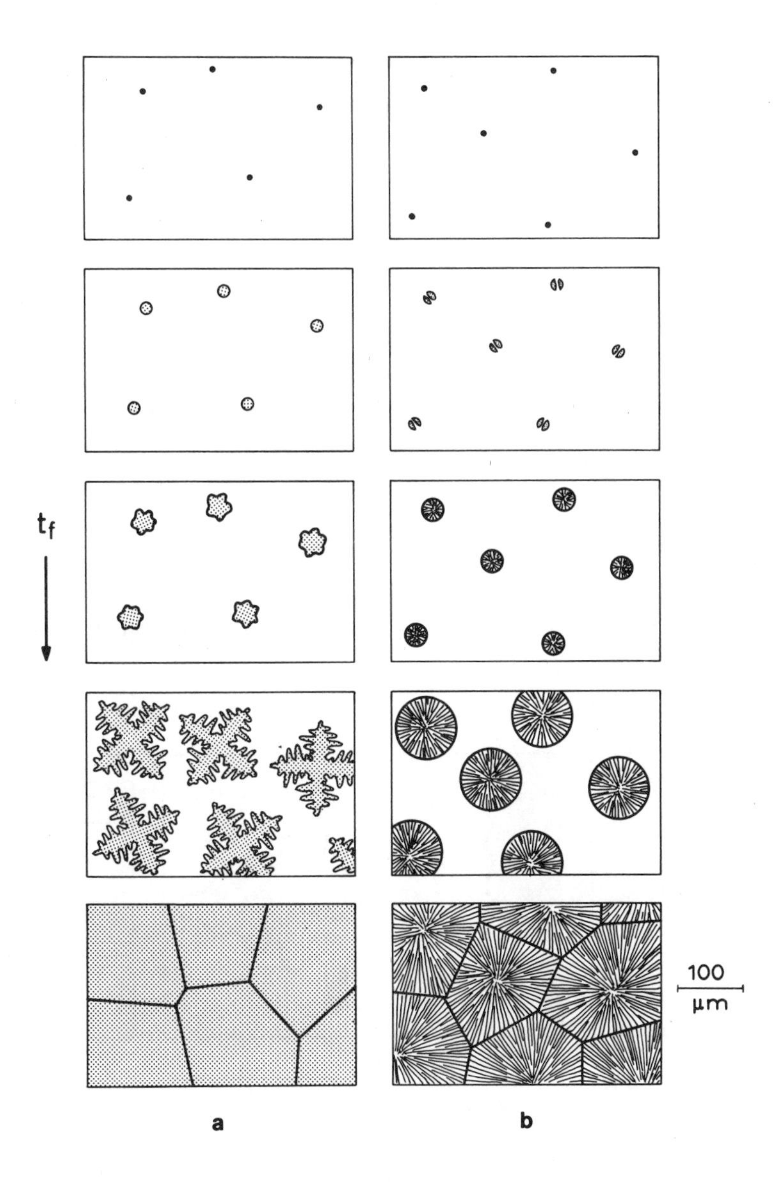

Figure 1.9 Process of Equiaxed Solidification of Dendrites and Eutectic

In each case, single-phase nuclei form initially. In pure metals or single-phase alloys (a), the nuclei then grow into spherical crystals which rapidly become unstable and dendritic in form. These dendrites grow freely in the melt and finally impinge on one another. In a pure metal after solidification, no trace of the dendrites themselves will remain, although their points of impingement will be visible as the grain boundaries. In an alloy, the dendrites will remain visible after etching due to local composition differences (microsegregation). In an eutectic alloy (b), a second phase will soon nucleate on the initial, single-phase nucleus. The eutectic grains then continue to grow in an essentially spherical form. In a casting, both growth forms, dendritic and eutectic, often develop together. Note that each grain originates from a single nucleus.

Note that the curvature, K, and the Gibbs-Thomson coefficient, Γ, are here defined so that a positive undercooling (decrease in equilibrium melting point) is associated with a portion of solid/liquid interface which is convex towards the liquid phase. The curvature can be expressed as (appendix 3):

$$K = \frac{dA}{dv} = \frac{1}{r_1} + \frac{1}{r_2} \qquad [1.6]$$

where r_1 and r_2 are the principal radii of curvature#. Thus, the total curvature of a sphere is $2/r$ and that of a cylindrical surface is $1/r$. The Gibbs-Thomson coefficient is given by:

$$\Gamma = \frac{\sigma}{\Delta s_f} \qquad [1.7]$$

For most metals, Γ is of the order of 10^{-7}Km. Hence, the effect of the solid/liquid interface energy, σ, only becomes important for morphologies which have a radius that is less than about 10 μm. These include nuclei, interface perturbations, dendrite tips, and eutectic phases (Fig. 1.10).

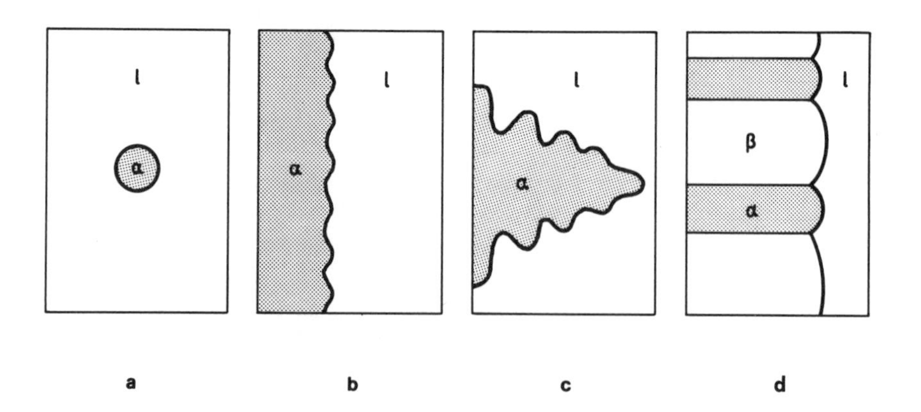

Figure 1.10 The Scale of Various Solid/Liquid Interface Morphologies

Solidification morphologies are determined by the interplay of two effects acting at the solid/liquid interface. These are the *diffusion* of solute (or heat), which tends to minimise the scale of the morphology (maximise curvature), and *capillarity* effects which tend to maximise the scale. The crystal morphologies actually observed are thus a compromise between these two tendencies, and this can be shown with respect to nucleation (a), interface instability (b), dendritic growth (c), and eutectic growth (d).

The two principal radii of curvature are the minimum and maximum values for a given surface. It can be shown that they lie in planes always perpendicular to each other.

1.5 Solute Redistribution

The creation of a crystal from an alloy melt causes a local change in the composition. This is due to the equilibrium condition for a binary system containing two phases:

$$\mu_l^A = \mu_s^A, \qquad \mu_l^B = \mu_s^B \tag{1.8}$$

(appendix 3). The difference in composition at the growing interface, assuming that local (i.e. at the interface) equilibrium exists in metals under normal solidification conditions, can be described by the distribution coefficient under isothermal and isobaric conditions (Fig. 1.11):

$$k = \left(\frac{C_s}{C_l}\right)_{T,P} \tag{1.9}$$

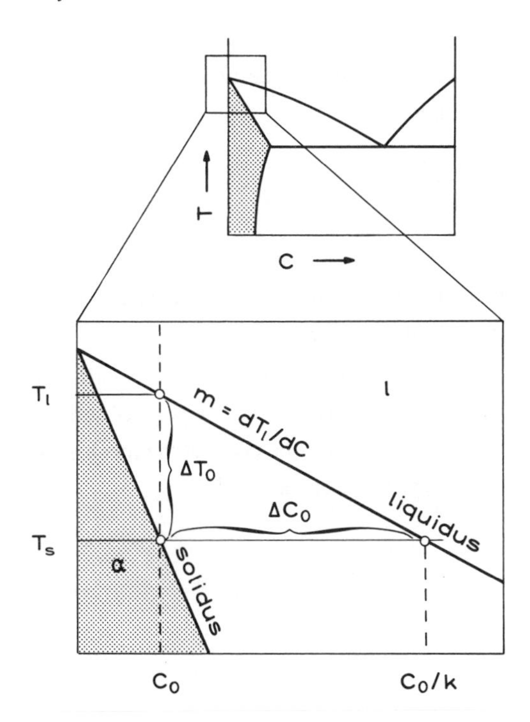

Figure 1.11 Solid/Liquid Equilibrium

In order to simplify the mathematical treatment of solidification processes, it is generally assumed that the liquidus and solidus lines of the phase diagram are straight, and therefore that the distribution coefficient, k, and the liquidus slope, m, are constant. The characteristic properties of the system are defined in the text (equations 1.9 to 1.11).

In most of the theoretical treatments to be presented later, the solidus and liquidus lines will be assumed to be straight. This means that k and m, the liquidus slope, are then constant. This violates the condition for thermodynamic equilibrium (equation 1.8) but often makes theoretical analyses more tractable. However, if the variation in composition is large, e.g. due to a large variation of growth rate, this assumption might lead to wrong results.

Throughout this book, m is defined so that the product, $(k-1)m$, is positive. That is, m is defined to be positive when k is greater than unity, and to be negative when k is less than unity. Two other important parameters of an alloy system are shown in figure 1.11. These are the liquidus-solidus temperature interval for an alloy of composition, C_0:

$$\Delta T_0 = -m\Delta C_0 = (T_l - T_s) \qquad [1.10]$$

and the concentration difference between the liquid and solid solute contents at the solidus temperature of the alloy:

$$\Delta C_0 = \frac{C_0(1-k)}{k} \qquad [1.11]$$

Under rapid solidification conditions equation 1.8 may no longer be satisfied, and k then becomes a function of V. This so-called non-equilibrium solidification can lead to a highly supersaturated crystal (see chapter 7).

In later chapters it will be shown how the above parameters influence the solidification microstructure. Meanwhile, the starting point of solidification (nucleation) will be considered briefly, and a look will be taken at the mechanisms by which atoms in the melt become part of the growing crystal.

Bibliography

Solidification Microstructure and Properties

M.C.Flemings, *Solidification Processing*, McGraw-Hill, New York, 1974, p. 328.

W.Kurz, P.R.Sahm, *Gerichtet erstarrte eutektische Werkstoffe*, Springer, Berlin, 1975.

G.F.Bolling, in *Solidification*, American Society for Metals, Metals Park, Ohio, 1971, p. 341.

J.F.Burke, M.C.Flemings, A.E.Gorum, *Solidification Technology*, Brook Hill, 1974.

Analytical Solutions to Heat Flow Problems in Solidification

H.S.Carslaw & J.C.Jaeger, *Conduction of Heat in Solids*, 2nd Edition, Oxford University Press, London, 1959.

G.H.Geiger, D.R.Poirier, *Transport Phenomena in Metallurgy*, Addison-Wesley, 1973.

J.Szekely, N.J.Themelis, *Rate Phenomena in Process Metallurgy*, Wiley - Interscience, New York, 1971.

Capillarity Effects

W.W.Mullins, in *Metal Surfaces - Structure, Energetics, and Kinetics*, ASM, Metals Park, Ohio, 1963, p. 17.

R.Trivedi, in *Lectures on the Theory of Phase Transformations* (Edited by H.I.Aaronson), TMS of AIME, New York, 1975, p. 51.

Thermodynamics of Solidification

J.C.Baker, J.W.Cahn, in *Solidification*, American Society for Metals, Metals Park, Ohio, 1971, p. 23.

M.C.Flemings, *Solidification Processing*, McGraw-Hill, 1974, p. 263.

J.S.Kirkaldy, in *Energetics in Metallurgical Phenomena* - Volume IV, (Edited by W.M.Mueller), Gordon & Breach, New York, 1968, p. 197.

M.Hillert, in *Lectures on the Theory of Phase Transformations* (Edited by H.I.Aaronson), TMS of AIME, New York, 1975, p. 1.

Casting Techniques

R.Flinn, in *Techniques of Metals Research* - Volume I, (Edited by R.F.Bunshah), Wiley, New York, 1968.

F.L.Versnyder, M.E.Shank, Materials Science and Engineering **6** (1970) 213.

T.F.Bower, D.A.Granger, J.Keverian, in *Solidification*, American Society for Metals, Metals Park, Ohio, 1971, p. 385.

Phase Diagrams

M.Hansen, *Constitution of Binary Alloys*, McGraw-Hill, New York, 1958.

T.B.Massalski et al. (Eds.), *Binary Alloy Phase Diagrams*, ASM, Metals Park, Ohio, 1986.

Metals Handbook - Volume 8, ASM, Metals Park, Ohio.

Rapid Solidification

J.C.Baker, J.W.Cahn, in *Solidification*, ASM, Metals Park, Ohio, 1971, p. 23.

W.J.Boettinger, S.R.Coriell, R.F.Sekerka, Material Science and Engineering **65** (1984) 27.

H.Jones, *Rapid Solidification of Metals and Alloys*, The Institution of Metallurgists, London, 1982.

Exercises

1.1 Discuss the shape of the upper surface of the ingot in figure 1.6. What would happen if the solidifying material was one of the following substances: water, Ge, Si, Bi?

1.2 From a consideration of the volume element in the mushy zone of figure 1.5 (upper part), define the local solidification time, t_f, in terms of the dendrite growth rate, V, and the length, a, of the mushy zone.

1.3 Sketch η–τ diagrams for the crystallisation of a pure metal and for an alloy, and comment on their significance in each case.

1.4 Equiaxed dendrites are developing freely in an undercooled melt. Discuss the direction of movement of the equiaxed dendrites in a quiescent melt. Where would most of them be found in solidifying melts of a) steel, b) Bi?

1.5 Sketch two different phase diagrams having a positive and a negative value of m. Show that the product, $m(k - 1)$, is always positive.

1.6 Using data from "Constitution of Binary Alloys" (M.Hansen and K.Anderko, McGraw-Hill, New York, 1958) or other similar phase-diagram compilations, estimate the distribution coefficient, k, of S in Fe at temperatures of between 1500 and 988°C, and of Cu in Ni at temperatures of between 1400 and 1300°C. Discuss the validity of the assumption that k is constant in these systems.

1.7 A molten alloy, like any liquid which has local density variations, will tend to exhibit the motion known as natural convection. What is the origin of this convection in a) pure metals, b) alloys? Discuss your conclusions with regard to various alternative solidification processes such as upward (as opposed to downward) directional solidification (Bridgman - Fig. 1.4), and casting (Fig. 1.5).

1.8 Give possible reasons for the good mould-filling characteristics which are exhibited by pure metals and eutectic alloys during the casting of small sections. Discuss them with regard to the interface morphology shown in figure 1.7.

1.9 Figure 1.4 illustrates two directional solidification processes which differ with respect to their heat transfer characteristics. In one case, a steady-state behaviour is established after some transient changes. In the other case, changes continue to occur with the passage of time. One process is not limited with regard to the length of the product, but is limited by its diameter. The other process is not affected by the diameter, but rather by the specimen length. Sketch heat flux lines, and G and V values as a function of t for both cases and relate them to the described characteristics of the two processes. Note that, in directional solidification, the temperature gradient in the liquid at the solid/liquid interface must always be positive, as shown in figure 1.7a,c.

1.10 Illustrate the changes in the temperature distribution of a casting as a function of time between the moment of pouring of a pure superheated melt and the establishment of the situation shown in figures 1.7a and b. Discuss the fundamental differences between a and b.

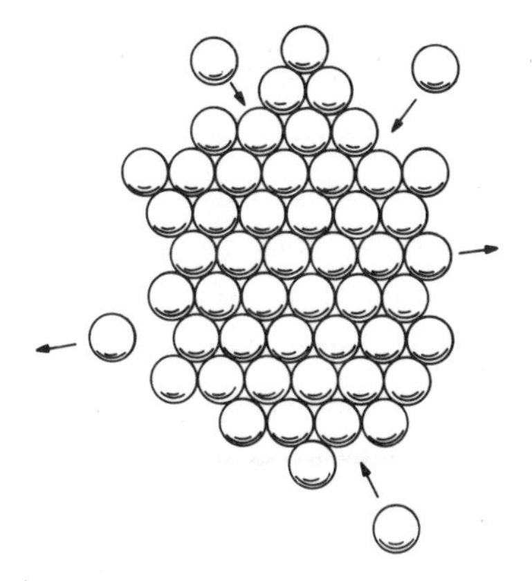

CHAPTER TWO

ATOM TRANSFER AT THE SOLID/LIQUID INTERFACE

From a thermodynamic point of view, solidification requires a heat flux from the system to the surroundings which changes the free energies, and therefore the relative thermodynamic stability, of the phases present. From the same point of view, thermodynamically stable phases are more likely to be observed, but the transformation of one phase into another requires rearrangement of the atoms. This may involve a relatively short-range (atomic) rearrangement to form a new crystal structure, as in the case of a pure substance. Alternatively, atomic movement may be required over much larger, but still microscopic, distances as in the case of alloy solidification where mass diffusion controls the transformation. Because of these atomic movements, solidification will always require some irreversible departure from equilibrium in order to drive the process.

Like chemical reactions, phase transformations are driven by thermal fluctuations and can only occur when the probability of transfer of atoms from the parent phase to the product phase is higher than that for the opposite process. However, before this stage is reached, it is necessary that some of the new phase, to which atoms of the parent phase can jump, should already exist. Therefore, stable regions of the new phase have to form. In liquid metals, random fluctuations may create minute crystalline regions (clusters, embryos) even at temperatures greater than the melting point, but these will not be stable. They continue to be metastable below the melting point because the relatively large excess energy required for surface creation tends to weight the 'energy balance' against their survival when they are small.

Once nucleation has occurred, atom transfer to the crystals has to continue in order to ensure their growth. The mechanisms involved during this second stage are discussed in section 2.3.

2.1 Conditions for Nucleation

It is inherently difficult to observe the process of nucleation because it involves such small clusters of atoms. Consequently, only extremely careful comparison of theoretical models and experimental results can clarify the very first stages of solidification. As demonstrated in figure 2.1 nucleation begins at some degree of undercooling, $\Delta T = \Delta T_n$ #, which for metals is generally very small in practical situations. The initially small grains which begin to grow do not appreciably modify the cooling rate imposed by the external heat flux, q_e. Increasing the undercooling has the effect of markedly increasing the nucleation rate, I, and also the growth rate, V, of the dendrites. The overall solidification rate approaches a maximum value when the internal heat flux (q_i), which is proportional to the latent heat of fusion and the volume rate of transformation, $\dot{f}_s(=df_s/dt)$, is equal to the external one (q_e) (equation 1.1). Here, $\dot{T} = 0$. During the first stage of equiaxed solidification, which is essentially nucleation-controlled, the volume fraction of solid is still very small. After some time, the temperature of the system has risen above the nucleation temperature and the second stage of solidification is growth-controlled. The number of grains present thus remains essentially constant and solidification proceeds first via the lengthening of dendrites, and then via dendrite arm thickening once the grains are in contact.

From this sort of consideration, it is possible to deduce that nucleation is the dominant process at the beginning of solidification and leads very rapidly to the

The undercooling, ΔT, is usually defined as the temperature difference between the equilibrium temperature of a system and its actual temperature. The latter is lower than the equilibrium temperature when the melt is undercooled. In this case, ΔT is greater than zero. The term, supercooling, is often used interchangeably with undercooling in the literature.

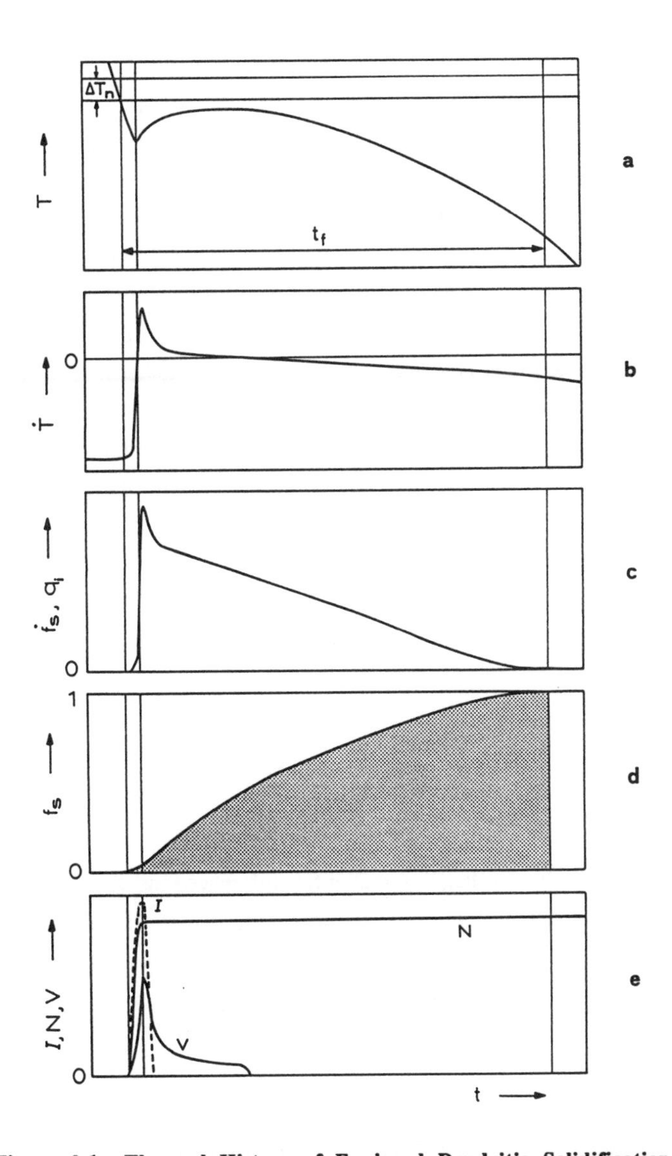

Figure 2.1 Thermal History of Equiaxed Dendritic Solidification

The above temperature-time curve is one which might well be obtained during a solidification sequence such as that pictured in figure 1.9a. The usual cooling curve (a) begins to deviate slightly at the undercooling where nucleation occurs, ΔT_n. At this point, the first fraction of solid, f_s, appears (d). With further cooling, the nucleation rate, I, rapidly increases to a maximum value (e). At the minimum in the temperature-time curve, the growth rate, V, of the grains (i.e. of the dendrite tips) is at its highest. The subsequent increase in temperature is due to the high internal heat flux, q_i, arising from the rate of transformation, $\dot{f}_s (= df_s/dt)$, and the latent heat released (c). (The maximum of the temperature can lie above the nucleation temperature.) Note that I is much more sensitive to temperature changes than is V (e). Most of the solidification which takes place after impingement of the grains involves dendrite arm coarsening at a tip growth rate, V, equal to zero. During this time interval, the number of grains, N, remains constant.

establishment of the final grain population, with each nucleus forming one equiaxed grain of the type shown in figure 1.7b or d. Note that even in the case of columnar solidification, the very first solid in a casting always appears in the form of equiaxed grains (Fig. 1.6). The conditions leading to nucleation are therefore of utmost importance in determining the characteristics of any cast microstructure.

In phase changes such as solidification, which are discontinuous, the transformation process cannot occur at any arbitrarily small undercooling. The reason for this arises from the large curvature of the interface associated with a crystal of atomic dimensions. This curvature markedly lowers the equilibrium temperature (appendix 3) so that, the smaller the crystal, the lower is its melting point. This occurs because the small radius of curvature creates a pressure difference between the two phases which is of the order of 100MPa (1kbar) for a crystal radius of 1nm. The equilibrium melting point of the system is thus lowered by an amount, ΔT_r. The critical size, r°, of a crystal, i.e. the size which allows equilibrium between the curved crystal and its melt, can be easily calculated. For a sphere (appendix 3) it is:

$$\Delta T_r = K\Gamma = \frac{2\Gamma}{r^{\circ}}$$

and

$$r^{\circ} = \frac{2\Gamma}{\Delta T_r} = \frac{2\sigma}{\Delta T_r \Delta s_f} \qquad [2.1]$$

This relationship indicates that, the smaller the difference (undercooling) between the melting point and the temperature of the melt, the larger will be the size of the equilibrium crystal. For nucleation of a spherical crystal of radius, r, to occur, a number of atoms, each of volume, v', given by:

$$n \cong \frac{4r^3\pi}{3v'} \qquad [2.2]$$

have to arrange themselves on the sites of the corresponding solid crystal lattice. It is evident that the probability of this event occurring is very small for large values of r°, i.e. at small undercoolings (equation 2.1).

As shown in figure 2.2, the critical condition for the nucleation of 1 mole is derived by summing the interface and volume terms for the Gibbs free energy:

$$\Delta G = \Delta G_i + \Delta G_v = \sigma A + \Delta g \cdot v \qquad [2.3]$$

where σ is the solid/liquid interface energy and Δg is the Gibbs free energy difference

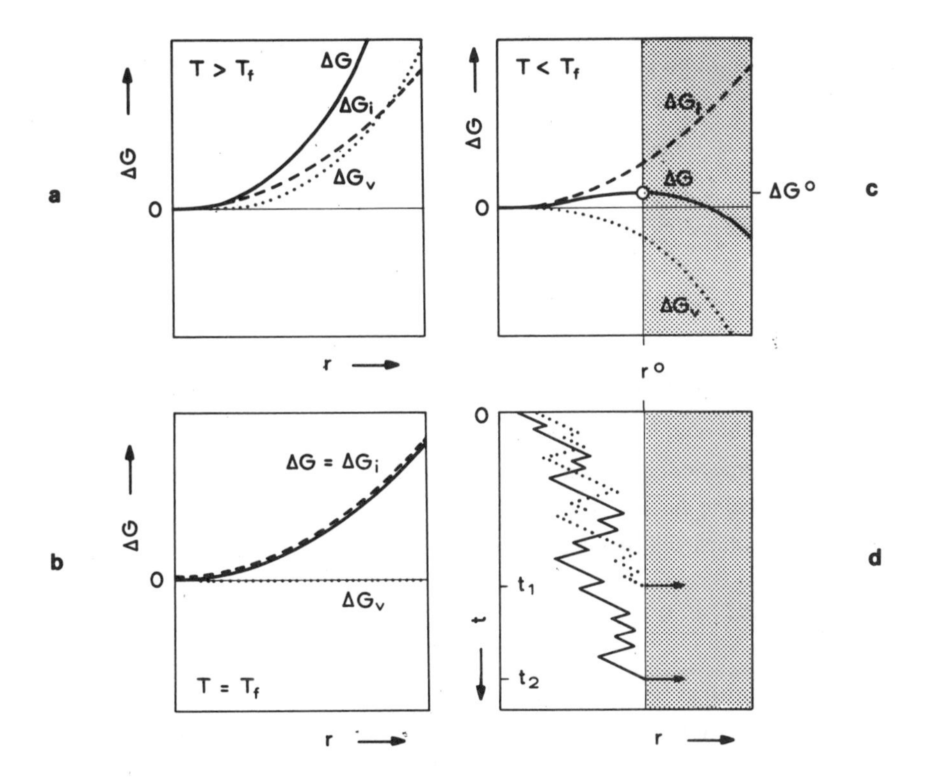

Figure 2.2 Free Energy of a Crystal Cluster as a Function of its Radius

The phenomenon of nucleation of a crystal from its melt depends mainly on two processes: thermal fluctuations which lead to the creation of variously sized crystal embryos (clusters), and creation of an interface between the liquid and the solid. The free energy change, ΔG_v, which is associated with the first process is proportional to the volume transformed. That is, it is proportional to the cube of the cluster radius. The free energy change, ΔG_i, which is associated with the second process is proportional to the area of solid/liquid interface formed. That is, it is proportional to the square of the cluster radius. At temperatures, T, greater than the melting point (a), both the volume free energy (ΔG_v) and the surface free energy (ΔG_i) increase monotonically with increasing radius, r. Therefore, the total free energy, ΔG, which is their sum, also increases monotonically. At the melting point (b), the value of ΔG_i still increases monotonically since it is only slightly temperature dependent. Because, by definition, thermodynamic equilibrium exists between the solid and liquid at the melting point, the value of ΔG_v is zero. Hence ΔG again increases monotonically with increasing radius. At a temperature below the equilibrium melting point (c), the sign of ΔG_v is reversed because the liquid is now metastable, while the behaviour of ΔG_i is still the same as in (a) and (b). However, ΔG_v has a 3rd-power dependence on the radius while ΔG_i has only a 2nd-power dependence. At small values of the radius, the absolute value of ΔG_v is less than that of ΔG_i, while at large values of r the cubic dependence of ΔG_v predominates. The value of ΔG therefore passes through a maximum at a critical radius, r°. Fluctuations may move the cluster backwards and forwards along the ΔG–r curve (c) due to the effect of random additions to, or removals of atoms from, the unstable nucleus (d). When a fluctuation causes the cluster to become larger than r°, growth will occur due to the resultant decrease in the total free energy. Thus, an embryo or cluster ($r<r^\circ$) becomes a nucleus ($r = r^\circ$) and eventually a grain ($r >> r^\circ$).

between the liquid and solid per unit volume. Again assuming a spherical form (minimum A/v ratio) for the nucleus,

$$\Delta G = \sigma 4\pi r^2 + \frac{\Delta g\, 4\pi r^3}{3} \qquad [2.4]$$

The Gibbs free energy per unit volume, Δg, is proportional to ΔT (appendix 3):

$$\Delta g = -\Delta s_f \Delta T \qquad [2.5]$$

The right-hand-side of equation 2.4 is composed of a quadratic and a cubic term. The value of σ is always positive whereas Δg depends upon ΔT, and is negative if ΔT is positive. This behaviour leads to the occurrence of a maximum in the value of ΔG when the melt is undercooled, i.e. when ΔT is positive (Fig. 2.2c). This maximum value can be regarded as being the activation energy which has to be overcome in order to form a crystal nucleus which will continue to grow. The criterion for the maximum is that:

$$\frac{d(\Delta G)}{dr} = 0 \qquad [2.6]$$

and can be regarded as being a condition for equilibrium between a liquid, and a solid with a curvature such that the driving *force* for solidification is equal to that for melting. Consequently, it is not surprising that setting the first derivative of equation 2.4 equal to zero should lead to equation 2.1.

Table 2.1 Critical Dimensions and Activation Energy for the Nucleation of a Spherical Nucleus in a Pure Melt ($\Delta g = \Delta s_f \Delta T$)

	Homogeneous Nucleation	Heterogeneous Nucleation
r^o	$-\frac{2\sigma}{\Delta g}$	$-\frac{2\sigma}{\Delta g}$
n^o	$-\left(\frac{32\pi}{3v'}\right)\left(\frac{\sigma}{\Delta g}\right)^3$	$-\left(\frac{32\pi}{3v'}\right)\left(\frac{\sigma}{\Delta g}\right)^3 f(\theta)$
ΔG_n^o	$\left(\frac{16\pi}{3}\right)\left(\frac{\sigma^3}{\Delta g^2}\right)$	$\left(\frac{16\pi}{3}\right)\left(\frac{\sigma^3}{\Delta g^2}\right) f(\theta)$

Table 2.2 Values of the Expression: $f(\theta) = (1/4)(2 + \cos\theta)(1 - \cos\theta)^2$

θ (°)	Type of Nucleation	$f(\theta)$
0 complete wetting	no nucleation barrier #)	0
10		0.00017
20		0.0027
30		0.013
40		0.038
50		0.084
70	heterogeneous	0.25
90		0.5
110		0.75
130		0.92
150		0.99
170		0.9998
180 no wetting	homogeneous	1

#) immediate growth can occur

Figure 2.2d demonstrates how fluctuations in a melt, corresponding to the conditions of figure 2.2c, will behave. At least one cluster which is as large as the critical nucleus (of radius, r°) must be formed before solidification can begin. The time which elapses before this occurs will be different (t_1, t_2,...) at different locations in the melt. In this case, fluctuations spontaneously create a small crystalline volume in an otherwise homogeneous melt (containing no solid phase). This is referred to as *homogeneous nucleation* because the occurrence of nucleation transforms an initially homogeneous system (consisting only of atoms in the liquid state) into a heterogeneous system (crystals plus liquid). Using equations 2.2 to 2.6, the critical parameters can be calculated and are given in table 2.1 where ΔG_n is equivalent to ΔG in equation 2.3, except that n (the number of atoms in the nucleus, equation 2.2) rather than the radius, r, has been used to describe the nucleus size.

As an example, suppose that an undercooling of 230K is required to cause homogeneous nucleation in small Cu droplets. From this value and the properties of the metal (appendix 14) it is estimated that $r^{\circ} = 1.28$nm and $n^{\circ} = 634$.

When the melt contains solid particles, or is in contact with a crystalline crucible or oxide layer, nucleation may be facilitated if the number of atoms, or the activation energy required for nucleation, are decreased. This is known as *heterogeneous nucleation*. A purely geometrical calculation shows that when the solid/liquid interface of the substance is partly replaced by an area of low-energy solid/solid interface between the crystal and a foreign solid, nucleation can be greatly facilitated. The magnitude of the effect can be calculated using the result derived in appendix 3:

$$f(\theta) = \frac{(2+\cos\theta)(1-\cos\theta)^2}{4} \qquad [2.7]$$

where θ is the wetting angle, in the presence of the melt, between a growing spherical cap of solid (nucleus) and a solid substrate (particle or mould wall). The n° and ΔG_n° values are decreased by small values of θ but the r° value is not.

Numerical values of $f(\theta)$, given by equation 2.7, are listed in table 2.2, and show that, under conditions of good solid/solid wetting (small θ) between the crystal nucleus and the foreign substrate in the melt, a large decrease in n° and ΔG° can be expected. This can have a dramatic effect on the nucleation rate and is used daily in foundries in the form of *inoculation*. Here, substances are added to the melt which are crystalline or form crystals at temperatures greater than the melting point. The effect is usually time-dependent since the added substances tend to dissolve in the melt. In the case of detaching of dendrite branches there is no nucleation problem at all since θ, and therefore ΔT, are zero. In this case, growth can commence immediately at $\Delta T \leq 0$.

The above arguments have been developed for pure metals with or without foreign particles. They can also be applied to alloys. In this case, the Gibbs free energy is not only a function of nucleus size (r or n), but also of composition. To a first approximation, the critical size and composition would be found in this case from the conditions, $d(\Delta G)/dn = 0$ and $d(\Delta G)/dC = 0$, which define a saddle point.

2.2 Rate of Nucleus Formation

In order to calculate the number of grains nucleated within a given melt volume and time (called the nucleation rate), the simplest case will be considered. This is an ideal mixture between an ensemble, N_n, of small crystalline clusters, each of which contain n atoms, and N_l atoms of the liquid. The equilibrium distribution (solubility) of these clusters can be calculated (appendix 4) leading to the result, for n (when $N_n \ll N_l$):

$$\frac{N_n}{N_l} = \exp\left(-\frac{\Delta G_n}{k_B T}\right) \qquad [2.8]$$

Equation 2.8 and figure 2.3 show that there are always crystal clusters in a melt, although they are not necessarily stable. Their number increases with decreasing value of ΔG_n. The number of clusters is shown schematically in figure 2.3 as a density of points. If the melt is superheated, $d(\Delta G_n)/dn$ is always positive and the *equilibrium* concentration of crystal nuclei is zero. In an undercooled melt, a maximum in ΔG_n, as a function of n exists, over which clusters can 'escape' and form the flux of nuclei, I. The maximum value, ΔG_n^o (table 2.1), varies with $1/\Delta T^2$. The value of N_n^o varies according to equation 2.8 and therefore:

$$N_n^o = K_1 \exp\left(-\frac{K_2}{T\Delta T^2}\right) \qquad [2.9]$$

where K_1 and K_2 are constants. If it is assumed here that the rate of cluster formation is so high or I is so low that the equilibrium concentration of critical clusters, N_n^o/N_l, will not change i.e. the source of nucleation will not be exhausted#, the steady-state nucleation rate is given by:

$$I = K_3 N_n^o$$

$$I = K_3 N_l \exp\left(-\frac{\Delta G_n^o}{k_B T}\right) \qquad [2.10]$$

where K_3 is a constant.

However, the formation of clusters will require the transfer of atoms from the liquid to the nuclei. An activation energy, ΔG_d, for transfer through the solid/liquid interface must therefore be added to equation 2.10, giving (appendix 4):

$$I = I_0 \exp\left(-\frac{\Delta G_n^o + \Delta G_d}{k_B T}\right) \qquad [2.11]$$

where I_0 is a pre-exponential factor. This important equation contains two exponential terms. One of these varies as $-1/T\Delta T^2$ (equation 2.9), while the other varies, like the diffusion coefficient, as $-1/T$. An increase in ΔT, giving more numerous and smaller nuclei of critical size, is accompanied by a decrease in T and fewer atoms are transferred from the liquid to the nuclei. These opposing tendencies lead to a maximum in the nucleation rate at a critical temperature, T_c, which is situated somewhere between the melting point ($\Delta T = 0$) and the point where there is no longer any thermal activation ($T = 0$K). This is illustrated by figure 2.4a. Note that I would exhibit a maximum value even in the absence of the diffusion term, ΔG_d. The presence of the latter term increases the temperature at which the maximum occurs.

Since, for a unit volume of the melt, the reciprocal of the nucleation rate is time, the

---

This assumption is a crude but useful simplification. For more details, the reader is referred to J.W.Christian, *The Theory of Transformations in Metals and Alloys*, Pergamon, Oxford, 2nd Edition, 1975, p. 418.

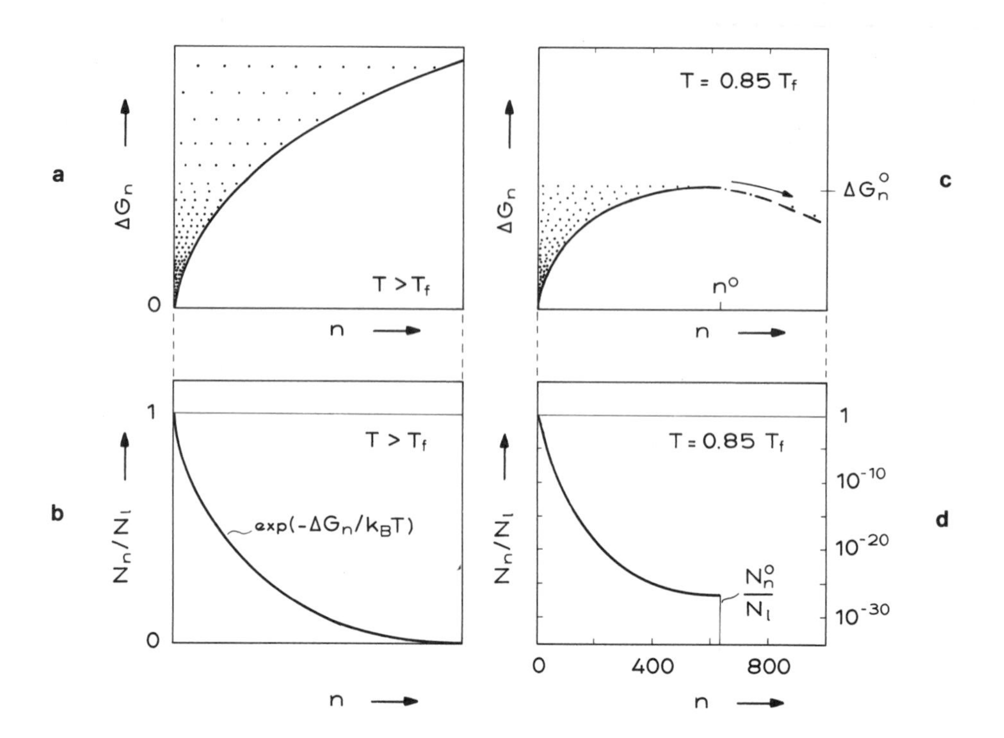

Figure 2.3 Dependence of Cluster-Size Distribution upon Temperature

Here, ΔG_n (a,c) is the free energy of a cluster containing n atoms, at two temperatures, and N_n (b,d) is the number of clusters containing n atoms, and N_l is the number of atoms in the liquid phase. There is an exponential relationship between ΔG_n and N_n. Thermal fluctuations are always creating small crystalline regions in the liquid, even at temperatures greater than the melting point (a). The number of clusters, N_n, divided by the number of atoms in the liquid, N_l, will be much smaller for large clusters (large r or large number of atoms, n), than for small ones (b). This variation in the distribution of cluster sizes is represented schematically (a) as a varying density of points. At temperatures below T_f (c), there will be a maximum, ΔG_n^o, in the free energy of the fluctuating system as is also shown in figure 2.2c. The clusters (nuclei) which reach this critical size will grow. The corresponding cluster-concentration, N_n^o/N_l, and cluster size, $n°$ (truncated minimum in figure d), are sensitive functions of the undercooling. The nucleation rate will depend on the number of clusters having the critical size, N_n^o/N_l.

I–T diagram can be easily transformed into a TTT-diagram (Fig. 2.4b) where the curve represents the beginning of the liquid-to-solid transformation. The effect of decreasing the wetting angle, θ, is felt mainly via its influence on the equilibrium concentration of nuclei and a decrease in ΔT, i.e. nucleation occurs closer to the melting point. At very high cooling rates, such as those encountered in rapid solidification processing, there may be insufficient time for the formation of even one nucleus, and a glassy

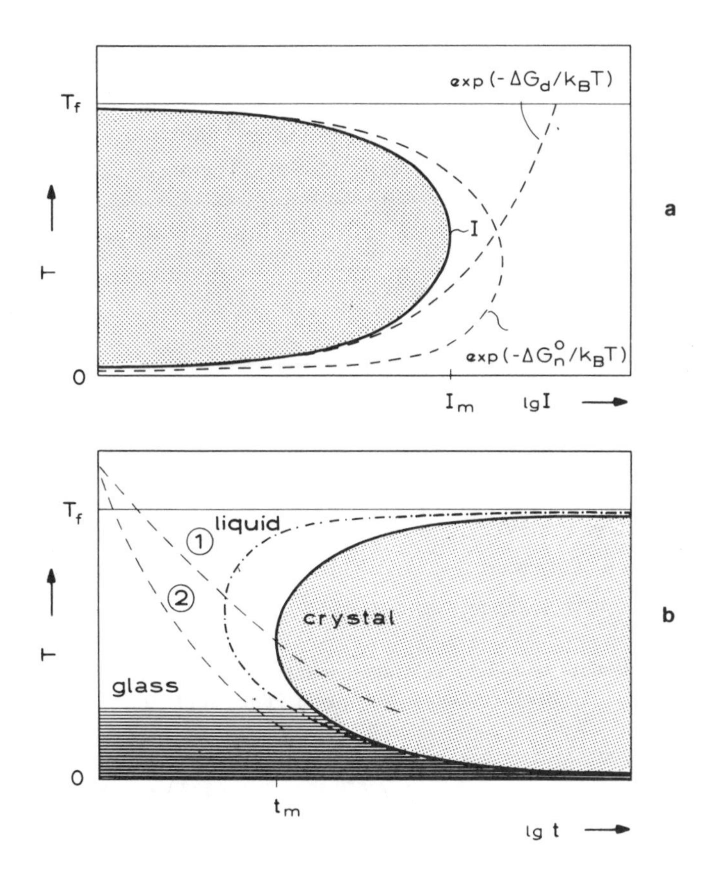

Figure 2.4 Nucleation Rate and Nucleation Time as a Function of Absolute Temperature

The overall nucleation rate, I (number of nuclei created per unit volume and time), is influenced both by the rate of cluster formation, which depends upon the nucleus concentration (N_n^o), and by the rate of atom transport to the nucleus. At low undercoolings, the energy barrier for nucleus formation is very high and the nucleation rate is very low. As the undercooling increases, the nucleus formation rate increases before again decreasing (a). The decrease in the overall nucleation rate, at large departures from the equilibrium melting point, is due to the decrease in the rate of atomic migration (diffusion) with decreasing temperature. A maximum in the nucleation rate, I_m, is the result. This information can be presented in the form of a TTT (time-temperature-transformation) diagram (b) which gives the time required for nucleation. This time is inversely proportional to the nucleation rate, and diagram (b) is therefore the inverse of diagram (a) for a given alloy volume. The diagram indicates that there is a minimum time for nucleation, t_m (proportional to $1/I_m$). This minimum value can be moved to higher temperatures and shorter times by decreasing the activation energy for nucleation, ΔG_n^o [dash-dot line in (b)]. When liquid metals are cooled by normal means, the cooling curve will generally cross the nucleation curve (curve 1). However, very high rates of heat removal (curve 2) can cause the cooling curve to miss the nucleation curve completely and an amorphous solid (hatched region, glass) is then formed via a continuous increase in viscosity (Fig. 1.1). Note that this figure relates to nucleation (start of transformation) only. The second curve of a TTT diagram which describes the end of the transformation, after growth has occurred, is not shown.

(amorphous) solid then results (cooling curve 2 in figure 2.4b).

It is interesting to calculate the effect of a slight change in ΔG_n^o, due perhaps to a change in $f(\theta)$ upon the nucleation rate. This can easily be done by approximating equation 2.11. At low values of ΔT, the $\exp[-\Delta G_d/k_BT]$ term is approximately equal to 0.01 and I_0 is approximately equal to $10^{41}m^{-3}s^{-1}$. The nucleation rate (in units of $m^{-3}s^{-1}$) therefore becomes:

$$I = 10^{39} \exp\left(-\frac{\Delta G_n^o}{k_BT}\right) \qquad [2.12]$$

A nucleation rate of one nucleus per cm^3 per second ($10^6m^{-3}s^{-1}$) occurs when the value of ($\Delta G_n^o/k_BT$) is about 76. Close to this value, changing the exponential term by a factor of two, from 50 to 100 for example, decreases the nucleation rate by a factor of 10^{22}. When $\Delta G_n^o/k_BT$ is equal to 50, 10^8 nuclei per litre of melt per microsecond are formed. If the latter term is equal to 100, only one nucleus will be formed per litre of melt over a period of 3.2 years (Fig. 2.5). This example shows that very slight changes in the solid/liquid interface energy can have striking effects.

Table 2.3 Absolute and Relative Undercoolings, Required to Give One Nucleus per Second per cm^3, as a Function of θ

θ (°)	$\Delta T/T_f$	$\Delta T(T_f = 1500K)$
180	0.33	495
90	0.23	345
60	0.13	195
40	0.064	96
20	0.017	25.5
10	0.004	6.5
5	0.001	1
0	0.0	0

Upon calculating the undercooling, for a constant value ($1/cm^3s$) of I, as a function of θ, another interesting result is revealed. This is illustrated by table 2.3 which reveals the change, in the undercooling for heterogeneous nucleation, as a function of the nucleus/substrate contact angle. If the substrate is highly dispersed, as in inoculation, the active surface area of the inoculant must also be taken into account in the pre-exponential factor (I_0), in equation 2.11. This effect is relatively small in comparison with that caused by a change in activation energy, and is therefore usually neglected.

However, the grain size will be inversely proportional to the particle density. When a fine grain size is required, it is clear that many finely dispersed particles should be introduced into the melt. More importantly, these particles should have a low interface energy when in contact with the solid which is to be nucleated. This is most likely to be so when the nuclei and the nucleated solid have similar atomic structures. In many casting situations, this is most effectively achieved by the detachment of dendrite arms by convection in the melt. This phenomenon is exploited in the continuous casting of steel by electromagnetically stirring the melt. The method is very effective since it produces nuclei (dendrite arms) which are free of any oxide film which might impair wetting. The presence of oxide films on inoculants is often a problem since they inhibit the action of the latter. For this reason, inoculation by chemical reaction or precipitation

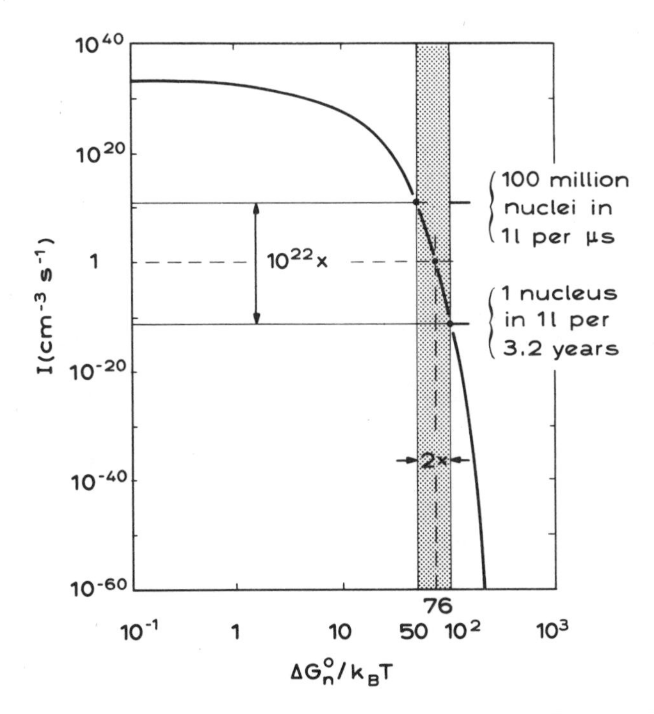

Figure 2.5 Nucleation Rate as a Function of Activation Energy, ΔG_n^o

Variations in the value of the term, $\Delta G_n^o/k_BT$, have a remarkable effect upon the rate of nucleation, I, due to the exponential relationship. If, for an observable rate of $I = 1/cm^3s$, $\Delta G_n^o/k_BT$ is changed by a factor of two, the resultant change in the nucleation rate is of the order of 10^{22}. Thus, changing the temperature or changing the value of ΔG_n^o can enormously increase or decrease the nucleation rate. The value of ΔG_n^o can be decreased by adding crystalline foreign particles which 'wet' the growing nucleus to the melt (inoculation), or by increasing the undercooling.

in the melt is favoured. Peritectic reactions are most effective in pure metals such as Al since precipitates having a higher melting point than that of the melt are formed and can promote nucleation before dissolution occurs.

2.3 Interface Structure

Once a nucleus is formed, it will continue to grow. Such growth will be limited by:

- the kinetics of atom attachment to the interface,
- capillarity,
- diffusion of heat and mass.

The relative importance of each of these factors depends upon the substance in question and upon the solidification conditions. This chapter will consider only the atomic attachment kinetics, and the other processes will be treated in chapters 3 to 5.

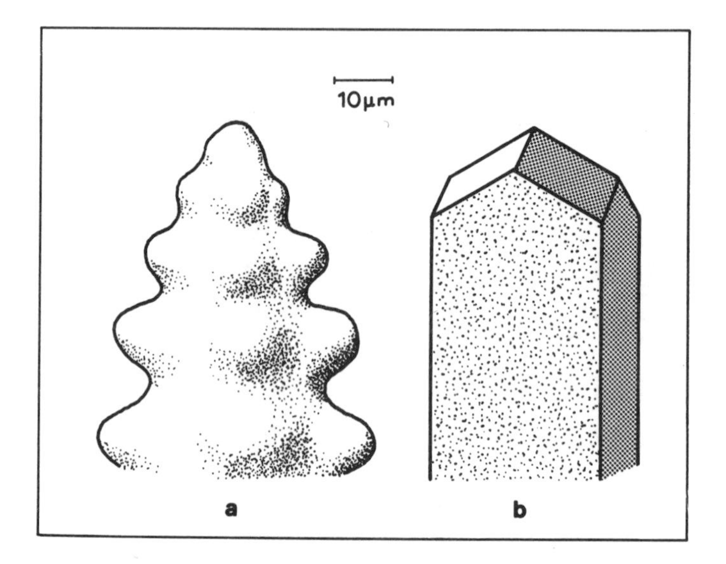

Figure 2.6 Non-Faceted and Faceted Growth Morphologies

After nucleation has occurred, further atoms must be added to the crystal in order that growth can continue. During this process, the solid/liquid interface takes on a specific structure at the atomic scale. Its nature depends upon the differences in structure and bonding between liquid and solid. During the solidification of a non-faceted material, such as a metal (a), atoms can be added easily to any point of the surface and the crystal shape is dictated mainly by the interplay of capillarity effects and diffusion (of heat and/or solute). Nevertheless, a remaining slight anisotropy in properties such as the interface energy leads to the growth of dendrite arms in specific crystallographic directions. In faceted materials, such as intermetallic compounds or minerals (b), the inherently rough, high-index planes accept added atoms readily and grow quickly. As a result, these planes disappear and the crystal remains bounded by the more slowly growing facets (low-index planes). The classes of non-faceted and faceted crystals can be distinguished on the basis of the higher entropy of fusion of the latter. This is due to the greater difference in structure and bonding between the solid and liquid phases as compared to metals, which exhibit only very small differences between the two phases.

The kinetics of atomic addition can play an important role in some substances. When the latter exhibits the 'non-faceted' growth morphology typical of a metal, it can be assumed that the kinetics of transfer of atoms from the liquid to the crystal are so rapid that they can be neglected. When the substance exhibits the faceted mode of growth typical of non-metals or intermetallic compounds, a large kinetic term may be involved. However, it is by no means certain that this term will dominate the growth process.

The classification of substances into faceted and non-faceted types is based upon their growth morphology (Fig. 2.6). Metals, and a special class of molecular compounds (plastic crystals), usually solidify with macroscopically smooth solid/liquid interfaces and exhibit no facets, despite their crystalline nature. This behaviour reflects an independence of the atomic attachment kinetics with respect to the crystal plane involved. A slight tendency to anisotropic growth remains and results from an anisotropy of the interface energy and the atomic attachment kinetics. This leads to the appearance of crystallographically determined dendrite trunk and arm directions of low-index type. On the other hand, substances exhibiting complex crystal structures and directional bonding form crystals having planar, angular surfaces (facets). Note that the faceted versus non-faceted classification also depends upon the growth process. A substance which exhibits non-faceted crystals when grown from the melt can give faceted crystals when grown from a solution or vapour.

In the present book, the classification will be applied to melt-grown crystals. This classification is of practical interest to metallurgists because of the importance of intermetallic phases and compounds in most alloys, and the large-scale industrial use of eutectic alloys (chapter 5) which contain a faceted phase as one component (e.g. Fe-C, Al-Si). From a theoretical point of view, the reason for this marked difference in morphology is worthy of mention here because of the light which it throws on the detailed structure of the solid/liquid interface at the atomic level. Finally, an understanding of atomic attachment kinetics aids the correct choice of transparent model systems which are often used in order to observe solidification phenomena directly (Fig. 4.3 and 4.16).

The growth rate of a crystal depends upon the net difference between the rates of attachment and detachment of atoms at the interface (appendix 5). The rate of attachment depends upon the rate of diffusion in the liquid, while the rate of detachment depends on the number of nearest neighbours binding the atom to the interface. The number of nearest neighbours depends upon the crystal face considered, i.e. upon the surface roughness at the atomic scale (number of unsaturated bonds). This is the simplest possible situation. In general, reorientation of a complicated molecule in the melt, surface diffusion, and other steps may be required.

Consider the essentially flat interface of a simple cubic crystal. Here, an atom in the bulk crystal has six nearest neighbours represented by the six faces of a cube (Fig. 2.7). There are five different positions at the interface, characterised by the number of nearest neighbours (1 - 5). In an undercooled system, where the crystal has a lower free energy, it is evident that an atom in position 5 will have a very much higher probability of remaining in the crystal than will an atom in position 1. In order to incorporate atom 1 into the crystal, a very large difference must exist between the force binding it to the crystal, and the force binding it to the liquid. In order to create such a difference, a large undercooling of the melt is required.

An *atomically flat* interface (Fig. 2.8a) will maximise the bonding between atoms in the crystal and those in the interface. Thus, such an interface will expose few bonds to atoms arriving via diffusion through the liquid. Such a crystal has a tendency to close up any gap in its solid/liquid interface at the atomic scale. This leads to crystals which are *faceted* at the microscopic scale and usually exhibit high undercoolings.

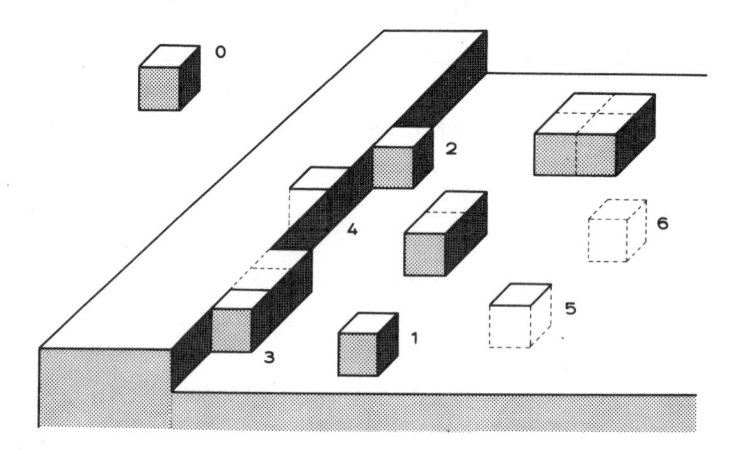

Figure 2.7 Variation in Bond Number at the Solid/Liquid Interface of a Simple-Cubic Crystal

In order to understand the two types of growth shown in figure 2.6, the various ways in which an atom can be adsorbed at the solid/liquid interface have to be considered. Growth is determined by the probability that an atom will reach the interface and remain adsorbed there until it has been fully incorporated into the crystal. This probability increases with an increasing number of nearest neighbours in the crystal. The possible arrangements of atoms on the crystal interface are indicated here, where the numbers specify the number of neighbouring atoms in the crystal (when the crystal coordination number is 6 as in a simple cubic crystal). The atoms of the liquid phase are not shown here. A special role is played by type 3 atoms in the growth of faceted crystals because, having three bonds, they can be considered to be situated half in the solid and half in the liquid. A likely growth sequence would be: addition of type 3 atoms until a row is complete; addition of a type 2 atom to start a new row; and so on until a layer is complete. Thereupon, nucleation of a new layer by the addition of a type 1 atom would be necessary. This is an unfavourable process and requires a very high undercooling. Therefore, other processes will play a role (Fig. 2.10).

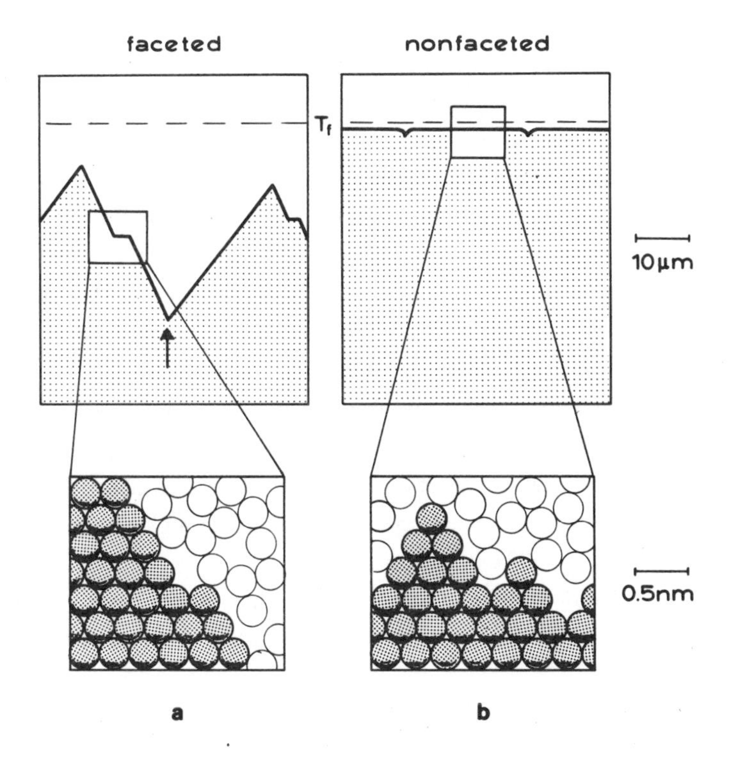

Figure 2.8 Form of Faceted (a) and Non-Faceted (b) Interfaces

A transparent organic substance, when observed under the microscope during directional solidification (upwards growth, bottom cool, top hot), can exhibit either of the growth forms shown in the upper diagrams. It is important to note that, during growth, a faceted interface (a) is jagged and faceted at the microscopic scale (upper diagram), but smooth at the atomic scale (lower diagram). On the other hand, a non-faceted interface (b) can be microscopically flat with some slight depressions due to grain boundaries (upper diagram), while at the atomic scale it is rough and uneven (lower diagram). This roughness causes the attachment of atoms to be easy and largely independent of the crystal orientation. Note also that the interface of a non-faceted material will grow at a temperature which is close to the melting point, T_f, while the interface of a faceted material might have a very high local undercooling. Such a point (arrow) is a re-entrant corner (Fig. 2.10), and is associated with an increased number of nearest neighbours. Thus, growth will tend to spread from here.

An *atomically rough* interface (Fig. 2.8b) always exposes a lot of favourable sites for the attachment of atoms from the liquid. Such an interface tends to remain rough and leads to smooth crystals which are *non-faceted* at the microscopic scale and exhibit low kinetic undercoolings.

Generally, it can be said that the greater the difference in structure and bonding between the solid and liquid phases, the narrower is the transition region over which these differences have to be accommodated. A sharp transition (i.e. an atomically flat, microscopically faceted interface) exhibits little tendency to incorporate newly arriving atoms into the crystal. Hence growth is more difficult and requires an additional (kinetic) undercooling. Also, because high-index crystallographic planes tend to be inherently rough and to contain many steps, a marked anisotropy in growth rates can develop with respect to low-index planes which are atomically flat. This, in turn, leads to the disappearance of the higher index planes, due to their more rapid growth, and leads to a characteristic crystal form which is bounded by the slowest growing faces (Fig. 2.9).

Metals, on the other hand, have a similar structure, density, and bonding in both the liquid and solid states. Hence, the transition from one phase to the other at the interface is very gradual and the interface becomes rough and diffuse (i.e. it exposes many suitable growth steps to atoms arriving from the melt).

The melting entropy is a convenient criterion for predicting this aspect of the crystallisation behaviour. Values of α (= $\Delta S_f/R$) which are less than ~2 can be taken to imply a tendency to non-faceted crystal growth, while higher α-values imply that faceted growth forms will be produced.

In table 2.4, a list of various substances is presented which compares their growth forms and their dimensionless entropies for crystallisation from a variety of media. The

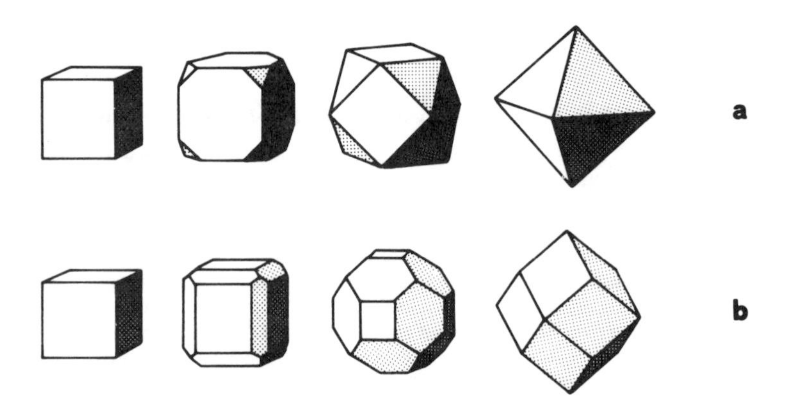

Figure 2.9 Development of Faceted Crystal Growth Morphologies

A growing cubic crystal which is originally bounded only by (100) planes (left hand side) will change its shape to an octahedral form (bounded by (111)) when the (100) planes grow more quickly than the (111) planes (a). Often, impurities change the growth behaviour of specific planes and this results in the appearance of different growth forms for the same crystal structure. If the (110) planes are the slowest-growing, this will lead to a rhombohedral dodecahedron (b). The slowest-growing planes (usually of low-index type) always dictate the growth habit of the crystal. The resultant minimum growth rate form is not the same as the equilibrium (non-growing) form, which is governed by minimisation of the total surface energy.

Table 2.4 Growth Morphologies and Crystallisation Entropies

Dimensionless Entropy ($\Delta S_f/R$)	Supersaturated Substance	Phase	Morphology
~1	metals	melt	non-faceted
~1	'plastic' crystals	melt	non-faceted
2-3	semiconductors	solution	nf/faceted
2-3	semimetals	solution	nf/faceted
~6	molecular crystals	solution	faceted
~10	metals	vapour	faceted
~20	complex molecules	melt	faceted
~100	polymers	melt	faceted

fact that metals and plastic crystals# fall into the same group is of great value in the study of solidification. The interior of metals can obviously not be observed using visible light, whereas the transparent, low-melting-point organic crystals can easily be studied in this way. Microscopic study of their interface morphology during growth has provided a good deal of useful information and continues to be an attractive research tool.

A high entropy of fusion increases the disparity in growth rates between the low-index planes and the faster-growing high-index planes. Table 2.5 presents some

Table 2.5 Growth Rate Coefficient, K_{hkl}, as a Function of Crystallisation Entropy for a Simple-Cubic Two-Dimensional Crystal (Jackson, 1968)

Dimensionless Entropy, α	K_{100}	K_{111}	K_{111}/K_{100}
1	0.2	0.1	0.5
5	0.007	0.01	1.4
10	0.000005	0.0001	20.0

Plastic crystals are a special class of molecular substance (organic or inorganic) whose molecular asymmetry is destroyed by rotational motion. That is, the molecules behave like spheres having surfaces defined by the envelope of all of the possible arrangements of the molecule's extremities. Since these 'spheres' no longer exhibit asymmetry or directional bonding, they can arrange themselves into simple crystal structures; particularly cubic ones, and many of their properties are analogous to those of metals. For example, the name, 'plastic', refers to their high malleability. However, they were first recognised as a distinct class due to their low entropies of fusion. This also indicates that their crystallisation behaviour will be analogous to that of metals.

numerical estimates for growth rate coefficients, K_{hkl}, as a function of the dimensionless entropy and crystallographic plane for a simple-cubic, two-dimensional crystal (appendix 5), used as a rough approximation of the behaviour of a real three-dimensional crystal. At small undercoolings, with respect to T_f, the actual growth rate of the faces can be obtained from the expression:

$$V = K_{hkl} \Delta T \quad [2.13]$$

This expression corresponds to a limited range of equation A5.7, and assumes that a simple growth mechanism is operating. It can already be seen from this simple example that a very marked anisotropy results in the case of a high melting entropy. That is, the ratio of $V_{[111]}$ to $V_{[100]}$ is increased, and the absolute growth rates are markedly decreased. As a consequence, a crystal of a substance having a large value of α will be bounded by (100) planes.

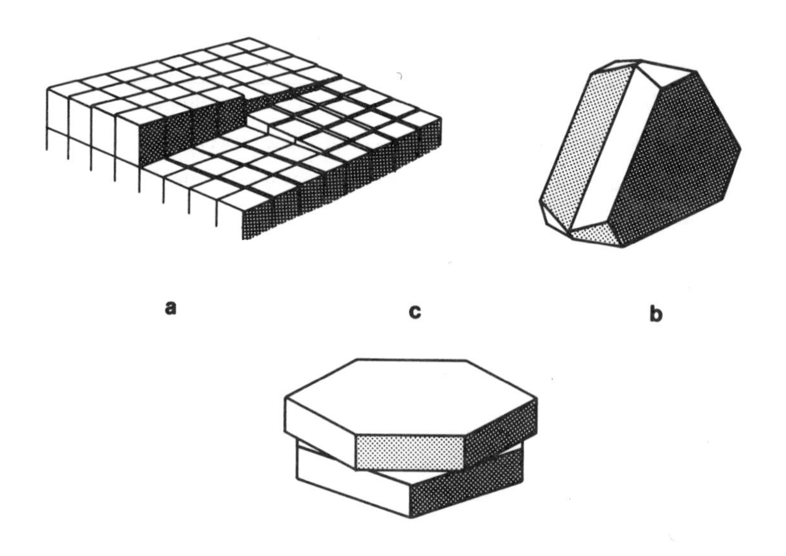

Figure 2.10 Repeatable Growth Defects in Faceted Crystals

As pointed out with reference to figure 2.7, steps in rows of atoms are favoured growth sites, but can easily be eliminated by the very growth which they promote. Several types of defect have been shown to provide steps and also to be impossible to eliminate by growth. These are the emergent screw dislocation (a) [F.C.Frank, *Discussions of the Faraday Society* **5** (1949) 48], which leads to the establishment of a spiral ramp; the re-entrant corner due to twinning (b) [D.R.Hamilton, R.G.Seidensticker, *Journal of Applied Physics* **31** (1960) 1165] which acts as a macroscopic step; and the twist (rotation) boundary which also provides effective steps (c). Depending upon the type of defect present, the faceted crystal can exhibit various morphologies: needles in the case of line defects (a), or plates in the case of planar defects - e.g. Si in Al-Si alloys (b) or graphite in cast iron (c) [I.Minkoff, in *The Solidification of Metals*, Iron and Steel Institute Publication 110, London, 1968]. The latter two defects are important in understanding the growth of irregular eutectic microstructures.

The observation of primary crystals growing from melts of various composition often permits the relative growth rates of different crystal faces to be deduced.

The widest range of morphologies arising from differences in the growth rates of various faces is probably that exhibited by hexagonal ice crystals growing from the vapour e.g. in the form of snowflakes. An example of more practical importance to metallurgists is that of cast iron. The faceted phase (graphite) grows at a different rate in different directions. When $V_{[0001]}$ is less than $V_{[1010]}$, plates delimited by (0001) planes appear, and the morphology is referred to as 'flake graphite'. When $V_{[1010]}$ is less than $V_{[0001]}$, hexagonal prisms form. Often, this is not immediately evident because the prisms tend to grow side by side in radial direction and the morphology is then called 'spherulitic'. Changes in the growth behaviour of graphite are obtained in practice by controlling the trace element concentration. For example, the presence of S leads to the appearance of flake graphite while the presence of Mg or Ce results in the formation of nodular cast iron.

Due to the greater difficulties experienced in attaching atoms to the surfaces of substances possessing a high entropy of fusion, surface defects are of particular importance in this case. Evidently, such defects do not greatly facilitate growth if the attachment of the atoms easily removes them. Therefore, only those defects which cannot be eliminated by growth are effective (Fig. 2.10). These include screw dislocations which emerge at the growth surface, twin boundaries, and rotation boundaries. Each of these supplies re-entrant corners (steps) which locally increase the number of bonds which an attached atom makes with the crystal (Fig. 2.7). This reduces the kinetic undercooling and leads to highly anisotropic growth. As a result, the morphology becomes sheet-like (planar defect in b and c) or whisker-like (line defect in a). This phenomenon has a marked effect upon the growth behaviour of the most widely used eutectic alloys, Fe-C and Al-Si (chapter 5).

Finally, it should be noted that under a sufficiently high driving force (e.g. high undercooling) the atomic interface of even a high melting entropy phase may become rough. Its growth behaviour then becomes more isotropic.

Bibliography

Nucleation Theory (pure substances)

M.Volmer, A.Weber, Zeitschrift für Physikalische Chemie **119** (1926) 277.

R.Becker, W.Döring, Annalen der Physik **24** (1935) 719.

D.Turnbull, J.C.Fisher, Journal of Chemical Physics **17** (1949) 71.

D.R.Uhlmann, B.Chalmers, Industrial and Engineering Chemistry **57** (1965) 19.

Nucleation Theory (alloys)

B.Cantor, R.D.Doherty, Acta Metallurgica **27** (1979) 33.

C.V.Thompson, F.Spaepen, Acta Metallurgica **31** (1983) 2021.

K.N.Ishihara, M.Maeda, P.H.Shingu, Acta Metallurgica **33** (1985) 2113.

Nucleation Experiments

D.Turnbull, R.E.Cech, Journal of Applied Physics **21** (1950) 804.

M.E.Glicksman, W.J.Childs, Acta Metallurgica **10** (1962) 925.

P.G.Boswell, G.A.Chadwick, Acta Metallurgica **28** (1980) 209.

J.H.Perepezko, J.S.Paik, in Proceedings of the Materials Research Society Symposium on *Rapidly Solidified Amorphous and Crystalline Alloys* (Edited by B.H.Kear and B.C.Giessen), North-Holland, New York, 1982.

J.H.Perepezko, Materials Science and Engineering **65** (1984) 125.

Inoculation

A.Cibula, *Grain Control*, University of Sussex, Institution of Metallurgists, London, 1969, p. 22.

I.Maxwell, A.Hellawell, Acta Metallurgica **23** (1975) 229.

T.W.Clyne, M.H.Robert, Metals Technology **7** (1980) 177.

J.H.Perepezko, S.E.LeBeau, in Proceedings of the 2nd International Symposium on *Al-Transformation Technology and its Applications*, Buenos Aires, August, 1981.

M.K.Hoffmeyer, J.H.Perepezko, Scripta Metallurgica **23** (1989) 315.

Solid/Liquid Interface Structure

K.A.Jackson, in *Liquid Metals and Solidification*, American Society for Metals, Cleveland, 1958, p. 174.

K.A.Jackson, Journal of Crystal Growth **3/4** (1968) 507.

F.Spaepen, Acta Metallurgica **23** (1975) 729.

G.H.Gilmer, K.A.Jackson, in *Crystal Growth and Materials* (Edited by E.Kaldis and H.J.Scheel), North-Holland, 1977.

D.Camel, G.Lesoult, N.Eustathopoulos, Journal of Crystal Growth **53** (1981) 327.

M.Elwenspoek, Journal of Crystal Growth **78** (1986) 353.

Crystal Growth Form

H.E.Cline, J.Walter, Metallurgical Transactions **1** (1970) 2907.

J.E.Walter, H.E.Cline, Metallurgical Transactions **4** (1973) 1775.

H.E.Cline, Journal of Crystal Growth **51** (1981) 97.

B.Lux, M.Grages, W.Kurz, Praktische Metallographie **5** (1968) 567.

I.Minkoff, S.Myron, Philosophical Magazine **19** (1969) 379.
H.Fredriksson, M.Hillert, N.Lange, Journal of the Institute of Metals **101** (1973) 285.

Exercises

2.1 Derive equation 2.1.

2.2 Compare the dA/dv ratio of spheres and cylinders with equation 1.5.

2.3 What is the meaning of the expression, $d(\Delta G)/dr = 0$, in terms of the forces acting upon the critical nucleus?

2.4 Develop an equation for ΔG, as a function of n, which is analogous to equation 2.4, and calculate ΔG_n° and n°.

2.5 Derive equation 2.5. What approximations are made? Why must caution be exercised when applying it to cases involving high undercooling?

2.6 For small undercoolings, the nucleation rate given by equation 2.11 can be written in the form, $I = K_1\exp[-K_2/T\Delta T^2]$. (a) Assuming heterogeneous nucleation, explain the origin of the term, ΔT^2, and give expressions for K_1 and K_2. (b) Determine by how much K_1 or K_2 must change in order to double the value of ΔT required to produce one nucleus per cubic metre per second when K_1 is initially equal to $10^{39}m^{-3}s^{-1}$.

2.7 In their classic experiments, D.Turnbull and R.E.Cech [*Journal of Applied Physics* **21** (1950) 804] divided a melt up into droplets which were only a few microns in diameter in order to measure the temperature at which homogeneous nucleation seemed to occur. Why did they use this method?

2.8 By measuring the undercooling required for homogeneous nucleation, it is possible to determine the solid/liquid interface energy, which is difficult to measure using other methods. Develop the equations needed for this, with the aid of equation 2.11. [In order to learn more about this technique, the reader should consult the papers of D.Turnbull: Journal of Applied Physics **21** (1950) 1022 and J.H.Perepezko et al. in *Solidification and Casting of Metals*, The Metals Society, London, 1979, p. 169].

2.9 Show that n° and ΔG° are functions of θ but r° is not. Why is this so? (See appendix 3, figure A3.7).

2.10 Determine the curvature and the melting point of a pure iron crystal in a wetted conical pore in the crucible surface ($\theta = 30°$), as illustrated below.

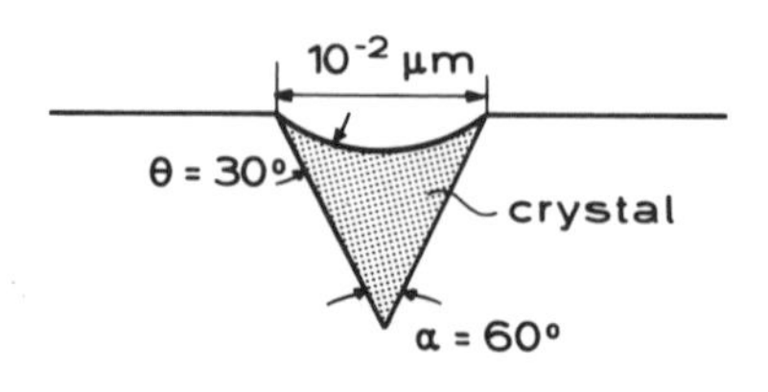

2.11 What will happen if one stirs 0.1wt% of Ti powder into a pure Al melt at 700°C just before casting? (Hint: consult the phase diagram). [For more details, see T.W.Clyne, M.H.Robert, *Metals Technology* **7** (1980) 177].

2.12 Water and Bi expand during solidification. From this fact alone, one can make predictions concerning the entropy of melting, and therefore the crystallisation behaviour, in terms of the faceted/non-faceted classification. What reasoning would you employ?

2.13 Show graphically in two dimensions why rapidly growing planes disappear during crystal growth and leave the slowest-growing ones.

2.14 For a given substance, what is the difference in the shape of an equilibrium crystal ($V = 0$), when σ has a pronounced minimum for $\{111\}$, and the shape arising from growth ($V > 0$) when V is a minimum for $\langle 100 \rangle$?

2.15 Explain why an atomically smooth interface is usually microscopically rough when growing in a temperature gradient (Fig. 2.8a - top).

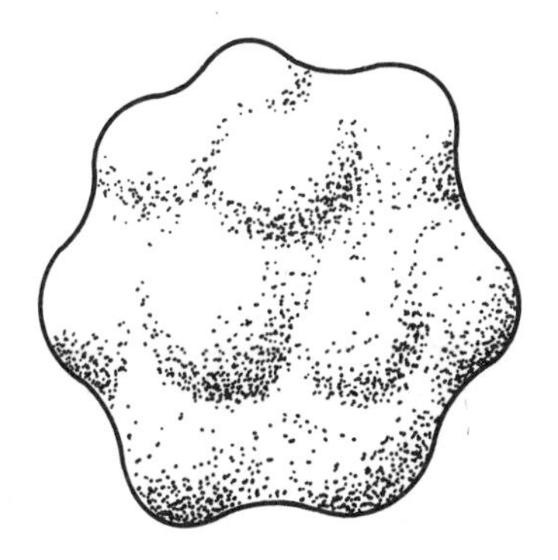

CHAPTER THREE

MORPHOLOGICAL INSTABILITY OF A SOLID/LIQUID INTERFACE

Classical thermodynamic definitions of stability are inapplicable to the determination of the morphology of a growing interface, and current extensions of equilibrium thermodynamics have not yet furnished a fully acceptable alternative. Meanwhile, in order to proceed with theoretical analyses of growth morphologies, it has been found necessary to use heuristically-based stability criteria. The simplest assumption made is that the morphology which appears is the one which has the maximum growth rate or minimum undercooling. An alternative method is the use of stability arguments. These involve perturbing (mathematically) the growing solid/liquid interface morphology in order to determine whether it is likely to change into another one. The interface is then said to be morphologically unstable if the perturbation is amplified with the passage of time and to be morphologically stable if it is damped out (Fig 3.1).

3.1 Interface Instability of Pure Substances

The conditions which lead to instability can easily be understood in the case of a pure substance. Figure 3.2 illustrates, in a schematic manner, the development of a perturbation during columnar and equiaxed growth (see also figure 1.7).

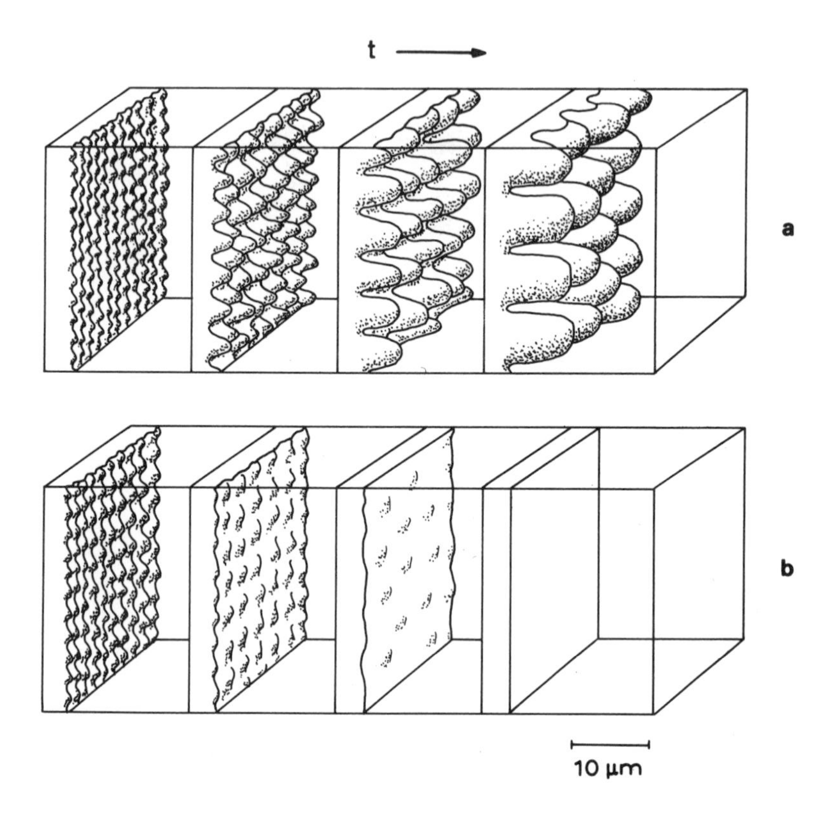

Figure 3.1 Initial Evolution of an Unstable (a) or Stable (b) Interface

Such an interface might be seen during the microscopic observation of transparent non-faceted organic substances. During growth, any interface will be subject to random disturbances caused by insoluble particles, temperature fluctuations, or grain boundaries. A stable interface is distinguished from an unstable interface by its response to such disturbances. It is imagined here that the interface is initially slightly distorted by a spatially regular disturbance. If the distorted interface is unstable (a), the projections may find themselves in a more advantageous situation for growth and therefore increase in prominence. In the case of a stable interface (b), the perturbations will be unfavourably situated and tend to disappear. During the casting of alloys, the solid/liquid interface is usually unstable. A stable interface is only obtained in special cases such as columnar solidification of pure metals (Fig. 1.7a) or directional solidification of alloys in a Bridgman-type furnace (Fig. 1.4a) under a sufficiently high temperature gradient, G. The indicated scale is typical for alloys under normal casting conditions.

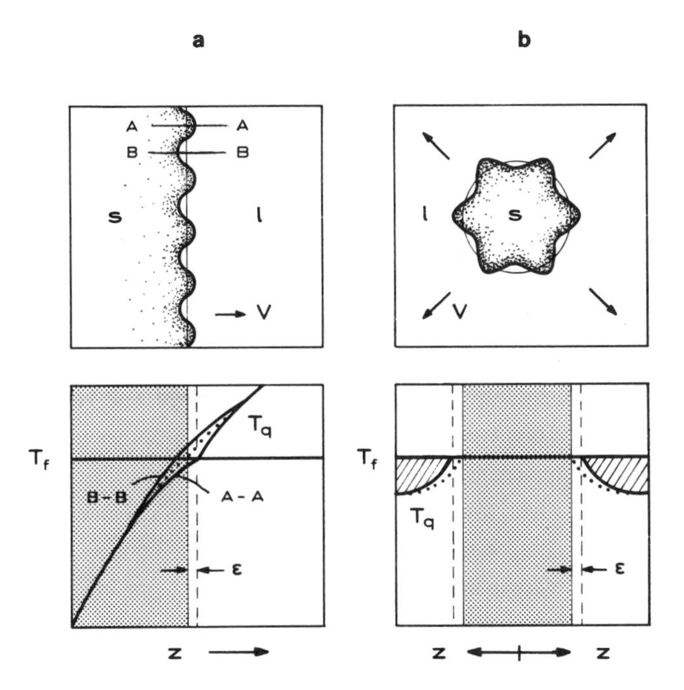

Figure 3.2 Columnar and Equiaxed Solidification of a Pure Substance

In a pure substance, stability depends on the direction of heat flow. In directional solidification, as in the columnar zone of a casting, the liquid temperature always increases ahead of the interface (a). Therefore, the heat flow direction is opposite to that of solidification. When a perturbation of amplitude,ε, forms at an initially smooth interface, the temperature gradient in the liquid increases while the gradient in the solid decreases (compare full and dotted lines along section, A-A). Since the heat flux is proportional to the gradient, more heat then flows into the tip of the perturbation and less flows out of it into the solid. As a result, the perturbation melts back and the planar interface is stabilised. In equiaxed solidification, the opposite situation is found (b). Here, the free crystals grow into an undercooled melt (cross-hatched region) and the latent heat produced during growth also flows down the negative temperature gradient in the liquid. A perturbation which forms on the sphere will make this gradient steeper (full line compared to the dotted line) and permit the tip to reject more heat. As a result, the local growth rate is increased and the interface is always morphologically unstable.

During columnar growth of a pure substance (Fig. 3.2a), the temperature given by the heat flux, T_q, increases in the z-direction (i.e. G is positive). The interface will be found at the isotherm where the temperature, T_q, imposed by the heat flux is equal to the equilibrium melting point, T_f#

If an interface perturbation is to remain at the melting point over its entire surface,

In general, a growing interface will always require some (kinetic) undercooling below the equilibrium melting point in advance (figure 2.8 and equation 2.13). In faceted substances, this undercooling may be appreciable (greater than 1K). In non-faceted substances, such as metals, its value will be negligibly small in most cases. Its effect can generally be ignored for these materials under normal solidification conditions.

and curvature effects on T_f are neglected, the temperature field (T_q) must be deformed so that the temperature gradient in the liquid at the tip (along line A-A) increases while the temperature gradient in the solid decreases. According to Fourier's first law this means that the heat flux from the liquid to the tip increases and that the flux in the solid decreases. Therefore, more heat will flow into the tip and less will flow out of it. Meanwhile, the reverse situation occurs in the depressions (line B-B). As a result, the perturbation tends to be damped out. Thus, the interface of a pure substance during columnar growth will always be stable.

Turning now to equiaxed solidification, the situation is completely different (Fig. 3.2b). In this case, the heat flux does not reach the mould wall via the solid but through the melt. Therefore, the melt must be undercooled in order to establish the necessary temperature gradient at the solid/liquid interface. The temperature gradient in the liquid will be negative, while the gradient in the solid is essentially zero. A perturbation will sense a higher negative gradient at its tip, leading to an increased heat flux, and a resultant increase in the growth rate of the tip. Thus, the interface of a pure substance (with negligible kinetic undercooling) will always be unstable under equiaxed solidification conditions. The result is that equiaxed grains in a pure metal always adopt a dendritic morphology (Fig. 1.7b). Because segregation is absent, the dendritic form will not be detectable in a cast, pure metal. Fortunately, the characteristics of dendritic growth in this case can be very closely observed in pure organic substances or in pure water.

It can be concluded that the solid/liquid interface of a pure metal will always be stable if the temperature gradient is positive, and unstable if the gradient is negative. Perturbations will grow if the temperature gradient at the interface, given by the heat flux, $G = dT_q/dz$, obeys the relationship:

$$G < 0 \qquad \text{(pure metals)} \qquad [3.1]$$

3.2 Solute Pile-Up at a Planar Solid/Liquid Interface

In alloys, the criterion for stable/unstable behaviour is more complicated because the local equilibrium melting point can vary along the solid/liquid interface. During the solidification of an alloy, solute will pile up ahead of the interface due to the smaller solubility of the solid when the distribution coefficient is less than unity (Fig. 1.11). The excess solute rejected from the solid will accumulate in an enriched boundary layer ahead of the interface. The fully developed diffusion boundary layer is illustrated in figure 3.3. It is established during a transient period before steady-state (time-independent) growth begins (Fig. 6.1). In this steady-state situation, where all of the concentrations are constant with respect to a reference frame moving with the solid/liquid interface, the solid forms at the solidus temperature. Therefore, the composition of the solid is equal to that, C_0, of the liquid far ahead of the interface (where the effect of the solute pile-up has not yet been felt). The solute concentration in the boundary layer decreases exponentially, from C_0/k to C_0, according to (appendix 2):

$$C_l = C_0 + \Delta C_0 \exp\left(-\frac{Vz}{D}\right) \quad [3.2]$$

Mathematically, the thickness, δ_c, of the boundary layer is infinite. However, for practical purposes it can be taken to be equal to the 'equivalent boundary layer' (appendix 2):

$$\delta_c = \frac{2D}{V} \quad [3.3]$$

This length is equal to the base-length of a right-angled triangle having a height which is

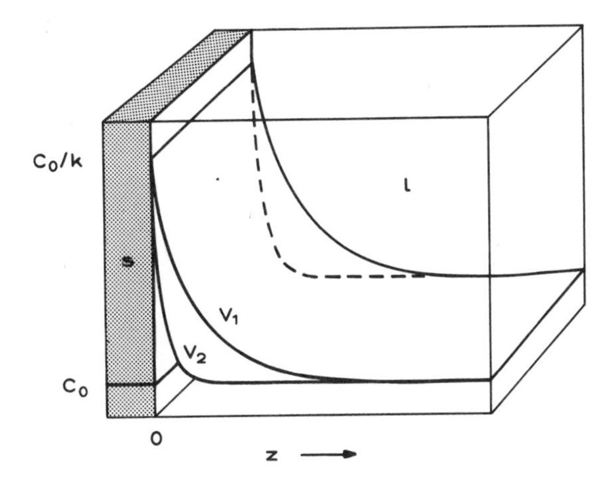

Figure 3.3 Steady-State Boundary Layer at a Planar Solid/Liquid Interface

In the case of an alloy, the stability arguments are similar to those advanced in the caption to figure 3.2, except that solute diffusion as well as heat flow effects must be considered. The former will be considered here. Firstly, as the interface advances, solute is rejected if the solubility of the solute in the solid is lower than that in the liquid (k less than unity). When the interface has been advancing for some time, the concentration distribution becomes time-independent (steady-state). If a planar interface is advancing under steady-state conditions, then the solute concentration in the solid is the same as that of the original melt (appendix 2). Under these circumstances, the concentration in the liquid decreases exponentially from the maximum composition, C_0/k, at the interface to the original composition, C_0, far from the interface. In general, the rate of rejection of solute at the interface is proportional to the growth rate. The rejected solute must be carried away by diffusion down the interfacial concentration gradient, and this therefore becomes steeper with increasing growth rate. In the figure, the boundary layers for two growth rates, $V_2 > V_1$, are shown.

equal to the excess solute concentration at the interface, and an area which is the same as that under the exponential curve. Equation 3.3 reveals that the equivalent boundary layer thickness is inversely proportional to the growth rate.

A simple flux balance shows that an interface of area, A, rejects J_s atoms per second:

$$J_s = A\left(\frac{dz'}{dt}\right)(C_l^* - C_s^*) \qquad [3.4]$$

where the term, $A(dz'/dt)$, represents the volume of liquid which is transformed to solid per unit time, and the second term represents the difference in the solute concentrations in the liquid and the solid at the interface. Under steady-state conditions, the resultant flux of rejected solute has to be balanced by an equal flux which takes solute away from the interface by diffusion. The flux in the liquid for a given cross-section, A, is:

$$J_l = -AD\left(\frac{dC_l}{dz}\right) \qquad [3.5]$$

Equating the fluxes and noting that, in the steady state, $C_l^* = C_0/k$, gives the important flux balance:

$$G_c = \left(\frac{dC_l}{dz}\right)_{z=0} = -\left(\frac{V}{D}\right)\Delta C_0 \qquad [3.6]$$

3.3 Interface Instability of Alloys

It is seen from the above that, during the solidification of an alloy, there is a substantial change in the concentration ahead of the interface. This change will affect the local equilibrium solidification temperature, T_l, of the liquid, which is related to the composition by:

$$T_l(C_0) - T_l = m(C_0 - C_l) \qquad [3.7]$$

where $T_l(C_0)$ is the liquidus temperature corresponding to the initial alloy composition. This relationship is shown in figure 3.4 and indicates that the concentration boundary layer can be converted, using the phase diagram, into a liquidus-temperature boundary layer. The liquidus temperature increases with increasing z, when the value of k is less than unity, because the value of m is then negative. It represents the local equilibrium temperature for the solidification of a corresponding volume element of the melt. In order to investigate stability, it is also necessary to determine the temperature, T_q, imposed by the heat flux. Both temperatures must be equal at the interface (see previous footnote). In the steady state growth of a planar interface, this will correspond to the solidus temperature for the composition, C_0, as shown in figure 3.4.

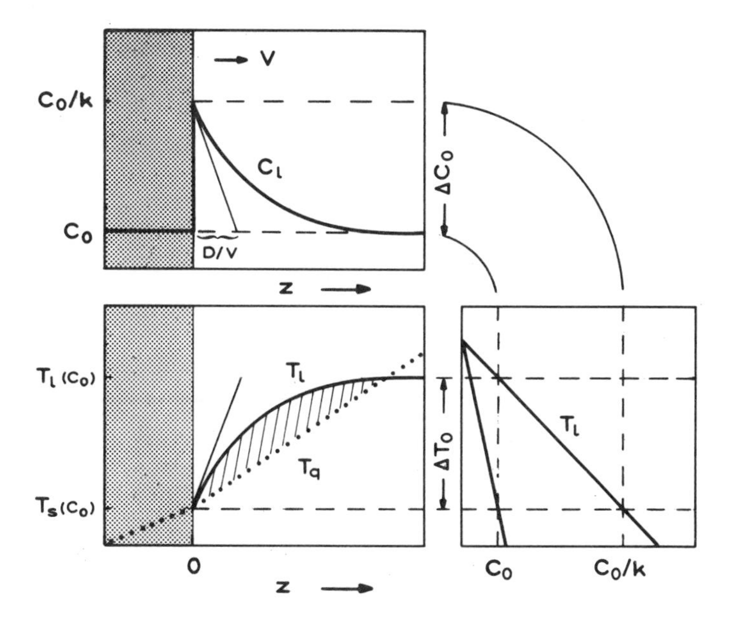

Figure 3.4 Constitutional Undercooling in Alloys

The steady-state diffusion boundary layer, shown in figure 3.3, is reproduced here (upper diagram) for a given growth rate. As the liquid concentration, C_l, decreases with distance, z, the liquidus temperature, T_l (i.e. the melting point), of the alloy will increase as indicated by the phase diagram. This means that if small volumes of liquid at various distances ahead of the solid/liquid interface were extracted by some means and solidified, their equilibrium freezing points would vary with position in the manner described by the heavy curve in the lower left-hand diagram. However, each volume element finds itself at a temperature, T_q, which is imposed by the temperature gradient arising from the heat flow occurring in the casting. Since, at the solid/liquid interface (z=0), T_q must be less than or equal to T_s in order to drive the atomic addition mechanism, there may exist a volume of liquid which is undercooled when the gradient of T_q is less than the gradient of T_l. This (cross-hatched) region is called the zone of *constitutional undercooling*. There exists a driving force for the development of perturbations in this volume as in the cross-hatched region of figure 3.2b.

Depending upon the temperature gradient,

$$G = \left(\frac{dT_q}{dz}\right)_{z=0} \qquad [3.8]$$

in the liquid at the solid/liquid interface (which is imposed by the external heat flux) there may, or may not, exist a zone of constitutional undercooling (Fig. 3.5). This zone is defined to be that volume of melt ahead of the interface within which the actual temperature, T_q, is lower than the local equilibrium solidification temperature, T_l. The melt in this zone is thus undercooled, i.e., in a metastable state. It is quickly seen that the condition required for the existence of such a constitutionally undercooled zone is

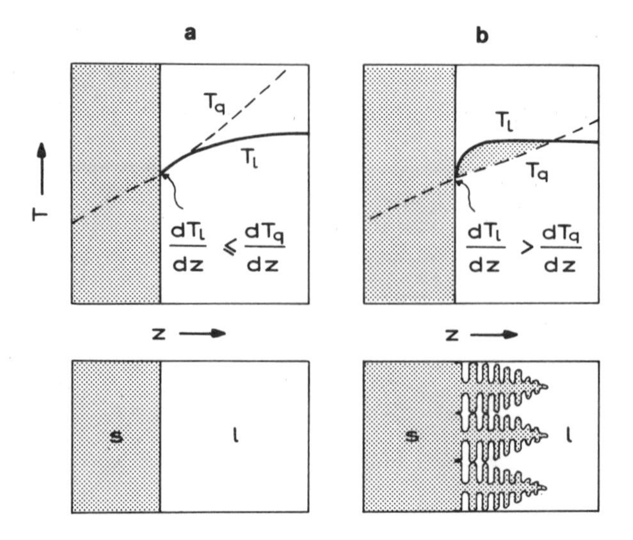

Figure 3.5 Condition for Constitutional Undercooling at the Solid/Liquid Interface, and the Resultant Structures

When the temperature gradient due to the heat flux is greater than the liquidus temperature gradient at the solid/liquid interface, the latter is stable (a). On the other hand, it can be seen that a driving force for interface change will be present whenever the slope of the local melting point curve (liquidus temperature) at the interface is greater than the slope of the actual temperature distribution. This is easily understood since the undercooling encountered by the tip of a perturbation advancing into the melt increases and therefore a planar interface is unstable (b). Note that the temperature profile in (b) is only hypothetical; after the dendritic microstructure shown in the lower figure has developed, the region of constitutional undercooling is largely eliminated. Only a much smaller undercooling remains at the tips of the dendrites as is shown in figure 4.7c.

that the temperature gradient, G, at the interface in the liquid should be lower than the gradient of liquidus temperature change in the melt. The latter gradient is obtained by multiplying the concentration gradient, G_C, by the liquidus slope, m. Therefore, the interface is constitutionally undercooled when:

$$G < mG_c \qquad \text{(alloys)} \qquad [3.9]$$

As before, consider the behaviour of a perturbation arising during directional solidification. Such a protuberance at the solid/liquid interface will increase the local temperature gradient in the melt. In the case of a pure melt (Fig. 3.2), this led to the disappearance of the perturbation. However, in an alloy melt (Fig. 3.4), the local concentration gradient will also become steeper and consequently the local gradient of the liquidus temperature will increase. Hence, the region of constitutional supercooling will tend to be preserved.

In the introduction to this chapter, it was stated that equilibrium thermodynamics principles could not properly be applied to solidification. However, it is interesting to see how far the use of such classical methods can be pursued in this non-equilibrium situation. In general, growth rates and undercoolings are found to be closely related by functions whose form depends upon the process which controls growth (atomic attachment, mass diffusion, or thermal diffusion). In each case, the growth rate increases with increasing undercooling. An interface perturbation can be imagined to experience a driving force, f', for accelerated growth which is given by the negative value of the first derivative of the Gibbs free energy with respect to distance:

$$f' = -\frac{d(\Delta G)}{dz} \quad [3.10]$$

For small undercoolings, $\Delta G = -\Delta S_f \Delta T$ and ΔS_f = constant. Thus,

$$f' = \Delta S_f \left(\frac{d\Delta T}{dz}\right) = \Delta S_f \phi \quad [3.11]$$

where $\Delta T = T_l - T_q$, and ϕ is the difference between the liquidus temperature gradient (mG_c) and the heat-flux-imposed temperature gradient at the interface (G):

$$\phi = \left(\frac{d\Delta T}{dz}\right)_{z=0} = \left(\frac{dT_l}{dz} - \frac{dT_q}{dz}\right)_{z=0} = mG_c - G \quad [3.12]$$

A positive driving force will exist which causes any perturbation to grow when ϕ is positive, i.e. when a zone of constitutional undercooling exists ahead of the interface. If its value is negative, G is greater than mG_c. The limiting condition for constitutional undercooling is therefore:

$$\phi = 0 \quad [3.13]$$

Thus, this pseudo-thermodynamic approach gives the same result as that deduced by considering the zone of constitutional undercooling. Since the concentration gradient at the interface is known, it is simple to derive the criterion for the existence of constitutional undercooling in another form. The interface will always become unstable if equation 3.9 is satisfied. Using equation 3.6, this can be written:

$$G < -\frac{mV\Delta C_0}{D} \quad [3.14]$$

Also, because $-m\,\Delta C_0 = \Delta T_0$ (equation 1.10), the limit of constitutional undercooling can be expressed in its usual form:

Table 3.1 Şummary of Stability Conditions in Pure Metals and Alloys

	Growth Conditions	
	Equiaxed ($G < 0$)	Columnar ($G > 0$)
pure metal	*unstable*	*stable*
alloy	*unstable*	*unstable* ($\phi > 0$)
		stable ($\phi < 0$)

$$\frac{G}{V} = \frac{\Delta T_0}{D} \qquad [3.15]$$

or

$$V_c = \frac{GD}{\Delta T_0}$$

where ΔT_0 can be replaced by the expression: $mC_0(k-1)/k$ (equations 1.10, 1.11). If G/V is smaller than $\Delta T_0/D$, instability will result. Use of the conductivity weighted mean temperature gradient $\bar{G} = (G_s\kappa_s + G_l\kappa_l)/(\kappa_s + \kappa_l)$ instead of the temperature gradient in the liquid gives a more precise value of this limit#.

Table 3.1 summarises the types of interface morphology to be expected under various conditions, and table 3.2 presents typical numerical values involved in constitutional undercooling. It is evident that, even for relatively low solute concentrations, high temperature gradients must be imposed in order to suppress interface instability and cellular or dendritic growth. Note that, in the preparation of single crystals, or in other directional growth processes, a gradient of 20 K/mm is considered to be high.

Table 3.2 Minimum Stabilising Temperature Gradient (K/mm) as a Function of Distribution Coefficient for D = 0.005 mm²/s, V = 0.01 mm/s, m = −10K/wt%

	C_0(wt %)		
k	10	1	0.01
0.5	200	20	0.2
0.1	1800	180	1.8

Note that an exact interface stability analysis leads to a slightly modified criterion (appendix 7).

So far, only the limit of stability has been estimated. Nothing has been said about the form and scale of the perturbations which will develop if the interface is unstable. Information concerning the dimensions of the initial, perturbed morphology is very important because this will influence the scale of the resultant growth morphologies. However, it must be remembered that the morphology which initially develops above the limit of stability is usually only a transient structure which disappears once the steady-state cellular or dendritic morphology is established (see for example, figure 4.3).

3.4 Perturbation Analyses

A drawback of the constitutional undercooling criterion is that it ignores the effect of the surface tension of the interface. It is reasonable to suppose that the latter should have a marked influence upon interface stability. In order to investigate this possibility, and to learn more about the morphological changes occurring near to the limit of stability, it is necessary to suppose that the interface has already been slightly disturbed and to study its development under the constraints of diffusion and capillarity. More details of this procedure are presented in appendix 7, and the present discussion is intended merely to point out some of the physical principles involved.

Firstly, it is assumed for simplicity that the purely mathematical perturbations (see figure 3.6), which have an infinitesimal amplitude, do not affect the thermal or solutal

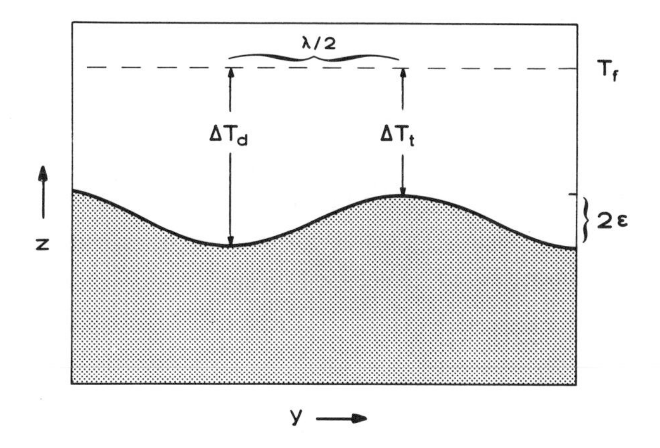

Figure 3.6 Interface Perturbations at a Solid/Liquid Interface

The existence of a zone of constitutional undercooling implies that a driving force for a change in the morphology is available, but gives no indication of the scale of the morphology which will appear. Experimental observations show that the initial form of the new morphology is periodic and may be approximated by a sinusoidal curve. Perturbation analysis permits the calculation of the wavelength of the instabilities which develop. The result is of great importance in the theory of dendrite growth.

diffusion fields. The perturbed interface can be described by a simple sine function:

$$z = \varepsilon \sin(\omega y) \qquad [3.16]$$

where ε is the amplitude, and ω (=2 π/λ) is the wave number. The temperature, T^*, of the interface can be deduced from the assumption of local equilibrium:

$$T^* = T_f + mC_l^* - \Gamma K^* \qquad [3.17]$$

This equation states that the difference between the melting point, T_f, and the interface temperature, T^*, is equal to the sum of the temperature differences due to the local interface composition and local interface curvature. Here, any temperature difference needed to drive atomic processes has been neglected. This is a reasonable assumption in the case of metals and other materials having a low entropy of fusion.

Taking just two points, at the tips (t) and depressions (d) of the interface and calculating the temperature difference gives:

$$T_t - T_d = m(C_t - C_d) - \Gamma(K_t - K_d) \qquad [3.18]$$

The curvatures at tips and depressions which are not too accentuated can be determined from the second derivative of the function which describes the interface shape (equation 3.16) at $y = \lambda/4$ and $3\lambda/4$:

$$K_t = -K_d = \frac{4\pi^2\varepsilon}{\lambda^2} \qquad [3.19]$$

(Note that K has the opposite sign to that which arises from the mathematical definition — equation 1.6). Since it is assumed that the temperature and concentration fields are unaffected by the presence of a very small perturbation, the temperature and concentration differences between the tips and depressions can be found from the gradients existing at the originally planar interface. (This corresponds to the assumption of small Péclet numbers, $\lambda V/2a < \lambda V/2D < 1$). Thus:

$$T_t - T_d = 2\varepsilon G \qquad [3.20]$$

and

$$C_t - C_d = 2\varepsilon G_c \qquad [3.21]$$

Substituting equations 3.19 to 3.21 into equation 3.18 leads to:

$$\lambda = 2\pi \left(\frac{\Gamma}{\phi}\right)^{1/2} \equiv \lambda_i \qquad [3.22]$$

where ϕ is the degree of constitutional undercooling as defined in equation 3.12. The wavelength, λ_i (given by equation 3.22), defines a critical perturbation which matches both the thermal and solutal diffusion fields. In this case, the wave-form will be stationary with respect to the unperturbed interface. That is, neither the tip nor the depression will grow.

In order to obtain a relationship between the time dependence of ε and λ, one can simplify equation A7.3 by assuming that the solid solubility in the alloy is essentially zero. Thus, k is approximately equal to zero and p is approximately equal to unity. Moreover, the differences in the thermal gradients in the liquid and solid, and the effect of latent heat, can be neglected:

$$\frac{\dot{\varepsilon}}{\varepsilon} = \left(\frac{V}{mG_c}\right)\left(b-\frac{V}{D}\right)\left(-\omega^2\Gamma - G + mG_c\right) \qquad [3.23]$$

where $\dot{\varepsilon} = d\varepsilon/dt$. This function ($\dot{\varepsilon}/\varepsilon - \lambda$) is plotted in figure 3.7 for an Al-Cu alloy, and exhibits a characteristic maximum. At λ values below the maximum,

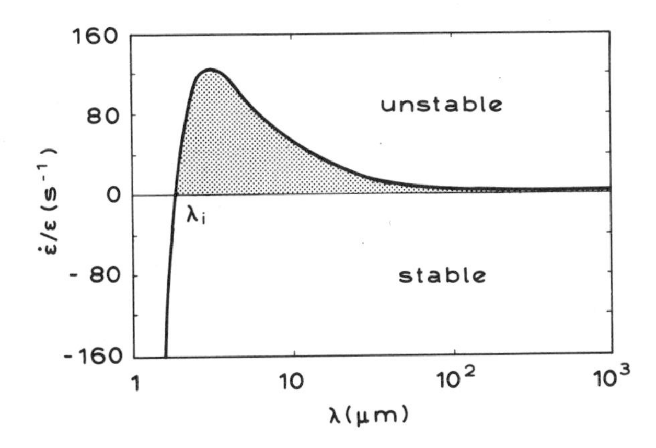

Figure 3.7 Rate of Development of a Perturbation at a Constitutionally Undercooled Interface

Here the parameter, $\dot{\varepsilon}/\varepsilon$, describes the relative rate of development of the amplitude of a small sinusoidal perturbation in the case of a specific alloy (Al-2*wt*%Cu) under given growth conditions (V = 0.1mm/s, G = 10K/mm). At very short wavelengths, the value of this parameter is negative due to curvature damping and the perturbation will tend to disappear (Fig. 3.1b). At wavelengths greater than λ_i and above, the sinusoidal shape will become more accentuated (instability - figure 3.1a). The wavelength having the highest rate of development is likely to become dominant. The reason for the tendency to stability at high λ-values is the difficulty of diffusional mass transfer over large distances. When the interface is completely stable, the curve will remain below the $\dot{\varepsilon}/\varepsilon = 0$ line for all wavelengths. This implies the disappearance of perturbations having any of these wavelengths (appendix 7).

the perturbations develop less quickly or disappear ($\dot{\varepsilon} < 0$) due to the effect of the high curvature. At λ values above the maximum the perturbations develop less quickly due to diffusion limitations. The wavelength, λ_i, describes the perturbed morphology which is at the limit of stability ($\dot{\varepsilon} = 0$) under conditions of constitutional undercooling. The first of the three terms on the right-hand-side of equation 3.23 is not zero, the second term becomes zero, upon substituting for b (see list of symbols) when:

$$\lambda = \infty$$

The third term vanishes when:

$$\omega^2 \Gamma = mG_c - G = \phi$$

or:

$$\lambda_i = 2\pi \left(\frac{\Gamma}{\phi}\right)^{1/2}$$

This is exactly the same as equation 3.22 which was derived above using a much simpler method. As the temperature gradients in liquid and solid, are not necessarily the same, G should be taken to be the mean temperature gradient

$$\overline{G} = (G_s \kappa_s + G_l \kappa_l) / (\kappa_s + \kappa_l)$$

The term, Γ/ϕ, is the ratio of the capillarity force to the driving force for instability. If ϕ tends to zero (the limit of constitutional undercooling), the minimum unstable wavelength approaches infinity. This is to be expected since, at the limit, only the planar interface should be observed. On the other hand, far from the limit of constitutional undercooling in the unstable regime:

$$G \ll mG_c = \frac{\Delta T_0 V}{D} \qquad [3.24]$$

Thus, the wavelength becomes, for $V >> GD/\Delta T_0$:

$$\lambda_i = 2\pi \left(\frac{D\Gamma}{V\Delta T_0}\right)^{1/2} \qquad [3.25]$$

The latter expression reveals that the wavelength of the unstable morphology is proportional to the geometric mean of a diffusion length (D/V) and a capillarity length ($\Gamma/\Delta T_0$). Increasing D or Γ, and decreasing V or ΔT_0 will increase the minimum unstable wavelength.

Bibliography

Constitutional Undercooling

W.A.Tiller, K.A.Jackson, J.W.Rutter, B.Chalmers, Acta Metallurgica **1**(1953) 428.

Stability Theory

W.W.Mullins, R.F.Sekerka, Journal of Applied Physics **34** (1963) 323.

W.W.Mullins, R.F.Sekerka, Journal of Applied Physics **35** (1964) 444.

R.F.Sekerka, Journal of Applied Physics **36** (1965) 264.

R.F.Sekerka, Journal of Crystal Growth **3** (1968) 71.

R.T.Delves, in *Crystal Growth* (Ed. B.R.Pamplin), Pergamon, Oxford, 1975, p. 40.

R.Trivedi, W.Kurz, Acta Metallurgica **34** (1986) 1663.

Interface Stability Experiments

S.C.Hardy, S.R.Coriell, Journal of Crystal Growth **7** (1970) 147.

R.J.Schaefer, M.E.Glicksman, Metallurgical Transactions **1** (1970) 1973.

K.Shibata, T.Sato, G.Ohira, Journal of Crystal Growth **44** (1978), 419.

Exercises

3.1 A solid/liquid interface becomes unstable when relation 3.1 (3.9) is obeyed in the case of a pure metal (alloy). Show that relation 3.1 is implied by relation 3.9. Discuss the differences.

3.2 Indicate why the constitutional undercooling criterion cannot yield the wavelength of the perturbed interface resulting from instability.

3.3 Discuss the advantage of the Bridgman method (Fig. 1.4a), over directional casting processes (Fig. 1.4b), with respect to the control of interface stability.

3.4 Determine the phase diagram from the information that the solute distribution ahead of the planar solid/liquid interface of an Al-Cu alloy has been found to be described by:

C [wt%] = $2(1 + [0.86/0.14]\exp[-Vz/D])$

under steady-state conditions. It has also been determined that the interface temperature is 624°, and it is known that the melting point of Al is 660°C. Give the values of k, m, ΔT_0, T_l , and ΔC_0.

3.5 What is the limit of stability, G/V, of the above alloy if it is given that $D = 3 \times 10^{-5}$ cm^2/s? Use the constitutional undercooling criterion.

3.6 Calculate the heat-flux required to produce a value of G which is sufficient to stabilise the planar front in question 3.5 at a growth rate, V, of 10 cm/h.

3.7 By analogy with figure 3.4, draw a diagram for $k > 1$ and discuss the constitutional undercooling criterion for this situation. Point out differences between this case and the case where $k < 1$.

3.8 Using the approach of Mullins and Sekerka (appendix 7), calculate the minimum value of G which is required to stabilise the planar interface of an Fe-0.09wt%S alloy during directional growth at a rate of 0.01mm/s. Calculate also the limit of absolute stability. (Let $D = 8 \times 10^{-9}$m^2/s, use k and m from exercise 1.6, and obtain other data from appendix 14.)

3.9 For the alloys, Al-2wt%Cu and Fe-1wt%Ni, determine the difference between the critical G/V ratio estimated according to the constitutional undercooling criterion (equation 3.15), the modified constitutional undercooling criterion (equation 3.15 with $G = \bar{G}$), and that estimated according to the Mullins-Sekerka criterion (equation A7.16 using the low thermal Péclet number approximation, i.e. $\xi_l = \xi_s = 1$ and equation A7.22c). It is assumed that the imposed temperature gradient, G, in the liquid is 10^5K/m. For Fe-1wt%Ni, use the m, k, and D values given in appendix 14 for Fe-10wt%Ni.

3.10 Calculate the ratio, λ_m / λ_i, for conditions of high constitutional undercooling and constant V and G values. Here, λ_m is the wavelength which corresponds to the maximum in figure 3.7. Note that b in equation 3.23 is a function of ω and is defined as (b_c in appendix 7, equation A7.4a) :

$$b = \left(\frac{V}{2D}\right) + \left[\left(\frac{V}{2D}\right)^2 + \omega^2\right]^{1/2}$$

Under certain conditions, this can be simplified by comparing the relative magnitudes of $V/2D$ and ω. Is that true of the case depicted in figure 3.7? Does the value of λ_m / λ_i depend upon the composition of the alloy?

3.11 Imagine that an experiment is carried out, on the alloy of question 3.4, in which the specimen is solidified using the Bridgman method with a temperature gradient, G, of 10K/mm and is maintained at the limit of constitutional undercooling until steady-state conditions are established. The growth rate is then doubled without changing the value of G. Calculate the minimum, and the most probable, wavelength which the resultant perturbation is likely to have.

3.12 For the case of pure Al at an undercooling of 1K, determine the order of magnitude of the radius of a growing sphere ($R \approx \lambda_i$) at which it will become unstable when growing in its undercooled melt. Assume, for the purpose of calculating G, that the steady-state solution for the temperature field around a sphere (analogous to the concentration field solution derived in appendix 2) is applicable.

CHAPTER FOUR

SOLIDIFICATION MICROSTRUCTURE: CELLS AND DENDRITES

Nearly all of the solidification microstructures which can be exhibited by a pure metal or an alloy can be divided into two groups: single-phase primary crystals and polyphase structures. The most important growth form, and the one to be discussed in this chapter, is the tree-like primary crystal, i.e. the dendrite. Polyphase structures (eutectics) will be described in the next chapter.

As shown in chapter 1, dendrites, eutectics, or combinations of these, make up the grains of any metallic microstructure after solidification. The growth of both morphologies can be described by analogous theoretical models; the development of which comprises two steps :

1. derivation of an *equation* which describes the general relationship between the scale of the microstructure, the undercooling, *and* the growth rate;

2. choice of a *criterion* which permits the definition of an unique relationship between the scale of the microstructure and the undercooling (in the case of equiaxed growth), *or* the growth rate (in the case of directional growth).

(This is the classical approach. More advanced treatments of the problem, developed in recent years, have shown that appropriate solutions can be obtained in quite different ways (see Langer 1986, Pelcé 1988)).

With regard to the first part of the problem, it is necessary to determine an expression

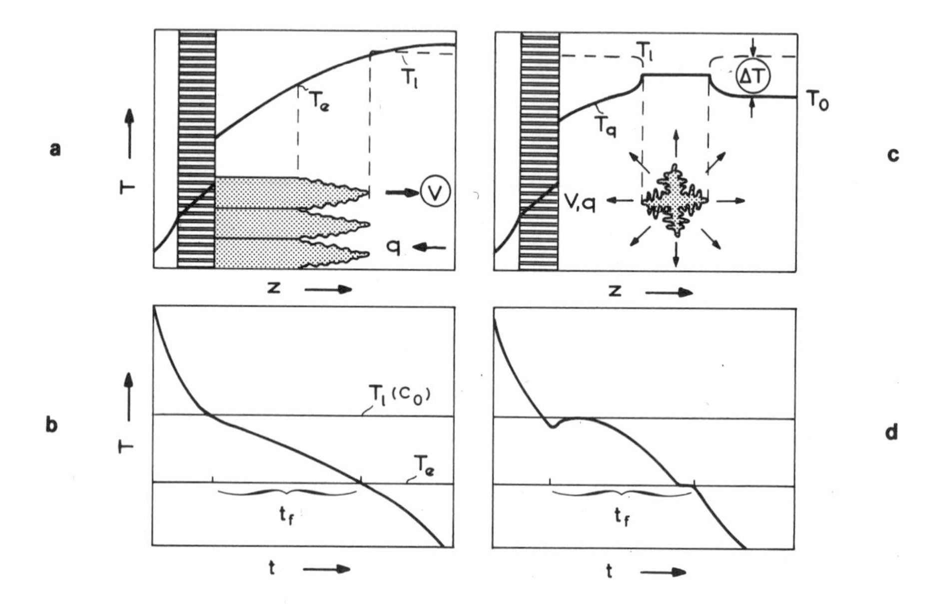

Figure 4.1 Thermal Fields and Cooling Curves of Alloy Dendrites

The upper part of the figure corresponds to the lower part of figure 1.7, with superimposed liquidus temperatures (broken lines). The case of directional (columnar) growth is illustrated in diagram (a) and that of equiaxed growth in diagram (c). If a thermocouple is placed at a fixed position in the melt and overgrown by the dendrites, different cooling curves will be recorded for directional growth (b) and equiaxed growth (d). This difference is essentially due to nucleation in the case of equiaxed solidification. Due to microsegregation (chapter 6) some eutectic will usually form in the last stages of solidification (i.e. at T_e). Note that, in the case of directional growth, the crystals are in contact with the mould and heat will be conducted through them in a direction which is parallel and opposite to that of their growth. Therefore, the melt is the hottest part of the system. In the case of equiaxed growth, the heat produced by solidification must be transported through the melt. Thus, in this case, the crystals are the hottest part and the heat flux, q, is radial and in the same direction as that of growth. The local solidification time, t_f, is the elapsed time between the beginning and the end of solidification at a fixed point in the system, i.e. from the tip to the root of the dendrite.

for the heat and/or solute distribution and to take account of the effects of capillarity. Step 2 above has been satisfied by using one of two alternative growth criteria. These are:

i growth at the extremum - i.e. at the maximum growth rate or minimum undercooling; this assumption has been justified on the basis of non-equilibrium thermodynamic concepts such as minimum entropy production;

ii growth at the limit of morphological stability; in the case of dendrite growth, use of this criterion leads to satisfactory agreement between theory and experiment.

4.1 Constrained and Unconstrained Growth

The situation in which the heat flow is opposite to the growth direction (i.e. directional or columnar solidification - figures 1.7a,c and 4.1a) is often referred to as *constrained* growth. That is, the rate of advance of the isotherms constrains the dendrites (which in this situation are *found only in alloy solidification*) to grow at a given velocity. This forces them to adopt the corresponding tip undercooling. The grain boundaries are parallel to the primary dendrite axes (trunks) and are continuous along the length of the solid. Each dendrite forms low-angle boundaries with its neighbours and many trunks, formed by repeated branching, together make up one grain (Fig. 4.2).

When the heat flows from the crystal into the melt (equiaxed solidification - figures 1.7b,d and 4.1c), the dendrites can grow freely, *in pure materials and alloys*, as rapidly as the imposed undercooling permits. The dendrites grow in a radial fashion until they impinge upon dendrites originating from other nuclei, and the grain boundaries form a continuous network throughout the solid (Fig. 1.9).

In the case of directional growth, most of the dendrites are arranged parallel to each other and a characteristic trunk spacing (λ_1) can be defined. In the case of equiaxed solidification, each dendrite forms a grain and the primary spacing, λ_1, is usually equal to the grain diameter. A secondary arm spacing, λ_2, can be defined for both columnar and equiaxed dendrites.

4.2 Morphology and Crystallography of Dendrites

The formation of a dendrite begins with the breakdown of an unstable planar solid/liquid interface. Perturbations are amplified until a marked difference in growth of the tips and depressions of the perturbed interface has occurred. As the tip can also reject solute in the lateral direction, it will tend to grow more rapidly than a depression, which tends to accumulate the excess solute rejected by the tips. Therefore, the form of

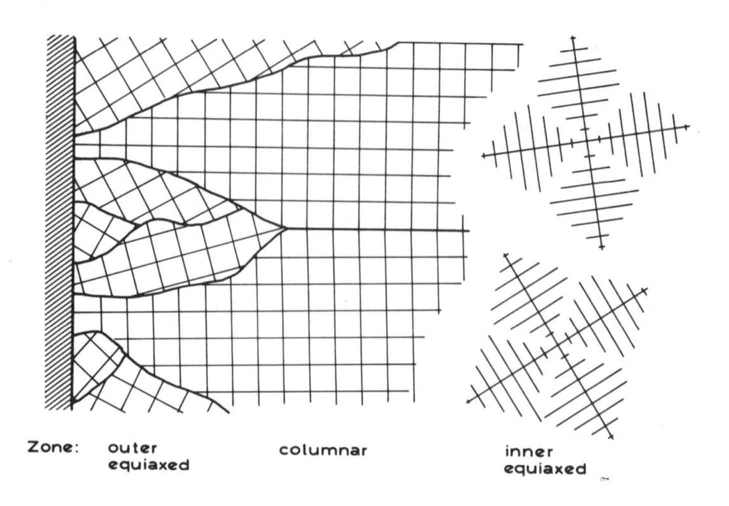

Figure 4.2 Formation of Columnar and Equiaxed Dendritic Microstructures

Initial competition of crystals nucleated at random at the mould wall, and their growth into the liquid, leads to grain selection. In cubic metals, columnar grains having one [001] axis close to the heat flux direction ultimately overgrow the others. Note that, in this section, differences in the rotation of the columnar grains about their growth axis is not visible. Large orientation differences are accommodated at high-angle grain boundaries. Branching of the dendrites leads to the creation of new trunks which are crystallographically related to the initial primary trunk and form a grain. The transition from columnar to equiaxed growth occurs when the melt has lost its superheat, becoming slightly undercooled, and detached dendrite arms growing in the melt form a barrier ahead of the columnar zone. (Compare with figure 1.6.)

the perturbation is no longer sinusoidal, but adopts the form of cells (Fig. 4.3). If the growth conditions are such as to lead to dendrite formation, the cells will rapidly change to dendrites, which exhibit secondary arms and crystallographically governed growth directions. Cell and primary dendrite spacings are much larger than the wavelength of the original perturbation which initiates the corresponding growth morphology (Fig. 4.3). It is important to note that, under normal solidification conditions, cells can only appear during the directional growth of alloys, i.e. for $G > 0$. For instance, cells can grow under conditions such as those indicated in figure 1.7c. In other cases (Fig. 1.7b,d), only dendrites will be observed. Figure 4.4 summarises the differences between cells and dendrites. Cells are usually a crystal morphology which grows anti-parallel to the heat flux direction. They grow under conditions which are close to the limit of constitutional undercooling of the corresponding planar interface. At high growth rates, close to conditions of absolute stability, cells can be also observed (see chapter 7). There are always many cells in one grain.

On the other hand, dendrites are crystalline forms which grow far from the limit of stability of the plane front and adopt an orientation which is as close as possible to the heat flux direction or opposite to it, but follows one of the preferred growth axes. These directions are crystallographically determined (table 4.1). Equiaxed dendrites grow along all of the available preferred directions when the heat extraction is isotropic, as is the case in an undercooled melt. In cubic crystals, the six [001] axes form the trunks and therefore the crystal orientation can easily be determined. In the case of a directionally solidified dendritic monocrystal (e.g. a turbine blade), all of the dendrites composing the monograin are aligned (Fig. 4.5), leading to improved high temperature properties.

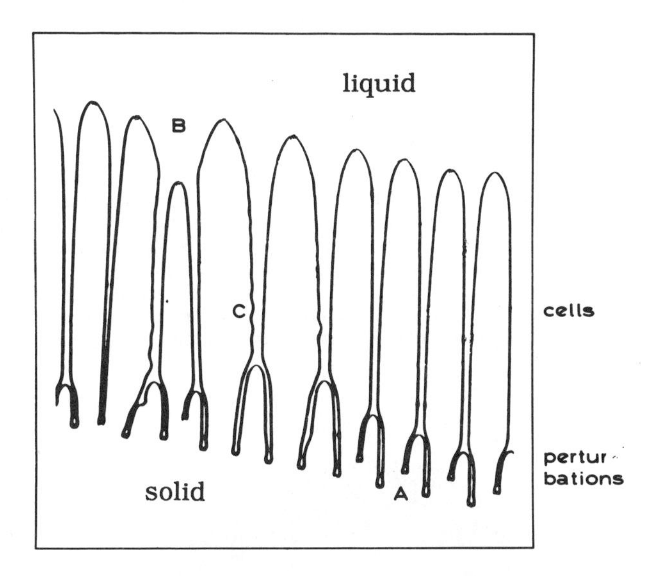

Figure 4.3 Breakdown of a Plane Solid/Liquid Interface to Give Cells

The development of perturbations at the constitutionally undercooled solid/liquid interface (lower part of figure) is only a transient phenomenon. The tips of the perturbations can readily reject solute while the depressed parts of the interface accumulate solute and advance much more slowly. The initial wavelength is too small for further rapid growth to occur, and the final result is the formation of a cellular structure. Note that the wavelength has approximately doubled between the initial perturbation and the final cells. Also, the spacing between the cells is not constant. The initial cellular morphology can adjust itself to give a more optimum growth form via the cessation of growth of some cells (B) in order to decrease their number, or by the division of cells in order to increase the number present. The division of cells is not shown here, but it resembles the change at point A, with two branches continuing to grow. Furthermore, the larger centre cells (C) have slightly perturbed surfaces and this suggests that, in the intercellular liquid, some driving force remains for further morphological change which might possibly lead to dendrite formation.

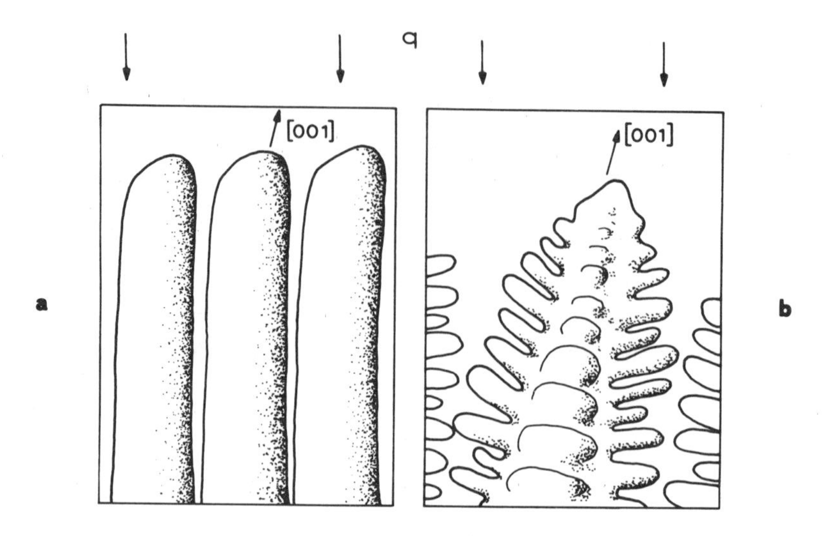

Figure 4.4 Cells and Dendrites

During directional growth in a positive temperature gradient, cells can exist as a stable growth form (a). Dendrites will be particularly common in conventional castings and are characterised by growth of the trunks and branches along preferred crystallographic directions, such as [001] in cubic crystals (b). Due to the anisotropy of properties such as the solid/liquid interface energy and the growth kinetics, dendrites will grow in that preferred crystallographic direction which is closest to the heat flow direction (q), whereas well-developed cells grow with their axes parallel to the heat flow direction; without regard to the crystal orientation. Only at the growing tip (a) may some trace of directionality exist. Between these two extremes, there is a range of intermediate forms (dendritic cell, cellular dendrite).

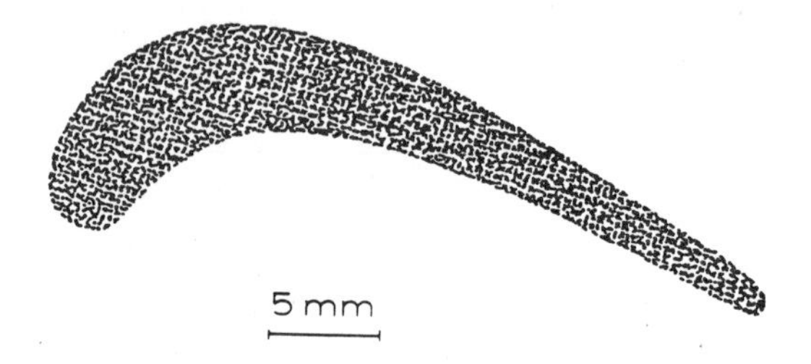

Figure 4.5 Transverse Section of a Monocrystalline Gas-Turbine Blade

In an alloy, the last (interdendritic) liquid to solidify always has a different composition compared to that of the first crystals (dendrite trunks) to form. Therefore, the dendritic structure can be revealed in a casting due to variations in etching behaviour. This is seen in the above transverse section of a directionally solidified monocrystalline turbine blade. The orientation of the dendrite trunks is parallel to the blade axis (and perpendicular to the plane of this section). The absence of high-angle grain boundaries improves the high-temperature strength and fatigue behaviour of superalloy castings used in jet aircraft. The formation of such a monocrystal can be understood with the aid of figure 4.2. During the initial stages of growth, when a columnar zone has progressed to some distance from a chill plate, the grains enter a restriction placed in the path of the growing crystals. As a result, only one grain can continue to grow and form the blade. This is the principle used in alloy single-crystal casting. (Photograph - Pratt and Whitney).

Table 4.1 Preferred Growth Directions of Dendrites of Various Materials

Structure	Dendrite Orientation	Example
face-centred cubic	<100>	Al
body-centred cubic	<100>	δ-Fe
body-centred tetragonal	<110>	Sn
hexagonal close-packed	$<10\bar{1}0>$	H_2O (snow)
	<0001>	$Co_{17}Sm_2(Cu)$

If a single dendrite could be extracted from the columnar zone of a casting during growth, it would resemble the one depicted schematically in figure 4.6. Behind a short paraboloid tip region, which often constitutes less than 1% of the length of the whole dendrite, perturbations appear on the initially smooth needle as in the case of the breakdown of a planar interface. These perturbations grow and form branches in the four [001] directions which are perpendicular to the trunk. If the primary spacing is sufficiently great, these cell-like secondary branches will develop into dendritic-type branches and lead to the formation of tertiary and higher-order arms. When the tips of the branches encounter the diffusion field of the branches of the neighbouring dendrites, they will stop growing and begin to ripen and thicken. Thus, the final secondary spacing will be very different to the initial one. Compare the upper and lower parts of figure 4.6. The final value of λ_2 is largely determined by the contact time between the branches and the liquid. This period is known as the local solidification time, t_f, and is given by the time required e.g. for a thermocouple placed at a fixed point to pass from the tip to the root of the growing dendrite (Fig. 4.1).

4.3 Diffusion Field at the Tip of a Needle-Like Crystal

The growth rate, as well as the dendrite morphology or spacing, are all largely dependent upon the behaviour of the tip region. During the growth of the tip, either heat (in the case of pure metals) or heat and solute (in alloys) are rejected (Fig. 4.7). These diffusion processes are driven by gradients in the liquid, and the latter are in turn due to differences in temperature (ΔT_t) and concentration (ΔC) ahead of the growing crystal. The concentration difference, ΔC, can be converted into a liquidus temperature difference, ΔT_c, via the phase diagram, as shown in figure A8.1. After adding the temperature difference, at the solid/liquid interface, caused by the curvature of the tip (ΔT_r), the coupling condition can be written:

$$\Delta T = \Delta T_c + \Delta T_t + \Delta T_r \qquad [4.1]$$

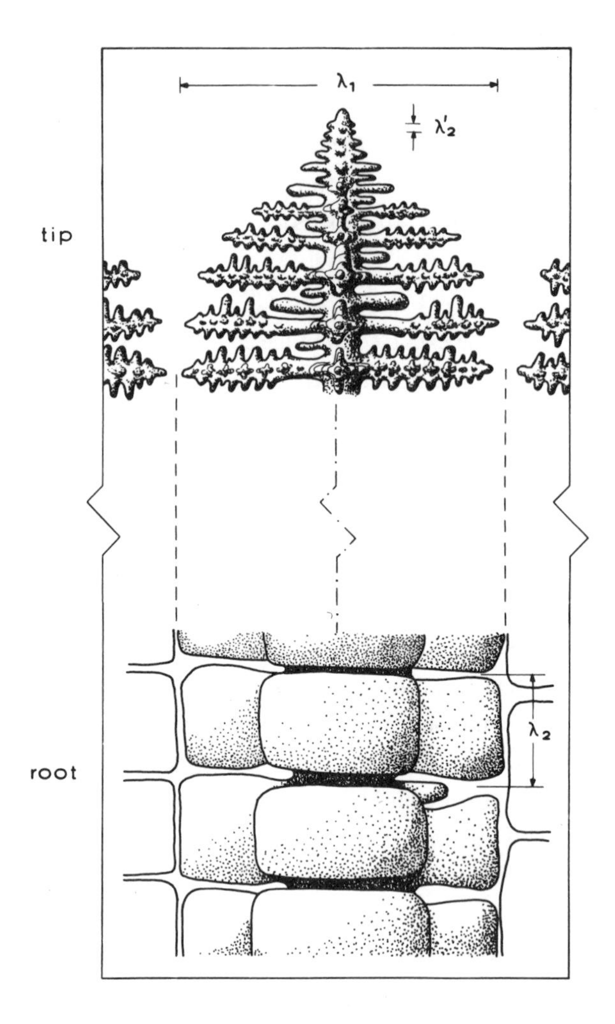

Figure 4.6 Growing Dendrite Tip and Dendrite Root in a Columnar Structure

Depending upon the directional growth conditions, the dendrite (from the Greek, dendron = tree) will develop arms of various orders. A dendritic form is usually characterised in terms of the primary (dendrite trunk) spacing, λ_1, and the secondary (dendrite arm) spacing, λ_2. Tertiary arms are also often observed close to the tip of the dendrite. It is important to note that the value of λ_1 measured in the solidified microstructure is the same as that existing during growth, whereas the secondary spacing is enormously increased due to the long contact time between the highly-curved, branched structure and the melt. The ripening process not only modifies the initial wavelength of the secondary perturbations, λ_2', to give the spacing which is finally observed, λ_2, but also often causes dissolution of the tertiary or higher order arms. The two parts of the figure are drawn at the *same scale*, refer to the *same dendrite*, and illustrate morphologies which exist at the *same time* but which are widely separated along the trunk length (by about 100 λ_1).

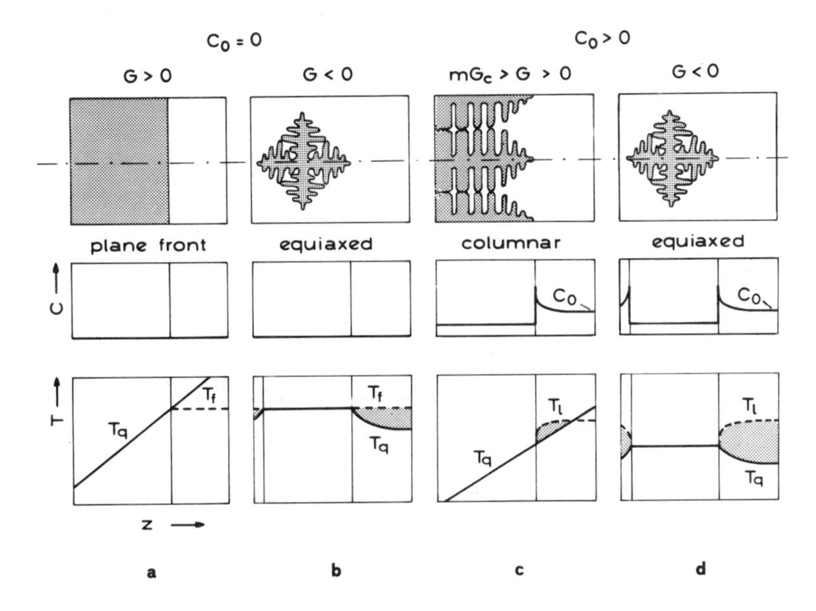

Figure 4.7 Concentration and Temperature Fields of Dendrites

These diagrams illustrate the heat and mass diffusion fields existing along the dendrite axis, and correspond to various possible cases (Fig. 1.7). In pure substances (a,b), there is no solute rejection and dendrites can form only in an undercooled melt. In the latter case (b), the heat rejection occurring during growth sets up a negative temperature gradient ahead of the interface. This leads to the establishment of the necessary conditions for the instability of thermal dendrites. In the case of alloys (c,d), dendrites can form regardless of the temperature gradient if the interface is constitutionally unstable. If G is greater than zero (c), the latent heat is transported, together with the unidirectional heat flux, into the solid. To a first approximation, therefore, solute rejection alone needs to be considered in the case of directionally solidified (solutal) dendrites. Equiaxed dendrites in alloys (d) reject both solute and heat.

In this equation, the possibility of a kinetic undercooling for atom attachment has been neglected. This is a reasonable assumption in the case of materials, such as metals, which exhibit a low entropy of melting under normal solidification conditions.

The current state of knowledge concerning the *equiaxed dendritic growth* of pure substances has been reviewed by Huang and Glicksman (1981). On the other hand, in the theory of the equiaxed dendritic growth of alloys, the problem of coupled heat- and mass-transport must be solved (appendix 9). In the case of the *directional growth of alloy dendrites*, the situation is somewhat simpler because, due to the imposed temperature gradient, the latent heat is transported through the solid and, to a first approximation, does not affect tip growth while solute is rejected ahead of the tips. In this case, only mass diffusion need be considered. This permits a simple solution of the problem of alloy dendrite growth in the columnar zone. This case is of great practical importance and will be considered here in greater detail.

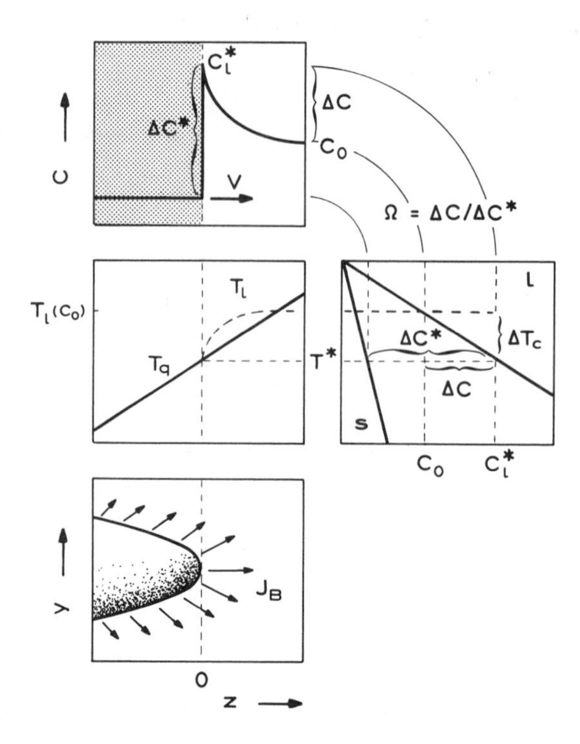

Figure 4.8 Solute Rejection at the Tip of an Isolated Dendrite

During directional solidification, where the isotherms move due to the imposed heat flux, a needle-like crystal can grow more quickly than a flat interface due to the more efficient solute redistribution: B-atoms rejected at the interface of a thin needle can diffuse outwards into a large volume of liquid. Thus, the solutal diffusion boundary layer, δ_c, of the needle is smaller than that of a planar interface. Also, because the interface is not planar, the solid formed does not have the same composition as the original liquid (as it does in the case of steady-state plane-front growth - Fig. 3.4). When a positive gradient is imposed, as in directional solidification, heat is extracted through the solid. If, furthermore, thermal diffusion is rapid (as in metals) the form of the isotherms will be affected only slightly by the interface morphology. Thus, in the case of directionally solidifying dendrites, *solute diffusion* alone will be the limiting factor. The growth temperature, T^*, of the tip will define a solute undercooling, ΔT_c, or, via the phase diagram, the degree of supersaturation, $\Omega = \Delta C/\Delta C^*$. The determination of Ω as a function of the other parameters requires the solution of the differential equation which describes the solute distribution. The simplest solution is obtained when the tip morphology is supposed to be hemispherical. Instead, the real form of the dendrite tip is closely represented by a paraboloid of revolution.

The rejection of solute changes the temperature of the solid/liquid interface at the tip (Fig. 4.8). The ratio of the change in concentration at the tip, ΔC, to the equilibrium concentration difference, ΔC^* ($= C_l^*[1-k]$: length of the tie-line at the temperature of the tip) is known as the supersaturation, Ω. This supersaturation (or the related undercooling, ΔT_c) represents the driving force for the diffusion of solute at the dendrite tip in an alloy. When the supersaturation is equal to zero the transformation rate will be zero. With increasing supersaturation, the growth rate of the new phase (the solid) will increase. The rejection rate, and therefore the growth rate, is influenced by the shape of the tip and, at the same time, the form of the tip is affected by the distribution of the rejected heat or solute. This interaction makes the development of an exact theory extremely complex. However, the dendritic shape can be described satisfactorily as a paraboloid of revolution; as first suggested by Papapetrou (1935). The mathematical solution of the diffusion problem for a paraboloid was derived by Ivantsov (1947) who deduced the following relationship between the supersaturation, Ω, the dendrite tip radius, R , and the growth rate, V:

$$\Omega = \mathrm{I}(P_c) \qquad [4.2]$$

where:

$$\mathrm{I}(P_c) = P_c \exp(P_c)\mathrm{E}_1(P_c) \qquad [4.3]$$

and the Péclet number for solute diffusion, $P_c = VR/2D = R/\delta_c$ #. Here, $\mathrm{E}_1(P)$ is the exponential integral function (appendix 8). The form of the Ivantsov function, $\mathrm{I}(P)$, is shown in figure A8.4. It is interesting to note that equation 4.3 can be approximated by a continued fraction of the type (Abramowitz & Stegun, 1965):

$$\mathrm{I}(P) = \cfrac{P}{P + \cfrac{1}{1 + \cfrac{1}{P + \cfrac{2}{1 + \cfrac{2}{P + \dots}}}}}$$

If only the first term is taken, one obtains:

$$\mathrm{I}(P) \cong P \qquad [4.4]$$

and insertion of this approximation into equation 4.2 gives:

$$\Omega \cong P_c \qquad [4.5]$$

This is the simple solution obtained for the diffusion field existing around a hemispherical cap (equation A8.11), and will be used in the following discussion

When the tip radius is related to the thermal diffusion length, δ_t, the corresponding dimensionless group is the thermal Péclet number, $P_t = VR/2a$ (see appendix 8).

because it permits a clearer insight into the physics of dendrite growth at low Péclet numbers than does the more complicated relationship of equation 4.3. Nevertheless, it should be kept in mind that equation 4.3 or more complicated ones (Trivedi, 1970) can easily be used in numerical calculations and this will then lead to more exact solutions. Equation 4.5 simply states that the response of the system characterised by the Péclet number is proportional to the driving force defined by the supersaturation. Note that P is the ratio of the tip radius, R, to the diffusion length, $2D/V$ (Fig. A2.5).

The solution to the diffusion problem, described by equation 4.5, is seen in figure 4.9, for an alloy dendrite (under directional solidification conditions), as a straight line at an angle of 45° in logarithmic coordinates. This indicates that, for a given Ω-value, R and V are not defined unambiguously. Thus, solution of the diffusion problem does not indicate whether the dendrite will grow quickly with small R or slowly with large R, but merely relates the sharpness of the tip to its rate of propagation.

The diffusion boundary layer around the tip is proportional to the tip radius (equation A2.26). Because the gradient of C is inversely proportional to the boundary layer thickness, a sharper tip has a steeper gradient. A sharper tip can grow more rapidly because it can reject solute (or heat in the case of a thermal dendrite) more efficiently; the flux being proportional to the gradient. However, there is a limit, to the possible sharpness of the dendrite tip, which is represented by the critical radius of nucleation, $R^o = r^o$ (table 2.1). At r^o, the growth rate is zero and therefore all of the supersaturation is used to create curvature and none remains to drive diffusive processes (Fig. 4.9).

4.4 Operating Point of the Needle Crystal - Tip Radius

The overall growth curve of a needle-like crystal, which reflects the sum of the capillarity and diffusion effects, follows the solid curve in figure 4.9, and exhibits a maximum close to R^o. Until recently, this maximum, R_e, was considered to be the radius at which the dendrite would actually grow. This so-called extremum criterion permitted the establishment of an unique solution, to the otherwise indeterminate growth problem, by setting the first derivative of the equation of growth equal to zero.

Langer and Müller-Krumbhaar (1977) have argued that a dendrite grows with a tip having a size at the limit of stability (marginal stability).# Thus, one can then immediately determine the expected tip radius by setting:

$$R_s = \lambda_i \qquad [4.6]$$

Since then, extensive theoretical modelling has shown that this criterion should be replaced by the solvability condition (Langer, 1986 and Pelcé, 1988). However, the fundamental relationship originally developed (equation 4.6) remains the same.

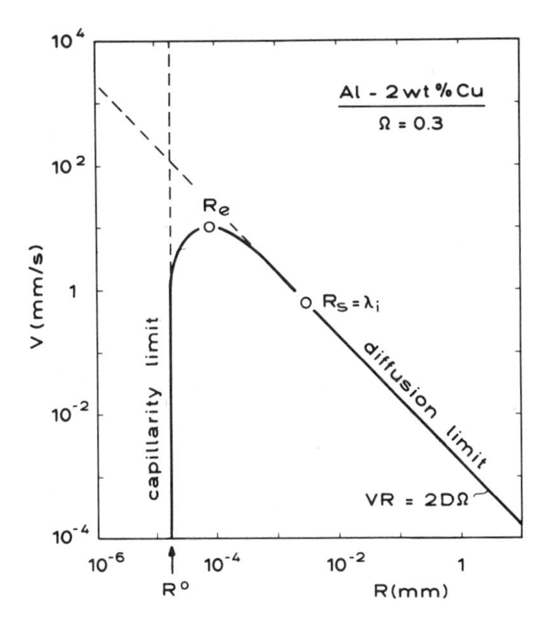

Figure 4.9 ***Unoptimised*** **Growth Rate of a Hemispherical Needle for Ω = Constant**

For a hemispherical needle crystal, the solution of the diffusion equation shows that the supersaturation, Ω, is equal to the ratio of the tip radius to the characteristic diffusion length. This dimensionless ratio is known as the Péclet number, P_c (= $RV/2D$). For a given supersaturation, the product, RV, is therefore constant and means that either a dendrite with a small radius will grow rapidly or one with a large radius will grow slowly (diagonal line). At small R-values, the diffusion limit is cut by the capillarity limit. The minimum radius, R^o, is given by the critical radius of nucleation, r^o (table 2.1). A maximum value of V therefore exists. Because it was reasoned that the fastest-growing dendrites would dominate steady-state growth, it was previously assumed that the radius chosen by the system would be the one which gave the highest growth rate (extremum value, $R = R_e$). However, experiment indicates that the radius of curvature of the dendrite is approximately equal to the lowest wavelength perturbation of the tip, which is close to λ_i (Fig. 3.7). This is referred to as growth at the limit of stability ($R = R_s$).

where λ_i is the shortest wavelength perturbation which would cause the dendrite tip to undergo morphological instability. To a first approximation, the wavelength of the marginally stable perturbation at a planar interface can be used. It is described by equation 3.22 and, for the condition $G << mG_c$ (which is true for dendrites) it is given by equation 3.25:

$$\lambda_i = 2\pi(\delta_c s)^{1/2} \qquad [4.7]$$

This wavelength is the geometric mean of a diffusion length, $\delta_c(\sim D/V)$ and a capillary length, s ($\sim \Gamma/\Delta T_0$). In figure 4.9 the value, R_s, at which R is at the limit of morphological stability is also indicated. It can be seen that this operating point is situated some distance from the extremum, and leads to the prediction of larger R-values. This prediction is consistent with experimental measurements and also with more recent analytical and numerical modelling.

Since R_s is considerably greater than R_e, the effect of curvature on the growth curve can be neglected. (The point, $R_s = \lambda_i$, is situated upon the fully diffusion-limited curve, therefore ΔT_r in equation 4.1 is almost zero.) The curvature then exerts its influence mainly through the λ_i-value, via the capillary length. Figure 4.10 illustrates how one obtains an unique solution by using the extremum criterion (dotted line) or with the aid of stability arguments (solid line). By using equation 4.6, Ω is eliminated from the V-R relationship. The final result for constrained growth ($V = f'[R_s]$) is given in figure 4.11 for growth rates, $V > V_c$, i.e. above the limit of constitutional undercooling of the plane front.

In order to understand these relationships more fully, the calculation (Kurz & Fisher 1981) of the tip radius of a dendrite growing under conditions of directional

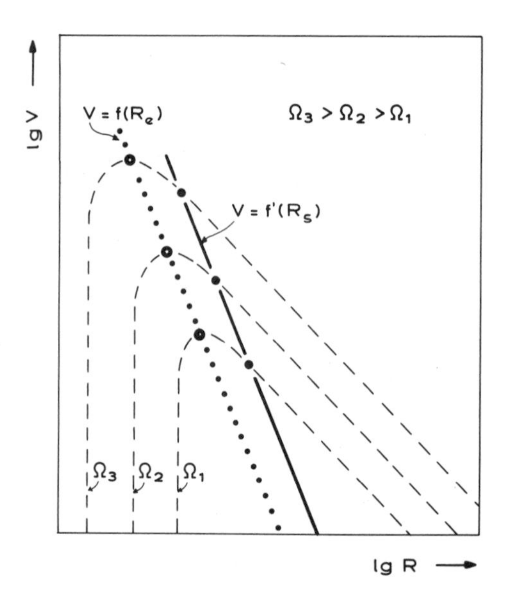

Figure 4.10 Growth Rate as a Function of Tip Radius for *Optimum* Growth

In this figure, it is demonstrated how, using different optimisation criteria, the observed V-R-values are obtained as a function of Ω. That is, for a given value of V, a given value of R is found. The unoptimised growth rate, V, of figure 4.9 is here shown as a function of the tip radius, R, for various supersaturations, $\Omega_3>\Omega_2>\Omega_1$ (broken lines). Two optimisation criteria are used here. One is that tip growth occurs at the extremum (dotted line) and the other is that it occurs at the limit of stability (solid line). The optimised curve deduced from the stability criterion (solid line) corresponds to the less steep portion of the solid line in figure 4.11. (Note that the coordinates there are inverted).

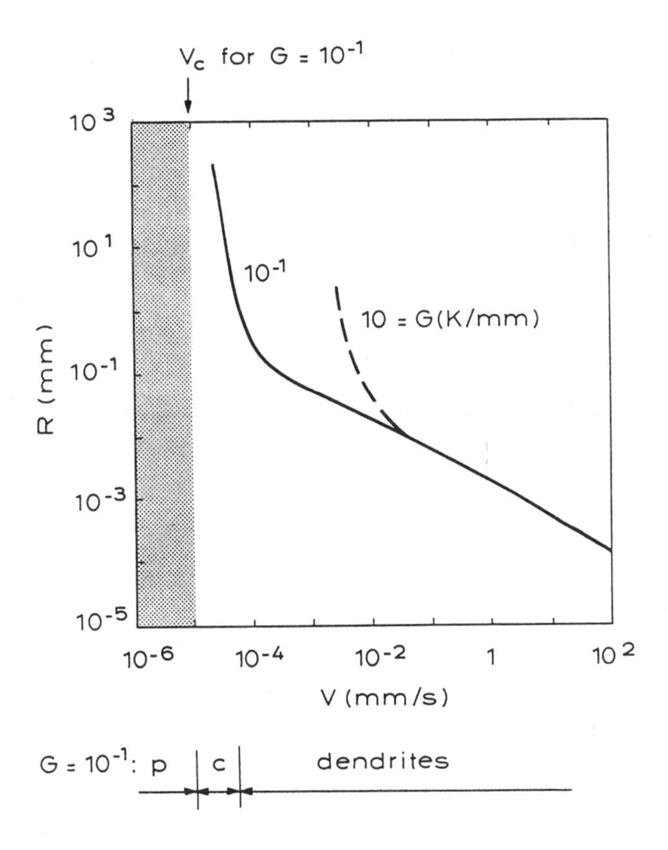

Figure 4.11 Optimised Dendrite Tip Radius as a Function of Growth Rate

If it is assumed that, in directional solidification, growth occurs with a tip radius which is equal to the minimum instability wavelength, λ_i, curves such as those above can be generated. They indicate the magnitude of the dendrite tip radius for a given growth rate and temperature gradient. Note the marked effect of the temperature gradient upon the radius of curvature at low growth rates (constrained growth regime or cellular regime e.g. for $G = 0.1$K/mm between $V = 10^{-4}$ and $V = 10^{-5}$mm/s). A sufficiently high gradient, or a sufficiently low growth rate ($V_c = GD/\Delta T_0$) will lead to the re-establishment of a planar interface (i.e. a 'dendrite' with an infinite radius of curvature).

solidification# and low Péclet numbers will be developed in more detail. This is the solutal case with an imposed temperature field, as in figure 4.8. The minimum wavelength of the instability of the tip is approximated by equation 3.22:

$$\lambda_i = 2\pi\left(\frac{\Gamma}{\phi}\right)^{1/2}$$

Note that directional solidification can be obtained by using a Bridgman-type furnace (G and V independent) or by directional casting (G and V coupled) - see exercise 1.9.

Using equation 4.6 and dropping the subscript for R gives:

$$R = 2\pi\left(\frac{\Gamma}{\phi}\right)^{1/2} \qquad [4.8]$$

where $\phi = mG_c - G$. In evaluating ϕ, it is assumed that $G_s = G_l = G$. All values except G_c, are known in equation 4.8, and G_c at the tip (in the steady state) can be deduced from a flux balance, i.e.

$$G_c = -VC_l^*p/D \qquad [4.9]$$

The unknown here is the tip concentration in the liquid, C_l^* which can be obtained by combining the definition of the supersaturation Ω, (equation A8.1),

$$C_l^* = \frac{C_0}{1-p\Omega}$$

with the diffusion solution (equation 4.2), leading to:

$$C_l^* = \frac{C_0}{1-p\,\mathrm{I}(P_c)} \qquad [4.10]$$

which can also be written as:

$$\frac{C_l^*}{C_0} = \mathrm{A}(P_c) \qquad [4.11]$$

where $\mathrm{A}(P_c) = [1 - p\cdot\mathrm{I}(P_c)]^{-1}$. From this, the tip concentration gradient can be obtained:

$$G_c = -VpC_0\mathrm{A}(P_c)/D \qquad [4.12]$$

Substituting this gradient into equation 4.8 leads to:

$$R = 2\pi\,\frac{\Gamma^{1/2}}{\left\{-\frac{mVpC_0\mathrm{A}(P_c)}{D} - G\right\}^{1/2}} \qquad [4.13]$$

and with the definition of the Péclet number ($R = 2P_cD/V$) to

$$V^2A' + VB' + G = 0 \qquad [4.14]$$

with

$$A' = \frac{\pi^2 \Gamma}{P_c^2 D^2}$$

and

$$B' = mC_0(1 - k)\mathrm{A}(P_c)/D$$

This quadratic equation can be easily solved and the value of R obtained as a function of V (Fig. 4.11)#. Noting that, for the dendritic growth regime, the imposed temperature gradient, G, has little effect one can rewrite equation 4.14 as:

$$V = \frac{-B'}{A'}$$

or

$$V = \frac{\Delta T_0 k \, \mathrm{A}(P_c) \, P_c^2 D}{\pi^2 \Gamma} \qquad [4.15]$$

with $\Delta T_0 k = -mC_0 p$. As for small Péclet numbers, $A(P_c)$ can be approximated by $1 + p\mathrm{I}(P_c)$ which, in this case, is close to unity. One obtains as an approximate solution

$$V = \frac{\Delta T_0 k P_c^2 D}{\pi^2 \Gamma}$$

or

$$R^2 V = \frac{4\pi^2 D \Gamma}{\Delta T_0 k} \qquad [4.16]$$

This result is the same as that which one obtains in using the hemispherical low Péclet number approximation of equation 4.5 (Kurz & Fisher, 1981). The final equations of the latter approach lead to simple relationships, which are useful for obtaining order of magnitude estimates, and are given in table 4.2.

Figure 4.12 shows the tip concentration for cells and dendrites. It is interesting to note that the composition is high in the cellular regime near to the plane front at low and high growth rates. In the steady state, the latter grows with the composition, C_0/k; leading to an homogeneous solid with the composition, C_0. The temperature is related

In using equation 4.14, one has to make sure that one always chooses the smaller radius corresponding to λ_i of the two λ values which are predicted by stability theory.

Table 4.2 Dendrite Growth Variables According to the Hemispherical Approximation $[(1-k)P_c<1]$

$R \cong 2\pi \left(\dfrac{D\Gamma}{\Delta T_0 kV}\right)^{1/2}$
$P_c \cong \pi \left(\dfrac{V\Gamma}{D\,\Delta T_0 k}\right)^{1/2}$
$\dfrac{C_l^*}{C_0} \cong \dfrac{1}{1-(1\text{-}k)P_c} \cong 1+(1\text{-}k)P_c$
$\Delta T = \Delta T_c = mC_0\left(1-\dfrac{1}{1-(1\text{-}k)P_c}\right)$

to the concentration via the liquidus line and shows, for a negative slope ($k < 1$), an inverse behaviour (Fig. 4.13).

The tip radius would not be so important if it did not influence other morphological parameters of the cell or dendrite. The theory presented here is useful in the case of isolated needles. Dendrites can be regarded as being isolated crystals, at least at their tips, but the applicability of this assumption to cellular growth is limited since cells always grow with their tips close together ($R \cong \lambda$) and their diffusion fields overlap. Furthermore, their tips differ from the parabolic shape of dendrites.

4.5 Primary Spacing of Dendrites after Directional Growth

The primary trunk spacing is an important characteristic of columnar dendrites and has a marked effect on the mechanical properties. On the other hand, in equiaxed growth in the absence of a primary spacing a corresponding parameter is the distance between the nuclei, or the grain size.

Returning to the primary spacing in the columnar (directional) growth situation, it is assumed that the cell or the dendrite envelope, representing the *mean cross-section* of the trunk and branches, can be described approximately by an ellipse (Fig. 4.14). The smaller radius of curvature of the ellipse is given by:

$$R = \frac{b^2}{a} \qquad [4.17]$$

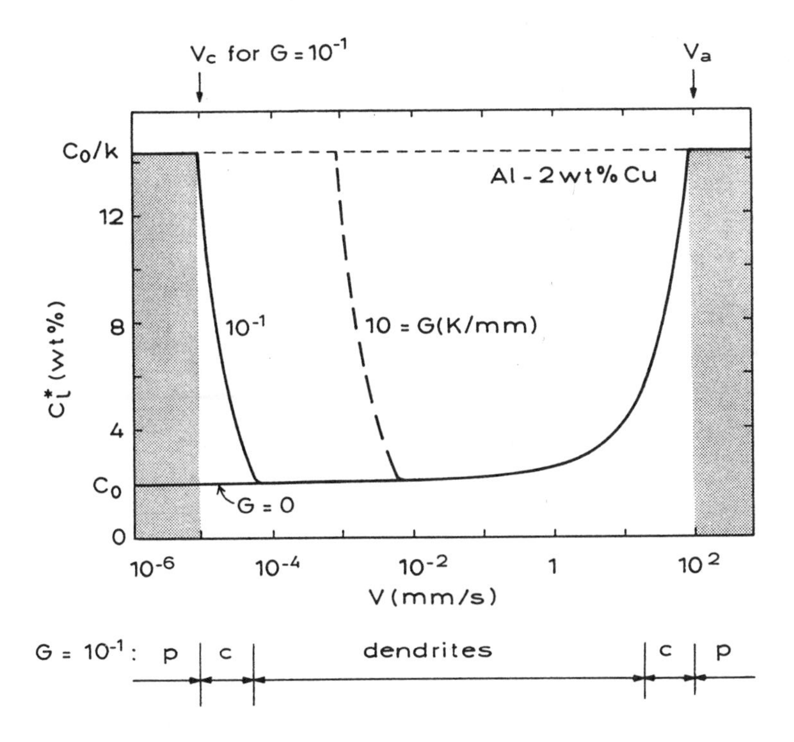

Figure 4.12 Liquid Concentration at the Tip of a Needle Crystal in Directional Growth

From the radius of curvature predicted by figure 4.11, the associated interface concentration can be derived. The blunt cell tips which exist close to the limit of constitutional undercooling, V_c, cannot easily dissipate the solute rejected there and the tip concentration will therefore be higher than that ahead of a sharp dendrite tip. If the temperature gradient is zero or negative, this will not occur as no cells are formed in this case. At very high growth rates, the interface concentration in the liquid will again increase to high values due to the increase in supersaturation necessary to drive the process. Again, at very high growth rates a crystal with a supersaturation of unity will grow which has the same composition, C_0, as the alloy (the composition of the liquid at the interface is then equal to C_0/k). Under these conditions, a planar solid/liquid interface will result; as in growth at low rates in a positive temperature gradient (grey regions). Note that the composition of the solid is related to C_l^* via the distribution coefficient, k. The latter becomes a function of V at growth rates of the order of 100mm/s or above (see chapter 7).

The semi-axis, b, is proportional to λ_1, where the proportionality constant depends upon the geometrical arrangement of the dendrites. An hexagonal arrangement is assumed in figure 4.14, where the last liquid is assumed to solidify at the centre of gravity of the equilateral triangle formed by three densely packed dendrites. This assumption leads to $b = 0.58\ \lambda_1$.

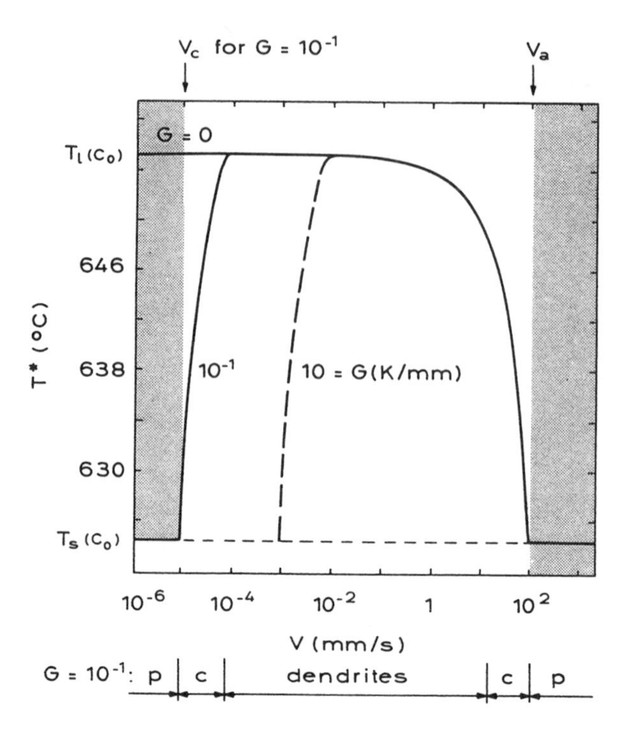

Figure 4.13 Interface Temperature of a Needle Crystal in Directional Growth

Use of the phase diagram and the assumption of the existence of local equilibrium at the solid/liquid interface permit the calculation of the temperature associated with the tip concentration (Fig. 4.12). At high and low growth rates, the tip temperature reaches the solidus temperature. In the low growth rate range, below the critical rate where the dendrite tip temperature and solidus temperature are equal (grey area on the left-hand side), the planar front is the more stable since it can grow at a higher temperature. (This is another way of interpreting the limit of constitutional undercooling). A similar argument applies at high growth rates.

The semi-axis length, a, is given by the difference between the tip temperature and the root temperature, divided by the mean temperature gradient in the mushy zone:

$$a = \frac{\Delta T'}{G} = \frac{T^* - T_s'}{G}$$

where, due to microsegregation, T_s' is often equal to the eutectic temperature if an eutectic exists in the system (chapter 6). Therefore, it can be assumed that the tip radius and the length of the interdendritic liquid zone together determine the primary spacing, due to purely geometrical requirements. From equation 4.17, one has $\lambda_1 \propto \sqrt{(Ra)}$, and:

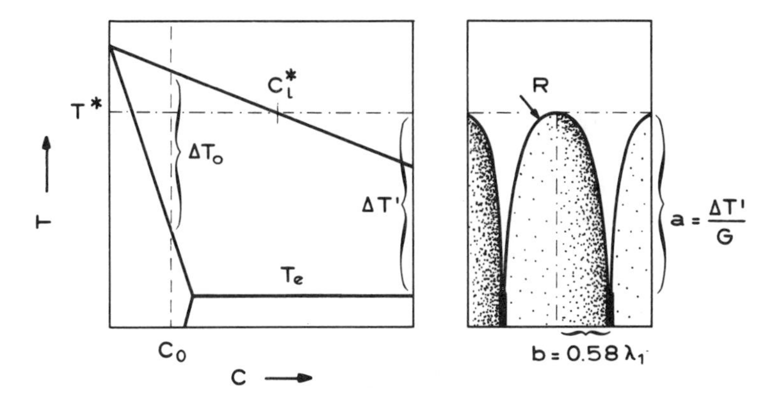

Figure 4.14 Estimation of the Primary Spacing in Directional Solidification

In practical applications, the dendrite tip radius is not as important a parameter as the primary spacing since it is very difficult to measure it directly. However, the tip radius has some influence on the primary spacing. In order to estimate the primary spacing, the dendrites are imagined to be elliptical in shape. The length of the major half-axis, a, of the ellipse is equal to $\Delta T'/G$. Here, $\Delta T'$ is the difference between the tip temperature and the melting point of the last interdendritic liquid. The primary spacing, λ_1, which is proportional to the minor half-axis, b, can be determined from simple geometrical considerations. The factor, 0.58, arises from the assumption that the dendrite trunk arrangement is close-packed hexagonal.

$$\lambda_1 = \left(\frac{3\Delta T' R}{G}\right)^{1/2} \qquad [4.18]$$

Knowing that $\Delta T'$ is strongly dependent upon the tip temperature, T^*, a sharp change in T^* at low and high V (Fig. 4.13) should also cause λ_1 to change sharply, as shown in figure 4.15. Cells will follow a different relationship for λ_1 as compared to that of dendrites. For the most important range of dendritic growth, at moderate growth rates, the following equation can be obtained by substituting equation 4.16 into equation 4.18 and assuming that, in this range, $\Delta T' \cong \Delta T_0$:

$$\lambda_1 = \frac{4.3(\Delta T_0 D \Gamma)^{0.25}}{k^{0.25} V^{0.25} G^{0.5}} \qquad [4.19]$$

This equation indicates that a variation in the growth rate has a smaller effect upon λ_1 than does a change in the temperature gradient. As in directional growth, the cooling rate, $\dot{T}$ is given by:

$$\dot{T} = -GV \qquad [4.20]$$

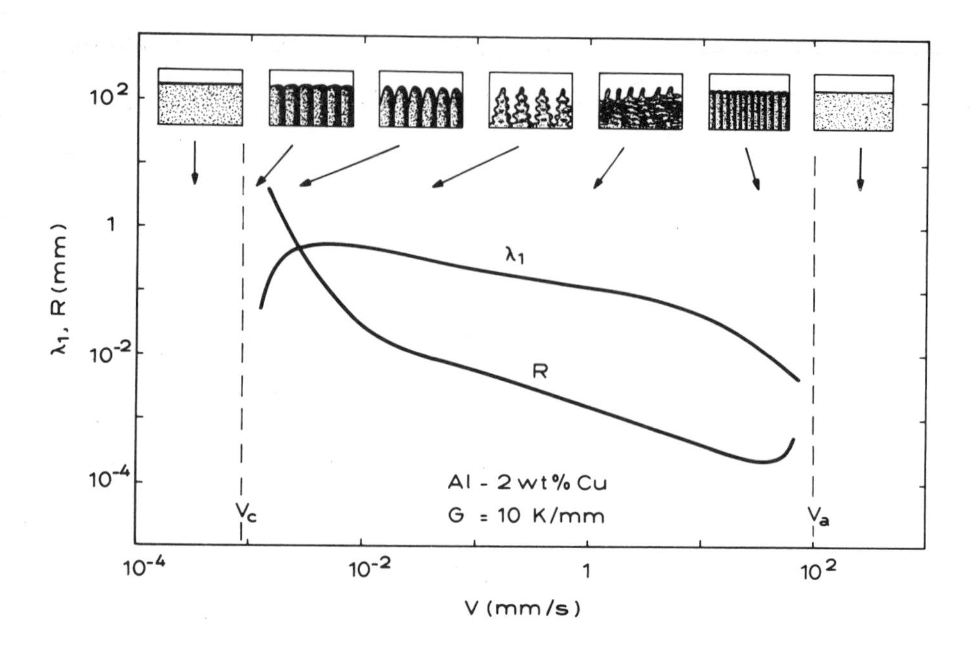

Figure 4.15 Morphology, Tip Radius, and Spacing of Cells and Dendrites

According to the dendrite model, the tip radius decreases from very large values at the limit of constitutional undercooling, V_c, to small values at high growth rates (as in figure 4.11). Over the range of dendritic growth, the primary spacing decreases approximately as the square-root of R (equation 4.18). The corresponding interface structures are also shown and vary from planar at growth rates less than V_c to cells and to dendrites which become finer and finer until they give rise to cellular structures again when close to the limit of absolute stability. At $V > V_a$, cell solidification structures disappear and again give a planar interface. (For more details on rapid V_a growth, see chapter 7).

It can be seen that λ_1 for dendrites will not obey simple relationships such as those often proposed in the literature e.g.:

$$\lambda_1 = K|\dot{T}|^n$$

Although equation 4.19 gives only a qualitative description of λ_1, due to the simplified model used, it nevertheless indicates that G and V have a different functional relationship to λ_1; a conclusion which is also borne out by Hunt's (1979) or Trivedi's (1984) analysis (see appendix 9)

Once the primary spacing is established, it will not change during, or after, solidification. This is not true of the secondary arms, which undergo a ripening process.

4.6 Secondary Spacing after Directional or Equiaxed Growth

As seen from figure 4.6, the secondary arms start very close to the tips. They initially appear as a sinusoidal perturbation of the paraboloid. As in the case of a planar solid/liquid interface which becomes unstable (Fig. 4.3), these perturbations grow, become cell-like, are sometimes eliminated by their neighbours, and a number of them finally become real secondary dendrites growing perpendicularly to the primary trunk (in the case of a cubic crystal). These secondary arms, with their higher-order branches, grow and eliminate each other as long as their length is less than $\lambda_1/2$. Once the diffusion fields of their tips come into contact with those of the branches growing from the neighbouring dendrites, they stop growing. A ripening process causes the highly-branched arms to change with time into coarser, less branched, and more widely-spaced ones (Fig. 4.6 and 4.16).

Careful inspection of the photographs in figure 4.16 suggests that one possible mechanism for the coarsening process is the melting of thinner secondary arms and an increase in the diameter of the thicker branches. This process is analogous to the Ostwald ripening of precipitates. The process is depicted schematically in the upper diagram of figure 4.16. Each time that a thin secondary arm melts, the local spacing is doubled. The driving force for the ripening process is the difference in chemical potential of crystals with differing interfacial energies due to differing curvatures. As in the ripening of precipitates, the spacing of the branches, λ_2, is proportional to the cube root of time (equation A9.55). Following Kattamis and Flemings (1965) and Feurer & Wunderlin (1977), one can write:

$$\lambda_2 = 5.5\,(Mt_f)^{1/3} \qquad [4.21]$$

with

$$M = \frac{\Gamma D \ln\left(\dfrac{C_l^m}{C_0}\right)}{m(1-k)(C_0 - C_l^m)} \qquad [4.22]$$

where C_l^m is often equal to C_e.

The value of M can easily vary by an order of magnitude. However, because its effect upon λ_2 is proportional only to the cube root, the differences will be relatively small when compared with the inevitable scatter to be expected in experimental measurements. Such results are presented in figure 4.17 for the secondary spacing of an Al-4.5wt%Cu alloy.

Measurements of the secondary spacing observed in directionally solidified or

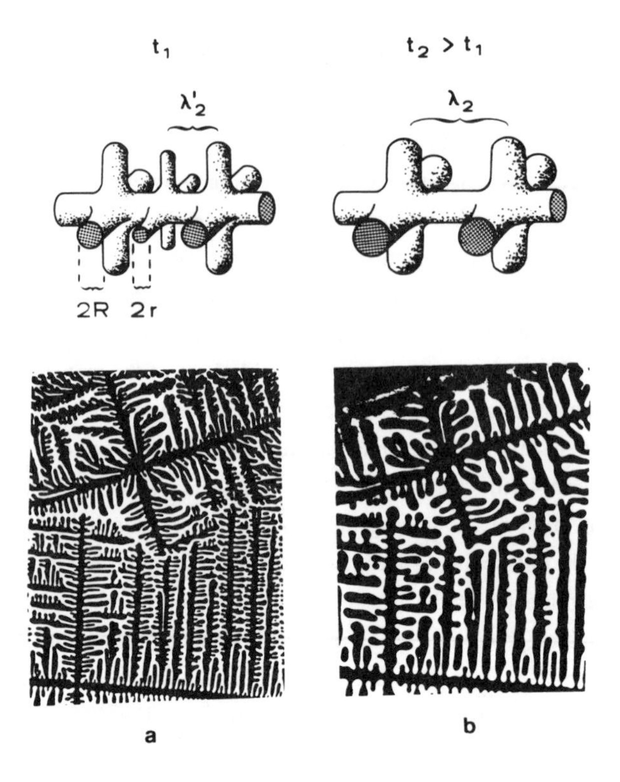

Figure 4.16 Establishment of the Secondary Dendrite Arm Spacing in Equiaxed Solidification

In contrast to the primary spacing, the secondary dendrite arm spacing, as measured in the solidified metal, is largely determined by annealing processes occurring during growth of the dendrites (Fig. 4.6). Due to a ripening phenomenon, smaller (higher curvature) features disappear and 'feed' the growth of the already larger features. The upper figures illustrate the model assumed in calculating the effect of these changes, while the lower photographs show equiaxed cyclohexane dendrites (a) just after solidification, and (b) 20 min later. In these photographs, the black areas correspond to the solid phase and the white areas to the liquid phase. Note that the primary spacing in an equiaxed structure is not well-defined and usually corresponds to the mean grain diameter. [Photographs: K.A.Jackson, J.D.Hunt, D.R.Uhlmann, T.P.Seward, Transactions of the Metallurgical Society of AIME **236** (1966) 149].

equiaxed structures give some indication of the local solidification conditions, and can be helpful when cooling curves are unavailable.

In the case of directional solidification, the local solidification time is given by:

$$t_f = \frac{\Delta T'}{|\dot{T}|} = \frac{\Delta T'}{|GV|} \qquad [4.23]$$

When there is no eutectic reaction in the system, the determination of $\Delta T'$ might be difficult. In chapter 6, it is shown that one can obtain an approximate value in this situation.

Figure 4.18 summarises the main points considered in this chapter. Thus, for one alloy, G and V are the main variables which determine the form and scale of the microstructures found after solidification of a given alloy. Specific G/V values (lines and bands running from the bottom left to the upper right) represent a constancy of microstructure (planar, cellular, columnar, and equiaxed dendritic). On the other hand, the various $G \cdot V$ ($= |\dot{T}|$) values (lines running from upper left to lower right) indicate a constant scale for these structures (e.g. λ_2). Thus, fine or coarse dendrites can be produced when G and V can be changed independently (see the insets in figure 4.18). Herein lies the value of the directional solidification (DS) method. That is, structures can be tailored to some extent so as to have optimum properties, as in the case of a turbine blade (Fig. 4.5). In a normal casting, G and V tend to be interrelated via the heat flux and the thermal properties of the metal. In a casting, therefore, only conditions close to the arrow in figure 4.18 can be exploited. When the growth rate is

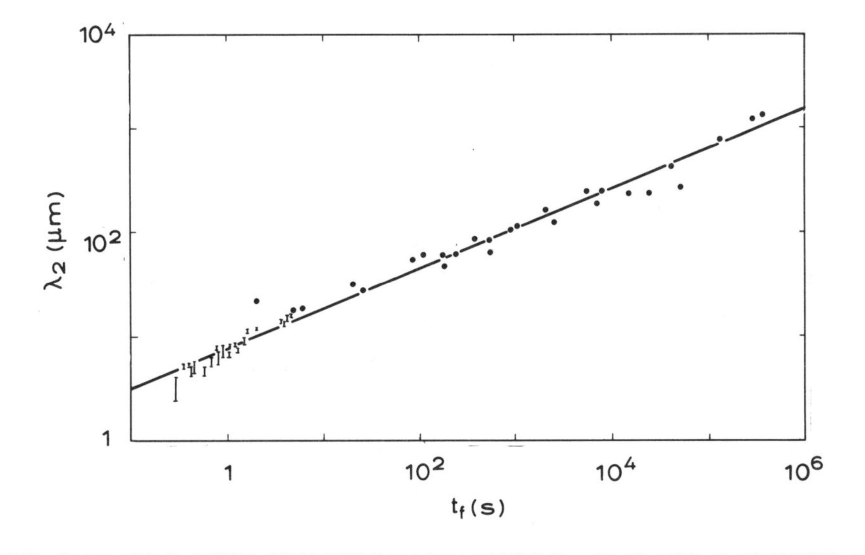

Figure 4.17 Secondary Spacing as a Function of Solidification Time

The best-fit curve to the experimental points for Al-4.5wt%Cu alloy over a wide range of solidification conditions shows that the secondary spacing varies approximately as the cube root of the local solidification time. The latter is defined as the time during which each arm is in contact with liquid (Fig. 4.1), and is therefore a function of the growth rate, the temperature gradient, and the alloy composition. The secondary spacing is important since, together with λ_1, it determines the spacing of precipitates or porosity and thus has a considerable effect upon the mechanical properties of as-solidified alloys (Fig. 1.2). [T.F.Bower, H.D.Brody, M.C.Flemings, Transactions of the Metallurgical Society of AIME **236** (1966) 624].

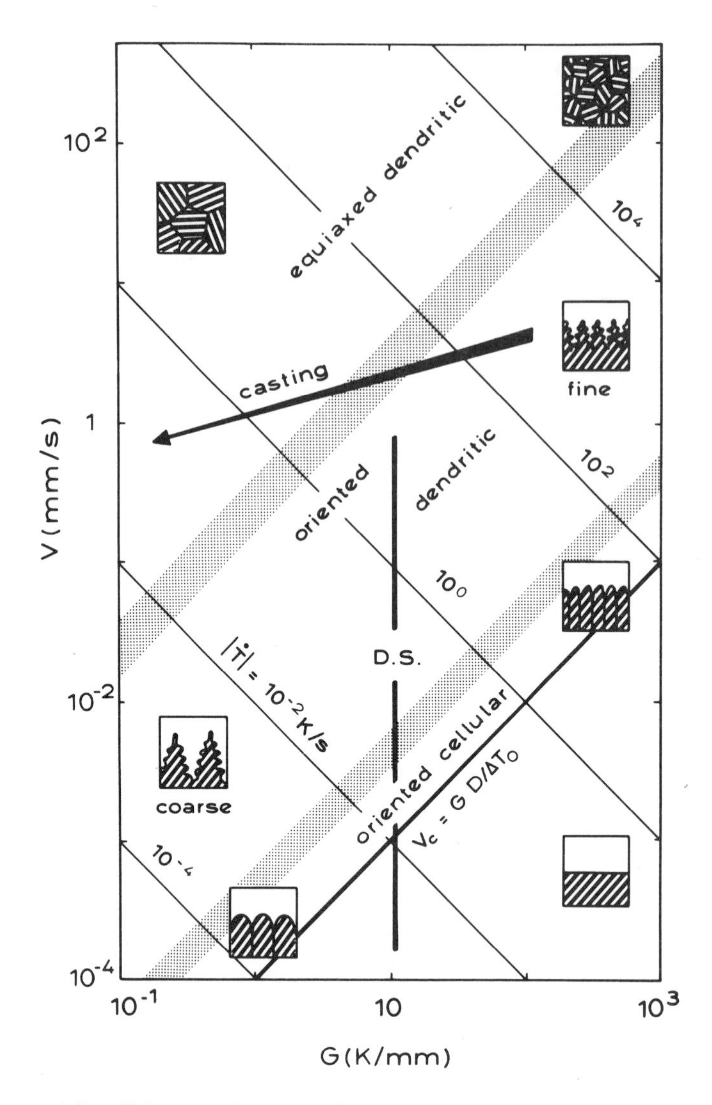

Figure 4.18 Schematic Summary of Single-Phase Solidification Morphologies

This diagram summarises the various microstructures which can be obtained using a typical alloy, with a melting range (ΔT_0) of 50K, when the imposed temperature gradient, G, or growth rate, V, are varied. Provided that a unidirectional heat flow is imposed, the product, $G \cdot V$, is equivalent to the cooling rate, $\dot{T}$, which controls the scale of the microstructures formed. The ratio, G/V, largely determines the growth morphology. Moving from the lower left to the upper right along the lines at 45° leads to a refinement of the structure without changing the morphology (G/V = constant). Crossing these lines by passing from the lower right to the upper left leads to changes in morphology (from planar, to cellular, to dendritic growth), and the scale of the microstructure remains essentially the same. The grey bands define the regions over which one structure changes into another. Superimposed on the diagram are typical conditions for two solidification processes, casting and directional solidification (D.S.). The conditions required to produce single-crystal turbine blades (Fig. 4.5) are those at the upper end of the thick vertical line marked D.S. Processes which produce perfect single crystals under conditions of plane front growth, such as those required for semiconductor-devices (silicon) are found at the bottom of the same vertical line. In a conventional casting, the growth conditions at the solid/liquid interface change with time approximately in the manner indicated upon following the slightly inclined arrow from right to left.

below the limit of constitutional undercooling, V_c (equation 3.15), plane front solidification is obtained and no solidification microstructure will develop. Note, however, that solidification with a planar interface does not alone ensure that a monocrystal will be obtained. To achieve this, elimination of all but one grain is necessary.

In castings there arises in the melt another important factor which will not be considered here. This is convective flow, which can markedly influence the transition from columnar to equiaxed dendritic growth. Generally, the presence of strong convection will decrease the length of the columnar dendritic zone and enhance equiaxed grain formation. This is due mainly to the melting-off of parts of dendrites and to the decreased value of G, and has a beneficial effect upon the internal quality of castings. In the continuous casting of steel, in particular, electromagnetic stirring has become a technologically important means of controlling the solidification structures.

Bibliography

Diffusion Field around Dendrite Tip

A.Papapetrou, Zeitschrift für Kristallographie **92** (1935) 89.

G.P.Ivantsov, Doklady Akademii Nauk SSSR **58** (1947) 567.

G.P.Ivantsov, *Growth of Crystals*, Consultants Bureau, NY, 1958, Vol. 1, p. 76.

G.Horvay, J.W.Cahn, Acta Metallurgica **9** (1961) 695.

Dendrite Tip Growth Theory (Free Growth)

R.Trivedi, Acta Metallurgica **18** (1970) 287.

R.Trivedi, Metallurgical Transactions **1** (1970) 921.

M.E.Glicksman, R.J.Schaeffer, and J.D.Ayers, Metallurgical Trans. **7A** (1976) 1747.

J.S.Langer, H.Müller-Krumbhaar, Journal of Crystal Growth **42** (1977) 11.

J.S.Langer, Reviews of Modern Physics **52** (1980) 1.

S.C.Huang, M.E.Glicksman, Acta Metallurgica **29** (1981) 701 and 717.

M.E.Glicksman, N.B.Singh, M.Chopra, in *Materials Processing in the Reduced Gravity Environment of Space* (Edited by G.E.Rindone), Elsevier, 1982, p. 461.

B.Cantor, A.Vogel, Journal of Crystal Growth **41** (1977) 109.

J.Lipton, M.E.Glicksman, W.Kurz, Metallurgical Transactions **18A** (1987) 341.

J.Lipton, W. Kurz, R.Trivedi, Acta Metallurgica **35** (1987) 957.

J.S.Langer, Physical Review **A33** (1986) 435.

P.Pelcé, ed., *Dynamics of Curved Fronts*, Academic Press, 1988.

Dendrite Tip Growth Theory (Constrained Growth)

M.H.Burden, J.D.Hunt, Journal of Crystal Growth **22** (1974) 109.

R.Trivedi, Journal of Crystal Growth **49** (1980) 219.

W.Kurz, D.J.Fisher, Acta Metallurgica **29** (1981) 11.

H.Esaka, W.Kurz, Journal of Crystal Growth **69** (1984) 362.

W.Kurz, B.Giovanola, R.Trivedi, Acta Metallurgica **34** (1986) 823.

Theory of Primary Dendrite Spacing

J.D.Hunt, in *Solidification and Casting of Metals*, The Met. Soc., London, 1979, p. 1.

W.Kurz, D.J.Fisher, Acta Metallurgica **29** (1981) 11.

R.Trivedi, Metallurgical Transactions **15A** (1984) 977.

Theory of Secondary Dendrite Spacing

T.Z.Kattamis, M.C.Flemings, Transactions of the Metallurgical Society of AIME **233** (1965) 992.

T.Z.Kattamis, J.C.Coughlin, M.C.Flemings, Transactions of the Metallurgical Society of AIME **239** (1967) 1504.

U.Feurer, R.Wunderlin, *Einfluss der Zusammensetzung und der Erstarrungsbedingungen auf die Dendritenmorphologie binärer Al-Legierungen*, Fachbericht der Deutschen Gesellschaft für Metallkunde, Oberursel, FRG, 1977.

P.W.Voorhees, M.E.Glicksman, Acta Metallurgica **32** (1984) 2001 and 2013.

Dendrite Spacing Measurements

K.P.Young, D.H.Kirkwood, Metallurgical Transactions **6A** (1975) 197.

T.Okamoto, K.Kishitake, I.Bessho, Journal of Crystal Growth **29** (1975) 131.

T.Edvardsson, H.Fredriksson, I.Svensson, Metal Science **10** (1976) 298.

H.Jacobi, K.Schwerdtfeger, Metallurgical Transactions **7A** (1976) 811.

G.M.Klaren, J.D.Verhoeven, R.Trivedi, Metallurgical Transactions **11A** (1980) 1853.

D.G.McCartney, J.D.Hunt, Acta Metallurgica **29** (1981) 1851.

M.A.Taha, H.Jacobi, M.Imagumbai, K.Schwerdtfeger, Metallurgical Transactions **13A** (1982) 2131.

K.Somboonsuk, J.T.Mason, R.Trivedi, Metallurgical Transactions **15A** (1984) 967.

R.Trivedi, K. Somboonsuk, Material Science Engineering **65** (1984) 65.

K.Somboonsuk, R.Trivedi, Acta Metallurgica **33** (1985) 1051.

H.Esaka, W.Kurz, Journal of Crystal Growth **72** (1985) 578.

Mathematical Functions

M.Abramowitz, I.A.Stegun, *Handbook of Mathematical Functions*, Dover, New York, 1965.

Exercises

4.1 What will happen when the angle of the dendrite trunk axis in figure 4.4b is at exactly 45° to the heat flow direction? Sketch a portion of such a dendrite growing under these conditions.

4.2 A microstructure such as that in figure 4.3 is being formed by directional solidification. The material is cubic and the cube axis is 20° away from the axis of the cells. What will happen when the growth rate is markedly increased?

4.3 What does monocrystalline mean in the case of figure 4.5 (no concentration variation in the solid, absence of low-angle boundaries, absence of high-angle boundaries)? Compare with the columnar zone of a casting (Fig. 4.2).

4.4 Design a mould which is suitable for the production of a dendritic monocrystalline casting, e.g. a gas-turbine blade.

4.5 Sketch a sequence of transverse sections of a dendrite which illustrates the region between the tip and the root of the dendrite in figure 4.6.

4.6 In appendix 8, the solution (equation A8.10) for solute diffusion around a hemispherical needle crystal is developed. Derive the equivalent solution for the case of heat diffusion. (Equiaxed dendrites of pure substances)

4.7 Imagine that the radius of curvature of a dendrite tip is changed during growth under a given set of conditions. What will happen to the concentration field of a columnar dendrite (Fig. 4.8) as the tip becomes sharper and sharper at a fixed growth rate. Note that $\Delta T = \Delta T_c + \Delta T_r$ in directional solidification. Sketch the concentration profile for the case where (a) $\Delta T > \Delta T_r$, (b) $\Delta T = \Delta T_r$, (c) $\Delta T < T_r$. Which situation corresponds to the critical nucleation radius? Which dendrite grows and which melts?

4.8 Calculate alloy undercooling for the case of figure 4.9 which represents an Al-2wt%Cu alloy solidifying under directional solidification conditions. From the value of $R°$, determine the Gibbs-Thomson parameter, Γ, used.

4.9 The unoptimised dendrite growth behaviour can be expressed as, $V = f(R)$ (Fig. 4.10) or as $\Delta T = g(R)$. Derive such an equation for the ΔT of a hemispherical needle crystal, and indicate the range of R values over which diffusion, or capillarity, is governing the growth. Calculate the extremum values for R and ΔT_c (start with equation A8.3).

4.10 When the results of more exact models for directionally solidifying dendrites are compared with those of the simple model given in table 4.2, the discrepancy between them is large at low and high growth rates. At both extremes, the Péclet number becomes large (P greater than unity). Discover which of the simplifications made is responsible for the unrealistic predictions of the simple model.

4.11 Calculate the tip temperature (table 4.2) of a dendritic growth front in Al-2%Cu alloy when $V = 0.1$mm/s and $G = 10$K/mm. Determine λ_1, λ_2, and the length of the mushy zone, assuming that G is constant in that region and that, due to microsegregation, the melting point of the last liquid is T_e. What is the value of the ratio, $\Delta T' / \Delta T_0$?

4.12 By using the lower limit, C_0/k, instead of C_l^m in equation 4.22, show how λ_2 varies with C_0 in a given alloy system. For many systems, this simplification is realistic.

4.13 In an Al-5wt%Si alloy casting, measurements of temperature and microstructure gave the results below. Compare these values to the theoretical ones and estimate the cooling rate (use the constants for Al-6wt%Si in appendix 14).

t_f(s)	λ_2(μm)
43	41 ± 3
330	81 ± 13
615	93 ± 3

4.14 In experiments involving strong, uniform flow of the melt perpendicular to the direction of growth of the solid/liquid interface (for example, in the electromagnetic stirring of steel during continuous casting), it is observed that columnar dendrites are inclined in a direction which is opposite to the flow direction. How would you explain this observation? Consider the way in which the boundary layers around the dendrite tip are changed. Would the same effect occur in pure metals?

4.15 Stirring of the melt during solidification is an efficient method for promoting the columnar-to-equiaxed transition in a casting. The reason for this transition in stirred castings is that the melt becomes more rapidly cooled and, at the same time, many dendrite branches are detached from the interface. With the aid of figure 4.1, indicate the two limiting temperatures at which the melt must be in order to make this transition possible.

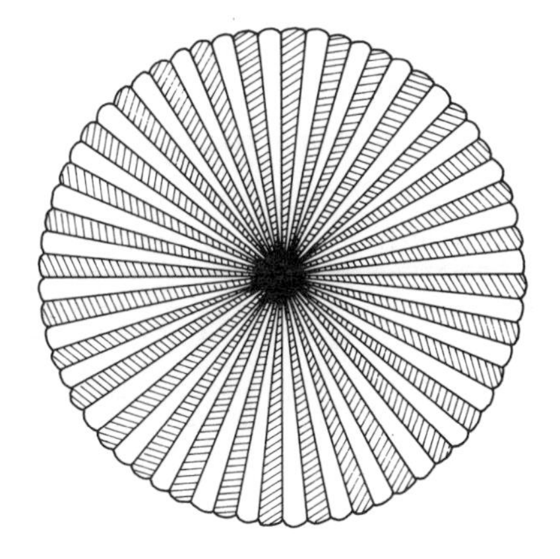

CHAPTER FIVE

SOLIDIFICATION MICROSTRUCTURE: EUTECTIC AND PERITECTIC

As shown in chapter 4, the solid which forms during solidification can adopt various morphological forms, and these can be present in a wide range of sizes. Dendrites make up the bulk of the microstructure of most alloys, but a number of important eutectic alloys are also used in practice. Eutectic morphologies are characterised by the simultaneous growth of two (or more) phases from the liquid. Due to their excellent casting behaviour, which is often similar to that of a pure metal, and the advantageous composite properties which are exhibited when solid, casting alloys are often of near-eutectic composition.

5.1 Regular and Irregular Eutectics

Due to the fact that they are composed of more than one phase, eutectics can exhibit a wide variety of geometrical arrangements. With regard to the number of phases present, as many as four phases have been observed to grow simultaneously. However, the vast majority of technologically useful eutectic alloys are composed of two phases. For this reason, only the latter type of eutectic will be considered here (Fig. 5.1).

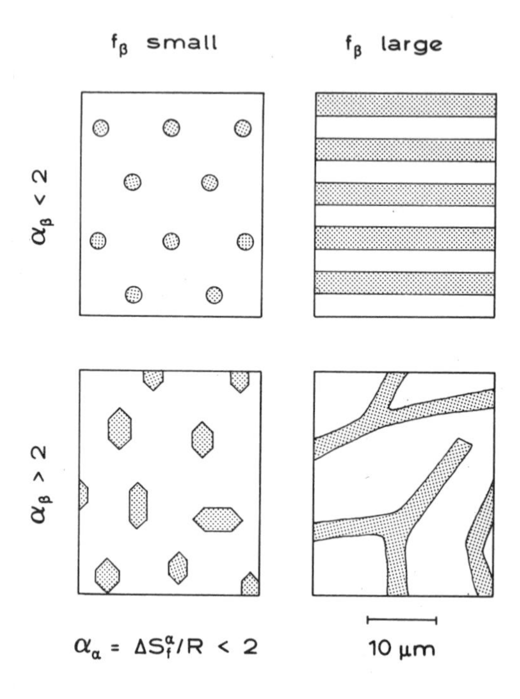

Figure 5.1 Types of Binary Eutectic Morphology

Seen in transverse cross-section, eutectic microstructures can be fibrous or lamellar, regular or irregular. The condition is that one phase (here the white α-phase) must always have a low entropy of fusion, so that $\alpha_\alpha < 2$. If both phases possess a low entropy of fusion, their growth is easy along all crystallographic directions (chapter 2.3) and the resultant structures are regular (non-faceted/non-faceted eutectic; upper part of the figure). When the low volume fraction phase possesses a high entropy of melting, as do semiconductors and intermetallic compounds for instance, the eutectics are of non-faceted/faceted type and the microstructures are usually irregular. The important eutectic casting alloys, Fe-C and Al-Si, belong to the latter class (lower right of the figure) in spite of the low volume fractions of C and Si. In general, fibres are the preferred growth form when a small volume fraction of one phase is present, especially in the case of non-faceted/non-faceted eutectics. This is so because, for λ = constant, the interface area, A, and therefore the total interface energy attributed to the surfaces between the phases decreases with decreasing volume fraction of the fibres [$A \propto (f_\beta)^{1/2}$], while the interface area is constant for lamellae. The interface area of fibres is lower than that for lamellae at volume fractions which are smaller than about 0.3. However, if the specific interface energy between the two phases is very anisotropic, lamellae may also be formed at a much lower volume fraction (as in Fe-C where $f_C = 0.07$).

At high volume fractions of both phases ($f \cong 0.5$); a situation which is encouraged by a symmetrical phase diagram, there is a marked preference for the formation of lamellar structures (e.g. Pb-Sn). On the other hand, if one phase is present in a small volume fraction, there is a tendency to the formation of fibres of that phase (e.g. Cr in NiAl-Cr). As a rule of thumb, one can suppose that when the volume fraction of one phase is between zero and 0.28, the eutectic will probably be fibrous; especially if both phases are of non-faceted type. If it is between 0.28 and 0.50, the eutectic will tend to be lamellar. If both phases possess a low entropy of fusion, the eutectic will exhibit a regular morphology. Fibres will become faceted if one phase has a high entropy of fusion or when interfaces having a minimum energy exist between the two phases. When faceting occurs, the eutectic morphology often becomes irregular and this is particularly true of the two eutectic alloys of greatest practical importance: Fe-C (cast iron) and Al-Si. These two eutectics are also generally close to lamellar structures even if the volume fractions of the second phases are much smaller than 0.28. Here, the strong anisotropy of growth of the faceted phase and of its interface energy plays an important role[#].

The regularity of an eutectic has a marked effect upon its mechanical properties, and this becomes especially important when it is required to control the orientation of the phases in order to obtain what are known as *in situ* composites. These are alloys where, using a controlled heat flux (Fig. 1.4), the eutectic phases can be caused to grow in a well-aligned manner. When the fibres are strong, as in the case of TaC in Ni-TaC eutectic, a marked increase in the creep strength of the alloy is found.

Because eutectic alloys exhibit small interphase spacings which are typically one tenth of the size of those of dendrite trunks under similar conditions of growth, a large

Table 5.1 Crystallography of Eutectic Alloys

Eutectic	Growth Directions	Parallel Interfaces
Ag-Cu	$[110]_{Ag}$, $[110]_{Cu}$	$(211)_{Ag}$, $(211)_{Cu}$
Ni-NiMo	$[1\bar{1}2]_{Ni}$, $[001]_{NiMo}$	$(110)_{Ni}$, $(100)_{NiMo}$
Pb-Sn	$[211]_{Pb}$, $[211]_{Sn}$	$(1\bar{1}\bar{1})_{Pb}$, $(0\bar{1}\bar{1})_{Sn}$
Ni-NiBe	$[112]_{Ni}$, $[110]_{NiBe}$	$(111)_{Ni}$, $(110)_{NiBe}$
Al-AlSb	$[110]_{Al}$, $[211]_{AlSb}$	$(111)_{Al}$, $(111)_{AlSb}$

[#] Nodular graphite in cast iron is an exception in non-faceted/faceted growth. In this case, the graphite grows as a primary faceted phase, together with austenite dendrites, and there is no eutectic-like growth of the two phases. Such a behaviour has been called divorced growth. If both phases are faceted, the types of coupled growth described in the present chapter are not observed.

total interfacial area exists between the two solid phases. For a 1 cm cube, this area is typically of the order of 1 m^2. Moreover, the specific energy of the interface is usually high and increases with increasing dissimilarity of the phases. As a result, there is a tendency for certain, lowest-energy, crystallographic orientations to develop between the phases and thereby minimise the interfacial energy. Extensive experimental studies have been made of the crystallography of eutectic alloys and the results have been reviewed by Hogan, Kraft and Lemkey (1971). Table 5.1 indicates the results for a number of systems. A list of most of the eutectics studied by means of directional solidification as well as their properties can be found in Kurz and Sahm (1975).

5.2 Diffusion-Coupled Growth

In order to determine the growth behaviour of the two eutectic phases, the simplest morphology for the solid/liquid interface will be assumed, i.e. that which exists during the growth of a regular, lamellar eutectic. For this case, the problem can be treated in two dimensions and, for reasons of symmetry (appendix 2), only half of a lamellae of each phase need be considered (Fig. 5.2). In this figure, the alloy is imagined to be growing in a crucible which is being moved vertically downwards at the rate, V'. In a steady-state thermal environment, this is equivalent to moving the solid/liquid interface upwards at a rate, $V = V'$. The alloy of eutectic composition is growing with its essentially isothermal interface at a temperature, $T^* = T_e - \Delta T$ below the equilibrium

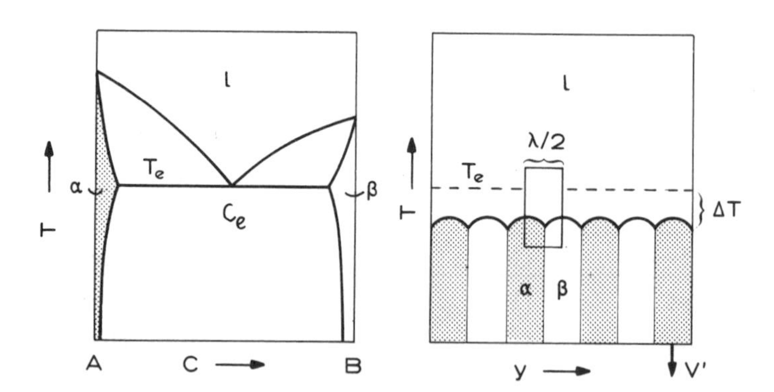

Figure 5.2 Phase Diagram and Regular Eutectic Structure

The figure shows an eutectic phase diagram and a regular lamellar two-phase eutectic morphology growing unidirectionally in a positive temperature gradient. The α- and β- lamellae grow side by side and are perpendicular to the solid/liquid interface. The form of the junctions where the three phases (α, β, liquid) meet is determined by the condition of mechanical equilibrium. In order to drive the growth front at a given rate, V, an undercooling, ΔT, is necessary. Due to the perfection and symmetry of the regular structures, only a small volume element of width, $\lambda/2$, need be considered in order to characterise the behaviour of the whole interface under steady-state conditions.

eutectic temperature. The α/β-phase interfaces are perpendicular to the solid/liquid interface and parallel to the growth direction. In order to proceed further, it is necessary to know more about the mass transport involved. It can be seen from the phase diagram that the two solid phases are of very different composition, while the melt composition, C_e, is intermediate in value. Obviously in the steady state, the mean composition of the solid must be equal to the composition of the melt. This makes it clear that eutectic growth is largely a question of diffusive mass transport.

Firstly, imagine that the two eutectic phases are growing separately from the eutectic melt with a plane solid/liquid interface (left-hand side of figure 5.3). During growth, the

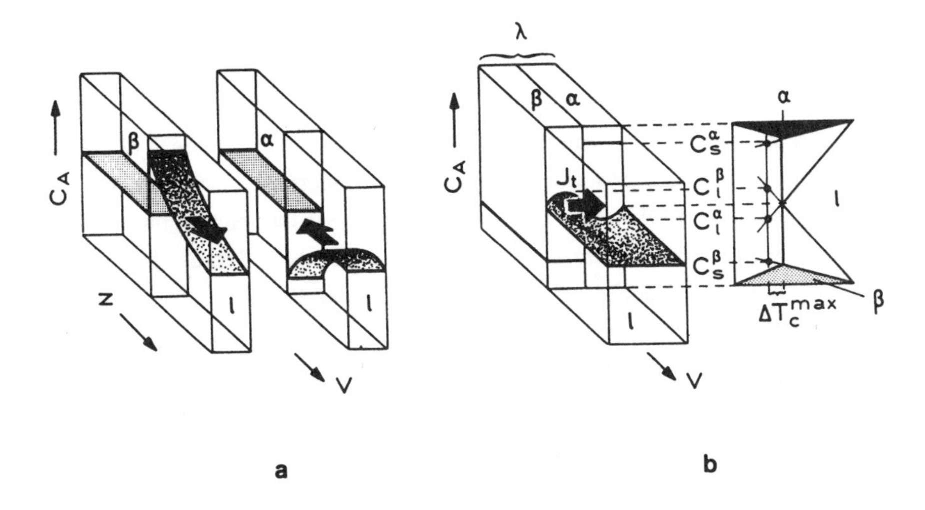

Figure 5.3 Eutectic Diffusion Field

If it is imagined that the two eutectic phases are growing from a melt of eutectic composition in separate adjacent containers (a), very large boundary layers, like that in figure 3.4, will be created. If both phases are constitutionally undercooled, their solid/liquid interfaces will break down to give dendrites, making solute rejection easier. If the two containers are now brought together and the intervening wall is removed (b), extensive lateral mixing will take place because of the concentration jump at the α/β interface. The large boundary layers of the planar interfaces of (a) (approximately equal to $2D/V$) are replaced by a very limited layer whose thickness is of the order of magnitude of the phase separation, λ. This marked change in the extent of the boundary layer is due to the diffusion flux which is established at, and parallel to, the eutectic solid/liquid interface and permits the rejection of solute by one phase to be balanced by incorporation of the solute into the other phase (diffusion coupling). The interface composition in the boundary layer oscillates, by a small amount, about the eutectic composition and the amplitude of the oscillation will decrease as λ decreases when V is constant. The lateral concentration gradients create free energy gradients which exert a 'compressive' force perpendicular to the α/β- interface and tend to decrease λ. The corresponding phase diagram has been placed next to the solid/liquid interface in such a way that the local phase equilibria can be determined. It can be seen that the amplitude of the concentration variation, $C_l^\beta - C_l^\alpha$, in the liquid at the solid/liquid interface is proportional to a maximum solute undercooling, $\Delta T_c{}^{max}$. (In this figure, curvature effects at the interface are not considered).

solid phases reject solute into the liquid. Thus, the α-phase will reject B-atoms into the melt, while the β-phase will reject A-atoms. Note here that, when the concentration is expressed as a fraction, $C_B = (1 - C_A)$. When the phases are supposed to be growing separately with a plane front, solute transport must occur in the direction of growth. This involves long-range diffusion and, in the steady-state, the solute distribution is described by the exponential decay discussed in chapter 3; with a boundary layer, $\delta_c = 2D/V$. Such a long-range diffusion field will involve a very large solute build-up and a correspondingly low (much lower than T_e) growth temperature at the interface. During steady-state growth, each phase would have the interface temperature indicated by the corresponding metastable solidus line when extended as far as the eutectic composition (see figure 3.4).

Imagine now that both phases are placed side-by-side and that the solid/liquid interfaces are at the same level (Fig. 5.3b). This situation is much more favourable since the solute which is rejected by one phase is needed for the growth of the other. Therefore, lateral diffusion along the solid/liquid interface, at right-angles to the lamellae, will become dominant and lead to a decrease in the solute build-up ahead of both phases. A periodic diffusion field will be established. Because the maximum concentration differences at the interface (compared to the eutectic composition) are much smaller than in the case of single-phase growth, the temperature of the growing interface will be close to the equilibrium eutectic temperature. The proximity of the lamellae, while making diffusion easier, also causes a departure from the equilibrium described by the phase diagram, due to capillarity effects (Fig. 5.4).

Both effects, diffusion and capillarity, are considered together in figure 5.5. In figure 5.5a, the diffusion paths (flux lines) at the interface are shown schematically. These are most densely packed (higher flux) at points near to the interface. They rapidly become less significant as the distance from the interface increases. The characteristic decay distance for the lateral diffusion is of the order of half the interphase spacing, λ. Note that the diffusion paths for the other species, in the opposite direction, are analogous. According to the phase diagram, the concentration variation at the solid/liquid interface (Fig. 5.5b) leads to a change in the liquidus temperature of the melt, T_l^*, in contact with the phases (thick line in figure 5.5c). The points where the liquid composition, C_B^* is equal to C_e are exactly at the eutectic temperature, while those points of the α-phase close to the α/β interface are at a higher liquidus temperature because the liquid in these regions has a lower content of B, as determined by the lateral diffusion field. On the other hand, the melt ahead of the β-phase is always richer in A than is the equilibrium eutectic composition. Therefore, its liquidus temperature is lower, compared to the equilibrium eutectic temperature, and decreases with increasing values of C_A. (These relationships can be understood with the aid of the phase diagram - see

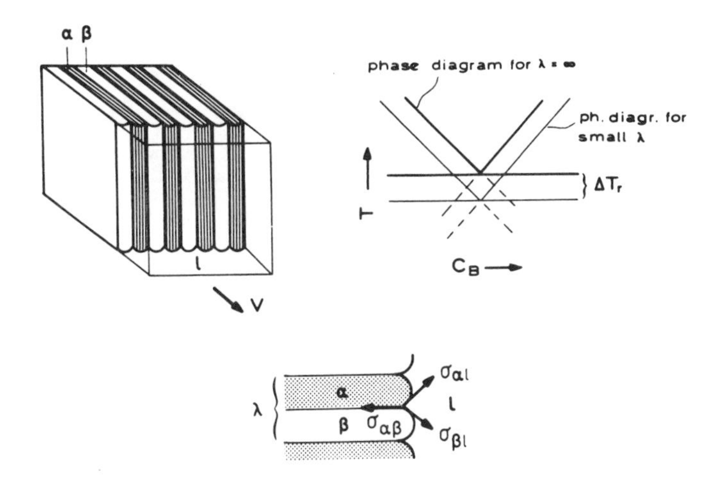

Figure 5.4 Curvature Effects at the Eutectic Interface

The diffusion field causes the λ-value of the structure to be minimised, and this leads to more rapid growth. There is an opposing effect which arises from the increased energy associated with the increased curvature of the solid/liquid interface as λ decreases. The latter can be expressed in terms of a curvature undercooling, ΔT_r, which depresses the liquidus lines of the equilibrium phase diagram as shown. The positive curvature of the solid phases in contact with the liquid arises from the condition of mechanical equilibrium of the interface forces at the three-phase junction (lower figure, see also appendix 3).

exercises 5.7 and 5.8). The capillarity undercooling, ΔT_r, shown also in figure 5.5 will be discussed in the next chapter.

In order to determine the solute distribution, a flux balance (appendix 2) must first be applied at the interface. In the present order-of-magnitude calculation it is assumed that the interface is planar and that the interface concentration variation of figure 5.5b can be approximated by using a saw-tooth waveform (constant concentration gradient) with amplitude, $\Delta C = C_l^\alpha - C_l^\beta$, and a diffusion distance in the y-direction of $\lambda/2$. In this way, the concentration gradient in the liquid at the solid/liquid interface is found to be: $(dC/dy)_{z=0} = -\Delta C/(\lambda/2)$. Note that this concentration variation decays very quickly ahead the interface, i.e. the amplitude $\Delta C \rightarrow 0$ when $z \sim \lambda/2$ and $(dC/dy)_{z=\lambda/2} \rightarrow 0$. Therefore, the mean concentration gradient leading to lateral solute diffusion in a narrow liquid band with a thickness of about $\lambda/2$, which is adjacent to the solid/liquid interface and perpendicular to the α/β interface, is half that of the interface value given above. That is, $\overline{dC/dy} = -\Delta C/\lambda$. The flux within this boundary layer is then:

$$J_t = D(\Delta C/\lambda)h\lambda/2 \qquad [5.1]$$

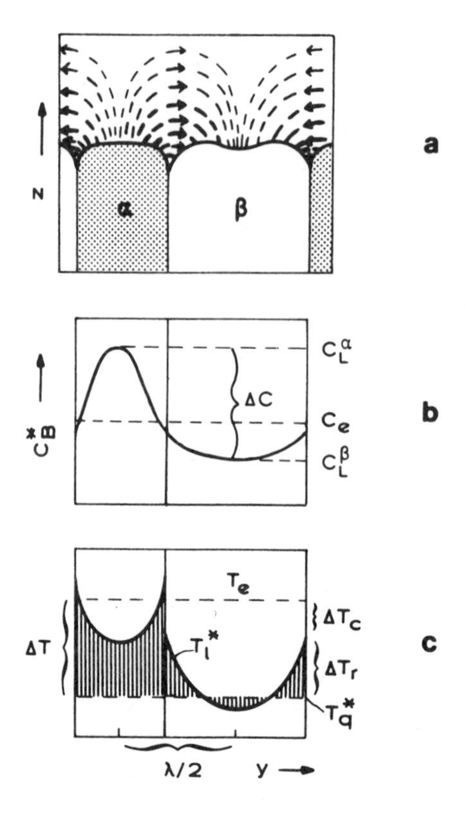

Figure 5.5 Eutectic Interface Concentration and Temperature

During growth, the diffusion paths of component B will be as shown in diagram (a). The concentration in the liquid at the interface will vary as in diagram (b). (Note that the eutectic composition is not necessarily found at the junction of the two phases, and that $C_B = 1 - C_A$). This sinusoidal concentration variation decays rapidly over one interphase spacing, in the direction perpendicular to the solid/liquid interface, as shown in figure 5.3b. The equilibrium between an attractive force arising from the diffusion field, and a repulsive force between the three-phase junctions arising from capillarity effects at small λ determines the eutectic spacing. The growing interface can be regarded as being in a state of local thermodynamic equilibrium. This means that the measurable temperature, T^{*}_{q}, of the interface which is constant along the solid/liquid interface (over $\lambda/2$) corresponds to equilibrium at all points of the interface. The latter is a function of the local concentration *and* curvature (c). The sum of the solute (ΔT_c) and curvature (ΔT_r) undercoolings must therefore equal the interface undercooling, ΔT. A negative curvature, as shown here at the centre of the β lamella, is required when the solute undercooling, ΔT_c, is higher than ΔT. The discontinuity in the solute undercooling, as the α/β-interface is crossed, is only a discontinuity in equilibrium temperature and not a real temperature discontinuity.

Here, the flux is weighted with the boundary layer cross-section, $h \cdot \lambda/2$, and has the dimensions [s^{-1}].

This flux, which is perpendicular to the growth direction, is needed in order to redistribute the solute which is rejected during the crystallisation of solid having a lower concentration than that of the liquid. The rejected flux (in the z-direction) for a symmetric phase diagram and equal volume fraction of both solid phases i.e. of one phase of half the width of the symmetry element and height, h, is:

$$J_r = VC_l^*(1 - k)h\lambda/4$$

Assuming that, under normal solidification conditions, the interface concentration deviates only slightly from the eutectic composition, i.e.ΔC is typically equal to a fraction of a percent and C_e is of the order of 50%, one can assume $C_l^* \cong C_e$ and the rejected flux becomes

$$J_r = VC_e(1 - k)h\lambda/4 \qquad [5.2]$$

Under steady-state conditions, the flux balance, $J_t = J_r$, can then be written:

$$\frac{\Delta C}{C_e(1 - k)} = \frac{\lambda V}{2D} \qquad [5.3]$$

which is, in fact, entirely analogous to the previously presented equation for the diffusional growth of a hemispherical needle. The left-hand-side of equation 5.3 corresponds to a supersaturation, while the right-hand-side is the Péclet number for eutectic growth. Therefore, one can also write equation 5.3 in the form:

$$\Omega_e = P_e \qquad [5.4]$$

The concentration difference, ΔC (= $\Delta C_\alpha + \Delta C_\beta$) which is required to drive solute diffusion during eutectic growth can be used to determine a temperature difference (undercooling) from the phase diagram (Fig. 5.6), via the liquidus slopes, $\Delta C_\alpha = \Delta T_c/(-m_\alpha)$, $\Delta C_\beta = \Delta T_c/m_\beta$ and $\Delta C = \Delta T_c\,[1/(-m_\alpha) + 1/m_\beta]$, leading, via equation 5.3, to a relationship of the form:

$$\Delta T_c = K_c \lambda V \qquad [5.5]$$

where K_c is a constant (see appendix 10).

From equation 5.4 or 5.5, it can be seen that this problem is not completely solved because, as in the case of dendritic growth, the above equations apply equally well to a fine eutectic growing at high rates or a coarse eutectic growing at low rates.

5.3 Capillarity Effects

Returning to the periodic concentration variation existing ahead of the solid/liquid interface (Fig. 5.5), it can be seen that the corresponding liquidus temperature varies from values greater than T_e, for certain regions of the α-phase, to values below the actual interface temperature, T_q^* for the central region of the β-phase. The difference (hatched region in figure 5.5c) has to be compensated by the local curvature in order to maintain local equilibrium at the interface. Thus, since T_q^* is constant due to the high thermal conductivity and small dimensions of the phases:

$$\Delta T = \Delta T_c + \Delta T_r = T_e - T_q^* = \text{constant}^{\#} \qquad [5.6]$$

A negative curvature (depression) may appear at the center of a lamella in order to compensate for a high local solute-controlled interface undercooling which is often associated with a large spacing, λ.

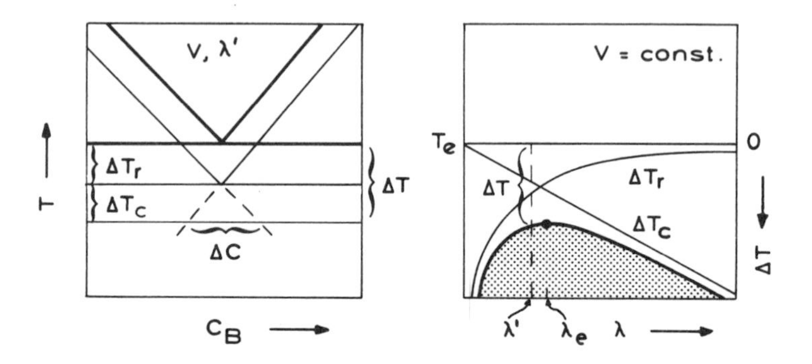

Figure 5.6 Contributions to the Total Undercooling in Eutectic Growth

Using figure 5.5c, a mean solute undercooling and a mean curvature undercooling can be defined. Both undercoolings vary in opposite senses when the spacing is changed. (The situation of figure 5.6a is shown for an arbitrary spacing, λ'). Here ΔT_c which is proportional to ΔC (driving diffusion) increases while ΔT_r decreases with increasing spacing (b). Note that ΔT is measured downwards with respect to T_e. The sum of the contributions exhibits a minimum in ΔT or a maximum in T with respect to λ. At smaller spacings, eutectic growth is controlled by capillarity effects ($\Delta T_r > \Delta T_c$). At larger spacings, diffusion is the limiting process. Generally, it is assumed that growth will occur at the extremum, λ_e. An increase in the growth rate increases the absolute value of the slope of the ΔT_c line, without influencing the ΔT_r curve, and displaces the maximum of the T-λ curve to smaller spacings.

At points close to the α/β interface, another condition must be imposed. That is, the α/β interface energy, $\sigma_{\alpha\beta}$, at the three-phase junction, has to be balanced by the sum of components of the α/l- and β/l-interface energies (Fig. 5.4). The angles at the three-phase junction are thereby determined by considerations of mechanical equilibrium (appendix 3).

The curvature of the α/l- or β/l- interface, which is necessary in order to match the

\# As in the case of directional dendrite growth, the thermal undercooling is zero during directional eutectic growth because heat is flowing into the solid ($G > 0$).

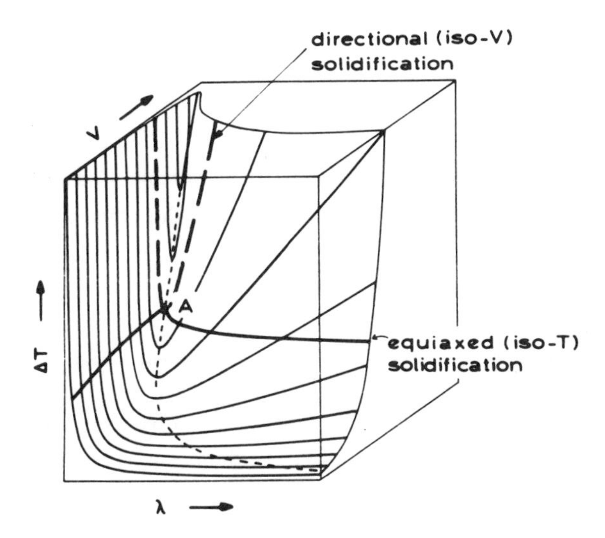

Figure 5.7 Optimisation of the Eutectic Spacing

A range of possible spacings exists, each of which satisfies local equilibrium requirements. This situation is described by the ΔT-V-λ surface of this figure which forms a valley running from the left rear to the front right of the diagram. (The ΔT axis is reversed here with respect to the equivalent diagram in figure 5.6b). It is seen that if the growth rate is constrained (V = constant), as in directional solidification, a minimum in the ΔT-λ curve (e.g. at point A) is obtained, corresponding to the maximum in the T-λ of figure 5.6b. If ΔT is maintained constant, as in equiaxed (isothermal) growth, the V-λ curve exhibits a maximum at point A. This curve for the plane-front growth of an eutectic is analogous to that in figure 4.9 for dendrites, where R replaces λ. The spacings corresponding to ΔT_{min} or V_{max} are called 'extremum' or 'optimum' spacings and, in regular eutectics, correspond closely to experimentally determined values. [After P.H.Shingu, Journal of Applied Physics, **50** (1979) 5743]

angles at the three-phase junction, changes the equilibrium temperature by an amount, ΔT_r, which is a function of y (Fig. 5.5c). By calculating the mean curvature over the α/l- and β/l- interfaces (appendix 10), the effect of capillarity can be related to a mean change in the liquidus temperature by (equation 1.5):

$$\Delta T_r = \Gamma K$$

Because the curvature, K, is proportional to $1/\lambda$:

$$K = \frac{K'}{\lambda} \qquad [5.7]$$

where K' is a constant. Use of equations 5.5 to 5.7 leads to the following relationships for the total solid/liquid interface undercooling:

$$\Delta T = K_c \lambda V + \frac{K_r}{\lambda} \quad [5.8]$$

Thus, a relationship is obtained which exhibits a maximum in the growth temperature (equivalent to a minimum in ΔT) as a function of λ (Fig. 5.6), where the growth rate is imposed and constant and ΔT is the dependent variable. Constant values of the growth rate are typical of directional solidification. On the other hand, the undercooling is imposed and constant in the initial stages of equiaxed solidification and results in a maximum in the V-λ-curve (Fig. 5.7).

5.4 Operating Range of Eutectics

Upon considering equation 5.8, it becomes clear that ΔT is not uniquely determined since it is a function of the product, λV. Therefore, another equation is required in order to determine the growth behaviour of an eutectic. This situation is analogous to that existing in dendrite growth. In the case of dendrite growth, the assumption of growth at the limit of morphological stability as the operating point has been found to correspond well with the experimental results. In the case of eutectic growth, both an operating point analogous to this one, and also the extremum point, have been found to explain various experimental results. Eutectic alloys which grow in a regular manner (e.g. Pb-Sn) can be described well by the use of the extremum criterion (see also the discussion in Seetharaman, Trivedi, 1988). Using this assumption, the first derivative of equation 5.8 is determined and set equal to zero:

$$\frac{d(\Delta T)}{d\lambda} = 0 \quad [5.9]$$

The condition that ΔT be a minimum implies, since $\Delta T = -\Delta G/\Delta S_f$, that $d(\Delta G)/d\lambda = 0$ and means that the driving force for spacing changes is zero. Insertion of the corresponding value of the spacing leads to the final result for growth at the extremum:

$$\lambda^2 V = \frac{K_r}{K_c} \quad [5.10]$$

$$\frac{\Delta T}{V^{1/2}} = 2(K_r K_c)^{1/2} \quad [5.11]$$

$$\Delta T \lambda = 2K_r \quad [5.12]$$

Equation 5.10 is illustrated in figure 5.8.

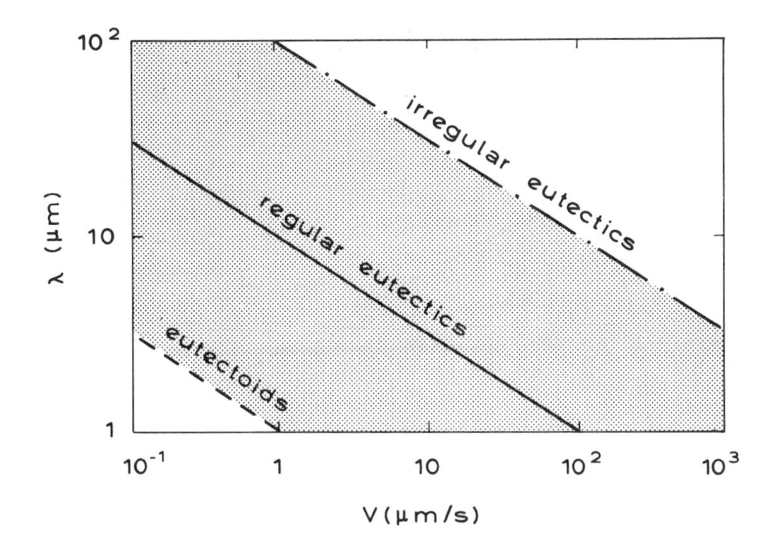

Figure 5.8 Eutectic and Eutectoid Spacings as a Function of Growth Rate

If the optimum spacings (Fig. 5.7) are determined for a range of growth rates (projection of the thin broken line on the λ-V surface in figure 5.7), curves such as the centre one above are found, where the spacing versus growth rate relationship can be described by $\lambda^2 V$ = constant. The eutectoid microstructure resembles that of the eutectic, as does its growth law, but diffusion occurs only through solid phases. Thus, there is a tendency to decrease the diffusion distances by decreasing the spacing, and the values of the latter are therefore smaller than those of regular eutectics. (Also, because of the slowness of solid-state diffusion, interface diffusion becomes important in eutectoid growth. In this case, the growth law changes.) Irregular eutectics, such as Fe-C and Al-Si, do not appear to grow at the extremum, but rather at larger values of λ and the larger spacings, with respect to regular eutectics, are explained by the branching difficulties of the faceted phase.

The situation is more complex when *irregular eutectics* are considered. In this case, the local spacing corresponding to the extremum value can be found, but the mean spacing is much larger (Fig. 5.8). Such large spacings can be explained by the difficulty which this class of eutectic experiences in branching. Branching is an essential mechanism which permits the eutectic to adapt its scale to the local growth conditions and to approach the extremum point. If one of the phases does not easily change direction during growth, and instead grows in a highly anisotropic manner (e.g. the faceted phases, C and Si, in Fe- and Al-alloys respectively), due to its atomic structure or planar defect growth mechanism, some lamellae of the faceted phase will diverge until one of them can branch. This behaviour can be understood with the aid of figure 5.9. When two adjacent lamellae diverge, the interface of the larger volume fraction phase will first become depressed because of the consequent increase in solute concentration at its centre (lower middle of figure 5.9). As the solute builds up more

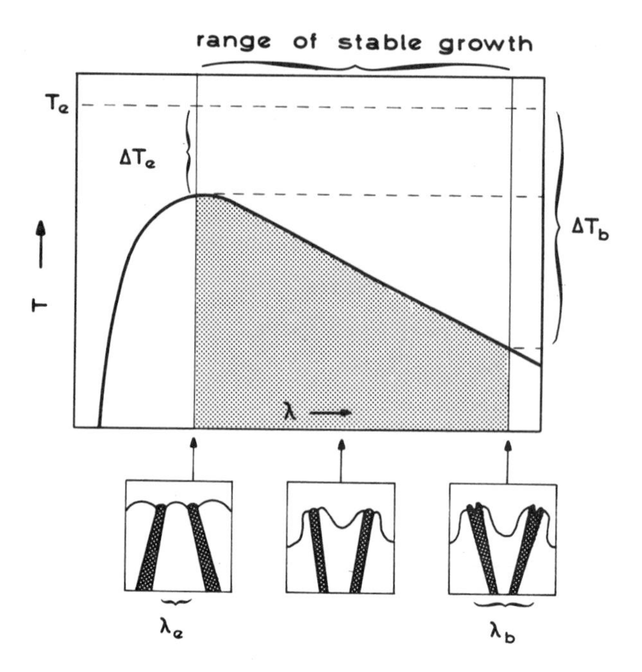

Figure 5.9 Spacing-Controlling Mechanisms in Irregular Eutectics

This figure illustrates the interface morphologies obtained due to convergent and divergent growth of the minor faceted phase of an irregular eutectic. In this case, the faceted phase can grow only along a planar growth defect (Fig. 2.10), making changes in the growth direction of the faceted phase, and therefore in the local spacing, very difficult. In convergent growth, the spacing will decrease during growth, thereby increasing the curvature of both phases. This will lead to the cessation of growth in that region because the interface temperature drops at the left of the extremum, λ_e, due to an increasing ΔT_r value. In divergent growth, the spacing becomes larger during growth, leading to increased solute pile-up ahead of both phases. This leads to the formation of depressions in the major phase at first (Fig. 5.5), and later in the minor phase. When the minor phase becomes depressed at λ_b, branching (formation of two lamellae from one) is possible and the spacing will be decreased again. This leads to zig-zag growth of the faceted phase between λ_e and λ_b.

and more at the interface of the diverging phases, the growth undercooling increases until the curvature can no longer compensate for the change and the interface becomes non-isothermal. Finally, the diverging phases will reach a spacing which is so large that even the low volume fraction phase will exhibit depressions at its solid/liquid interface. Under these conditions, the single lamella may branch into two. When a new lamella has been created, it will usually diverge from the other one of the pair and tend to converge on neighbouring lamellae at the interface (Fig.5.10). But, since the phase separation is decreasing, the interface temperature will increase, due to the decreasing solute build-up, and eventually reach the maximum in temperature (or minimum in undercooling). As the faceted phase cannot easily change its growth direction, its growth will decrease the local spacing to a value which is below the extremum value. However, smaller values will soon decrease the temperature appreciably due to the steep slope of the curve in this region. As a result, any spacing smaller than the extremum value will tend to be increased by the cessation of growth of one of the neighbouring lamellae (a termination appears).

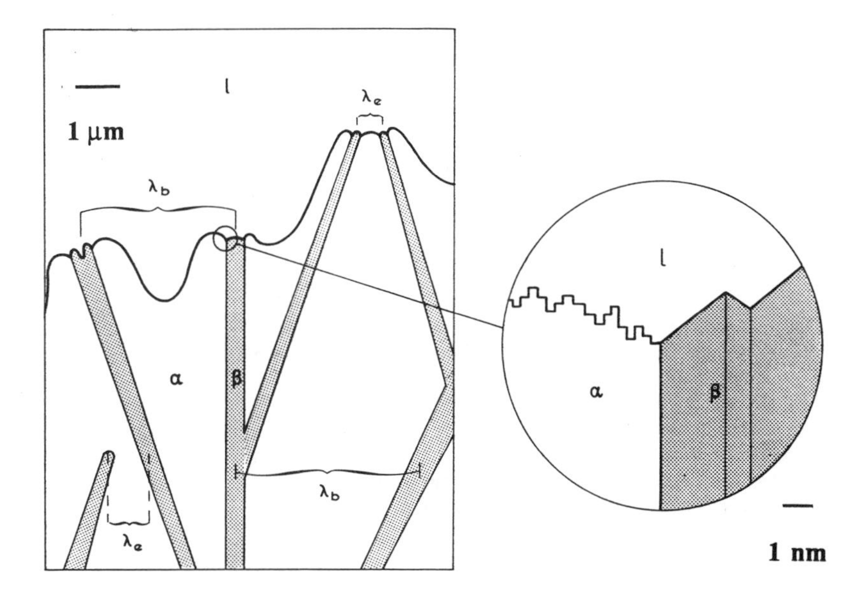

Figure 5.10 Growth of Irregular Eutectics

A difficulty (stiffness) in smoothly changing the growth direction of certain faceted phases results in the zig-zag growth structure described in figure 5.9. Here, the corresponding spacings are defined as well as the non-isothermal character of the solid/liquid interface. The resultant microstructures are irregular, and common examples are the two eutectics, Fe-C and Al-Si, upon which most commercial casting alloys are based. The magnified inset showing atomic processes represents the atomically diffuse, easily growing interface of the non-faceted α-phase and the planar defect growth mechanism for the faceted β-phase (e.g. Si, figure 2.10 b). The latter growth defects needed for easy growth of β make this phase grow with a marked "stiffness".

Thus, the range of stable eutectic growth is located between the extremum value, λ_e, and the branching spacing, λ_b. Only those eutectics with branching difficulties will exploit the whole range and this explains their coarse spacing (Fig. 5.8), large undercoolings, and large spacing variations (= irregularity).

The branching point can be calculated by using a stability analysis which is analogous to the criterion used in the case of dendritic growth. The main difficulty

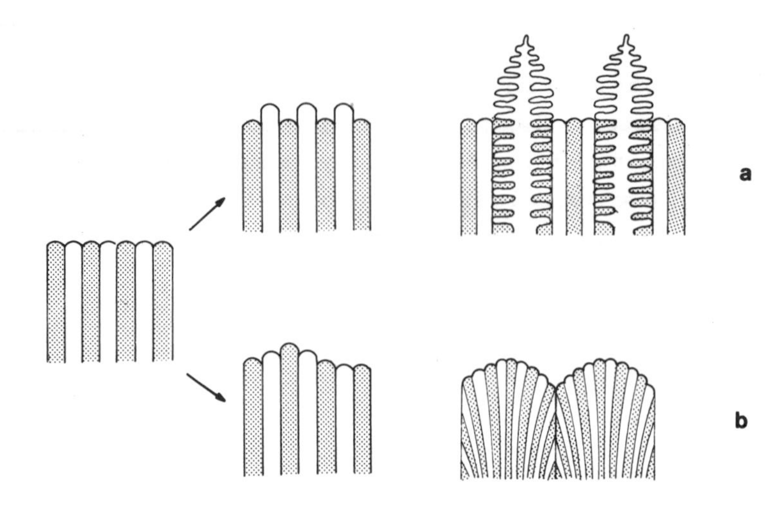

Figure 5.11 Types of Eutectic Interface Instability

The planar eutectic solid/liquid interface can become unstable just as in the case of a single-phase interface. Here, there are two different ways in which an instability can develop; instability of one phase (a), or instability of both phases (b). The former leads to the appearance of dendrites of one phase (plus interdendritic eutectic) and is mainly seen in off-eutectic alloys in binary systems. Alternatively, a third (impurity) element may destabilise the morphology as a whole because a long-range diffusion boundary layer is established ahead of the composite solid/liquid interface. [Recall that the eutectic tie-line of a binary system degenerates to an eutectic three-phase ($l+\alpha+\beta$) region in a ternary system]. This can lead to the appearance of two-phase eutectic cells or dendrites (b).

involved in using this approach is the estimation of the concentration gradient at the non-isothermal interface (Fisher & Kurz, 1980 and Magnin & Kurz, 1987).

5.5 Competitive Growth of Dendrites and Eutectic

As shown in figure 5.11, binary eutectics can undergo two types of morphological instability; single-phase or two-phase. The latter is analogous to the morphological instability of a planar single-phase interface (chapter 3) but, due to the very complex behaviour involved, quantitative analysis is difficult. In general, it can be said that a third alloying element which is similarly partitioned between both solid phases will lead to two-phase instability and the appearance of eutectic cells (Fig. 5.11b) or even eutectic dendrites.

During off-eutectic growth of a pure binary eutectic, single-phase instability can occur and result in the appearance of mixed structures, that is, dendrites of one phase and interdendritic two-phase eutectic (Fig. 5.11a). The reason for the latter effect is that, due to the long-range boundary layer built up ahead of the solid/liquid interface in

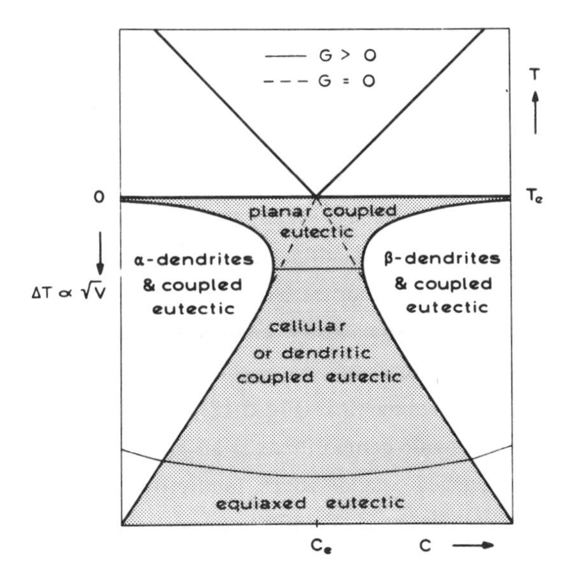

Figure 5.12 Coupled Zone of Eutectics

From consideration of the eutectic phase diagram alone, it might be thought that microstructures consisting entirely of the eutectic can only be obtained at the exact eutectic composition. In fact, due to the growth characteristics of dendrites and eutectics, at compositions close to the eutectic, the latter can often grow more rapidly than the dendrites and therefore outgrow them over a *range* of growth conditions. This occurs during directional growth, in a Bridgman furnace for instance when the dendrite tip temperature is low. When $G > 0$ this can happen at both low and high growth rates (Fig. 4.13). The coupled zone (grey region) represents the growth temperature/composition region where the eutectic grows more rapidly (or at a lower undercooling) than do dendrites of the α or β phases. This zone, corresponding to an entirely eutectic microstructure, may take the form of an anvil, where the upper widening is detected in experiments carried out at low growth rates and high (positive) temperature gradients, and the lower widening is found for growth at high growth rates; corresponding to high undercoolings. Within the coupled zone, an increased growth rate (decreased temperature) will destabilise the solid/liquid interface, due to the presence of impurities. This leads firstly to the formation of two-phase cells and later to the formation of two-phase dendrites (Fig. 5.11b). Outside of the coupled zone, primary dendrites and interdendritic eutectic will grow simultaneously (Fig. 5.11a). For a discussion of the transition from eutectic to glass formation, the reader is referred to chapter 7.

this case, one phase becomes heavily constitutionally undercooled. This can be deduced from the fact that, in an off-eutectic composition, the alloy liquidus is always higher than the eutectic temperature (Fig. 5.12). That is, the corresponding primary phase will be more highly undercooled and tend to grow faster than the eutectic. This case is of considerable importance because the properties of a casting can be appreciably impaired or enhanced when single-phase dendrites appear. Under some circumstances, dendrites can be observed in alloys having an exactly eutectic composition if the growth rate is sufficiently high. The reason for this behaviour is qualitatively explained by figure 5.13. The undercooling of the eutectic interface as a function of V is described by equation 5.11, while the undercooling of the dendrite tips obeys an analogous relationship when the temperature gradient is equal to zero. When growth occurs in a positive temperature gradient, the temperature-velocity curve exhibits a maximum (Fig. 4.13) and the eutectic curve (which is usually unaffected by G) may be cut at both high and low growth rates. When the dendrite curve is below the eutectic curve, only eutectic will be observed. When the dendrite curve is higher, both dendrites and eutectic are observed. If the undercoolings which lead to entirely eutectic growth are determined for a range of compositions, they make up what is known as the coupled zone.

When the coupled zone is symmetrical (Fig. 5.12), the eutectic morphology will obviously be obtained for eutectic compositions at all growth rates (undercoolings). However, in the case of skewed coupled zones, high growth rates may lead to the formation of α-dendrites even on the β side of the eutectic composition. Such skewed zones are usually associated with eutectics which contain one phase having anisotropic growth characteristics. Thus, a skewed zone is normally associated with irregularity of the eutectic morphology (e.g. Al-Si or Fe-C). These assumptions hold only for normal growth conditions. In the case of *rapid solidification processing* (for example: fibre-

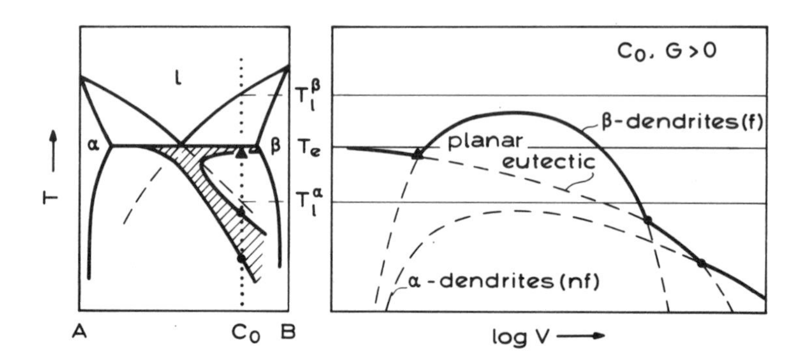

Figure 5.13 Origin of Skewed Coupled Zones

Symmetrical coupled zones (Fig. 5.12) are associated with regular eutectics and reflect the similar undercoolings of the two primary dendrite types. When the eutectic is irregular, the associated high undercoolings at high growth rates of the eutectic and the faceted primary β-phase, compared to that of the non-faceted α-phase, lead to the establishment of a skewed zone. The most important practical effect of this is that a fully eutectic microstructure may not be obtained when an alloy of eutectic composition is rapidly solidified. Because the zone is skewed towards the phase which has growth problems (such as Si in the Al-Si system), it is necessary to use a starting composition which is richer than C_e in the faceted element in order to obtain dendrite-free eutectic microstructures.

spinning, strip-casting, laser surface remelting), new phenomena may occur due to the extremely high undercooling reached. These have been summarised in an interesting paper by Boettinger (1982) and will be treated in more detail in chapter 7.

5.6 Peritectic Growth

In peritectic alloys, steady-state coupled growth, analogous to that of eutectics has not been observed. Some evidence exists which suggests that coupled growth may be possible over short distances but, in general, dendrites of one phase are formed. At a point behind the dendrite tip, partial reaction with the interdendritic liquid begins and gives the peritectic phase (Fig. 5.14). The microstructure of these alloys is therefore mostly dendritic, with the peritectic phase, and possibly an eutectic, forming the interdendritic precipitates. (Glardon & Kurz, 1981). The rate of this type of phase transformation has been analysed by Hillert (1979) and St John & Hogan (1987).

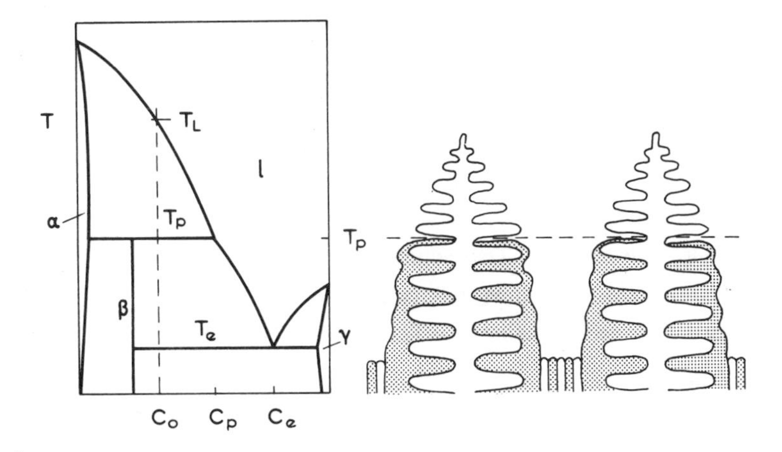

Figure 5.14 Directional Peritectic Growth

This is an important reaction which occurs in steels, bronzes, and other alloys. It has been suggested that conditions similar to those used in off-eutectic solidification might suppress primary dendrite growth and lead to the formation of an eutectic-like microstructure, at least over short distances. This has not been clearly demonstrated, and the usual sequence of events is that primary dendrite trunks form close to T_l and begin to react with the liquid to form β-phase close to T_p. Because this involves diffusion through the solid β-phase, the reaction is very slow and is rarely completed before the remaining, un-reacted liquid disappears or undergoes an eutectic reaction at T_e (if there is an eutectic in the phase diagram). Normal solidification conditions will tend to produce a microstructure which consists of primary α-dendrites, with a surface layer of β-phase, separated by interdendritic $\alpha+\gamma$ eutectic. Dissolution of the interdendritic peritectic phase by preferential chemical etching reveals the form of the dendrites (as on the cover of this book).

Bibliography

Theory of Eutectic Growth

C.Zener, Transactions of the Metallurgical Society of AIME **167** (1946) 550.

M.Hillert, Jernkontorets Annaler **141** (1957) 757.

W.A.Tiller, in *Liquid Metals and Solidification*, American Society for Metals, Cleveland, Ohio, 1958, p. 276.

K.A.Jackson, J.D.Hunt, Transactions of the Metallurgical Society of AIME **236** (1966) 1129.

T.Sato, Y.Sayama, Journal of Crystal Growth **22** (1974) 259.

D.J.Fisher, W.Kurz, Acta Metallurgica **28** (1980) 777.

R.Trivedi, P.Magnin, W.Kurz, Acta Metallurgica **35** (1987) 971.

P.Magnin, W.Kurz, Acta Metallurgica **35** (1987) 1119.

Eutectic Growth Experiments

B.Lux, W.Kurz, M.Grages, Praktische Metallographie **6** (1969) 464.

J.D.Livingston, H.E.Cline, E.F.Koch, R.R.Russell, Acta Metallurgica **18** (1970) 399.

A.Hellawell, Progress in Materials Science **15** (1970) 1.

H.A.H.Steen, A.Hellawell, Acta Metallurgica **20** (1972) 363.

R.M.Jordan, J.D.Hunt, Metallurgical Transactions **3** (1972) 1385.

Y.Sayama, T.Sato, G.Ohira, Journal of Crystal Growth **22** (1974) 272.

B.Lux, A.Vendl, H.Hahn, Radex Rundschau (1980) No.1/2, p. 30.

H.Jones, W.Kurz, Metallurgical Transactions **11A** (1980) 1265.

H.Jones, W.Kurz, Zeitschrift für Metallkunde **72** (1981) 792.

V.Seetharaman, R.Trivedi, Metallurgical Transactions **19A** (1988) 2955.

Coupled Zone

G.Tammann, A.A.Botschwar, Zeitschrift für Anorganische Chemie **157** (1926) 26.

A.Kofler, Zeitschrift für Metallkunde **41** (1950) 221.

E.Scheil, Giesserei, technisch-wissenschaftliche Beihefte **24** (1959) 1313.

M.Tassa, J.D.Hunt, Journal of Crystal Growth **34** (1976) 38.

W.Kurz, D.J.Fisher, International Metals Reviews **5/6** (1979) 177.

W.J.Boettinger, in *Rapidly Solidified Amorphous and Crystalline Alloys* (Eds. B.H.Kear, B.C.Giessen), Elsevier, New York, 1982.

R.Trivedi, W.Kurz, in *Solidification Processing of Eutectic Alloys* (Eds. D.M.Stefanescu, G.J.Abbaschian, R.J.Bayuzick), The Metallurgical Society, 1988, p. 3.

Directionally Solidified Eutectics

L.M.Hogan, R.W.Kraft, F.D.Lemkey, in *Advances in Materials Research*, (Ed. H.Hermann), 1971, Vol. 5, p. 83.

E.R.Thompson, F.D.Lemkey, in *Composite Materials*, Volume 4, (Ed. K.G.Kreider), Academic Press, New York, 1974.

W.Kurz, P.R.Sahm, *Gerichtet erstarrte eutektische Werkstoffe*, Springer, Berlin, 1975.

P.Magnin, W.Kurz, Metallurgical Transactions **19A** (1988) 1955 and 1965.

Peritectics

W.J.Boettinger, Metallurgical Transactions **5** (1974) 2023.

R.Glardon, W.Kurz, Journal of Crystal Growth **51** (1981) 283.

D.H.St John, L.M.Hogan, Acta Metallurgica **25** (1977) 77.

M.Hillert, in *Solidification and Casting of Metals*, The Metals Soc., London, 1979.

D.H.St John, L.M.Hogan, Acta Metallurgica **35** (1987) 171.

Cast Iron

B.Lux, Cast Metal Research Journal **18** (1972) 25 and 49.

I.Minkoff, *The Physical Metallurgy of Cast Iron*, Wiley, New York, 1983.

General

R.Elliott, *Eutectic Solidification Processing - Crystalline and Glassy Alloys*, Butterworth, London, 1983.

Exercises

5.1 Calculate the equilibrium volume fraction of graphite at T_e in an Fe-C alloy of eutectic composition. (Density of graphite = 2.15 x 10^3kg/m^3, density of C-saturated Fe = 7.2 x 10^3kg/m^3).

5.2 Calculate, on purely geometrical grounds, the volume fraction at which fibrous or lamellar structures have the lower total α/β interface energy. Assume that the α/β interface energy is isotropic and that the phase separation, λ, is equal and constant in both cases.

5.3 In figure 5.5b, the solute concentration in the liquid at the eutectic solid/liquid interface ($z = 0$) is given. Calculate the $C(y)$ function corresponding to $z = \lambda$ for an alloy of eutectic composition.

5.4 In order to demonstrate the potential value of producing 'in-situ' composites via directional eutectic solidification, calculate the total length of fibres contained in a cube, of such a composite material, having an edge length of 1cm and a fibre spacing, λ , of 1 μm.

5.5 An eutectic stores part of the transformation energy in the form of α/β interfaces (Fig. 5.4). By what amount, ΔT, will the melting point of a lamellar eutectic with $\lambda = 1$ µm and $\sigma_{\alpha\beta} = 5 \times 10^{-8}$J/mm^2 be lowered? As shown in appendix 14, the magnitude of Δs_f for metals is typically of the order of 10^6J/m^{3}K.

5.6 Find expressions for K_c and K_r (equation 5.8) for the simple case shown in the main text, and derive the solutions for extremum growth.

5.7 Using the phase diagram, explain the apparent discontinuity of the liquidus temperature, T_l^* at the α/β-junction shown in figure 5.5c.

5.8 Draw analogous diagrams to those in figure 5.5 for the case where both phases have a positive curvature, and for the case where both phases have depressions at their centres.

5.9 It can be easily shown that, at least over one half-spacing of the eutectic (Fig. 5.2), the solid/liquid interface of the eutectic must be very close to isothermal: an enormous heat flux in the y-direction would be required in order to change, even slightly, the interface temperature of the two phases. Using the properties of Al (appendix 14), calculate this lateral heat flux, assuming that ΔT_y = 0.1K and that $\lambda = 1$ µm.

5.10 Experiments performed on eutectic Fe-C alloys reveal the following relationship for the mean lamellar spacing: $\bar{\lambda}^2 V = 4 \times 10^{-7}$mm^3/s. Does this support the extremum criterion? Assume that $\theta_{Fe} = 20°$, and $\theta_C = 80°$. Use equation A10.31.

5.11 Is an effect of G upon the spacing of regular eutectic microstructures at a given growth rate to be expected? Explain your answer.

5.12 Draw a T-V diagram, similar to that in figure 5.13, for a symmetrical coupled zone under conditions where $G > 0$.

5.13 Repeat exercise 5.12 for the case when $G \leq 0$. Compare your results with figure 5.12. (First study figure 4.13 in detail).

5.14 Under constrained growth conditions ($G > 0$) at low growth rates, coupled eutectic growth without the appearance of dendrites is possible at off-eutectic compositions (Fig. 5.12). Calculate the limit of stability of a hypereutectic Al-Al_2Cu (θ) alloy containing 36wt%Cu at which θ–dendrites plus eutectic will appear as in figure 5.11a. (Hint: use the simple constitutional undercooling criterion and replace ΔT_0 by T_l - T_e). Show qualitatively to what point in figure 5.12 this situation corresponds. What will happen if G is doubled?

5.15 From the growth equations for dendrites (when $G \leq 0$) and eutectic: $\Delta T_d = K_d V^{1/2}$ and $\Delta T_e = K_e V^{1/2}$, determine the limiting growth rate and temperature of the coupled zone as a function of the composition. (Note that both growth morphologies must grow at the same rate in order to exist side-by-side under steady-state conditions. Assume that the constants, K_d and K_e, are independent of the composition and that ΔT_d and ΔT_e are defined as shown in the figure below.)

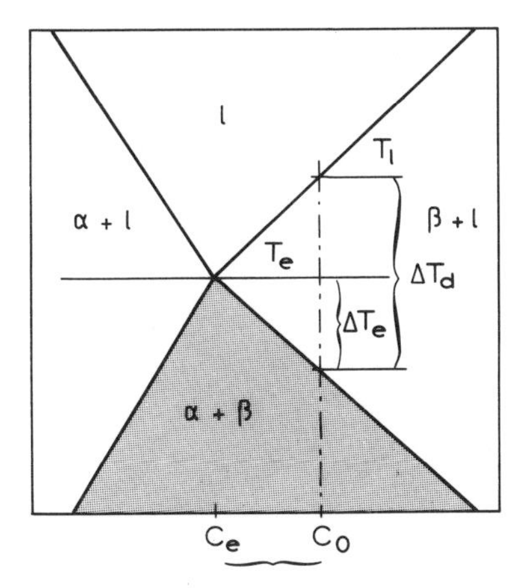

5.16 Sketch the solid/liquid interface of a unidirectionally and dendritically solidifying steel containing 0.2wt%C. Note that, due to the rapid solid-state diffusion of carbon in δ-Fe and γ-Fe, the liquid/solid and solid/solid transformations closely follow the behaviour to be expected on the basis of the equilibrium diagram.

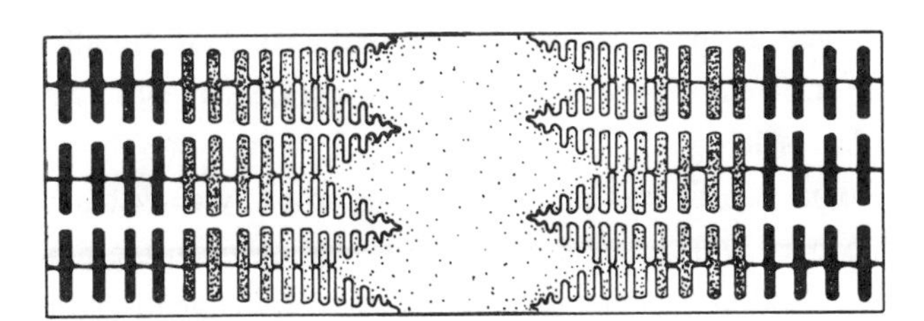

CHAPTER SIX

SOLUTE REDISTRIBUTION

It has been explained in some detail in chapter 3 that the solid/liquid interface rejects solute into the liquid when the solubility of the solute element in the solid is smaller than that in the liquid. In this case, the liquidus slope, m, is negative and the distribution coefficient, k, is smaller than 1. On the other hand, m is positive and k is greater than unity when the solubility is greater in the solid than in the liquid. In this case, solute will diffuse from the liquid to the solid. This leads to the creation of a depleted zone ahead of the solid/liquid interface.

As far as metals growing under normal solidification conditions are concerned, local equilibrium is assumed to hold at the solid/liquid interface. Then, at the interface, the solid concentration is related to the liquid concentration by the equilibrium distribution coefficient (equation 1.9):

$$C_s^* = kC_l^*$$

This compositional difference will always lead to concentration variations, in the solidified alloy, which are known as segregation. Note that the solute distribution in the liquid ahead of the solid/liquid interface leads to the appearance of the various growth morphologies and the latter in turn determine the solute distribution in the solid. Concentration differences over microscopic distances, interdendritic precipitates and porosity are the result.

Because solute can be transported by diffusion or by convection (or both), the segregation pattern will be quite different depending upon the process involved. Convection can lead to the transport of mass over very large distances, compared to those involved in diffusional processes, and may result in macrosegregation, that is, compositional differences over distances equal to the size of a large casting. As convection will not be treated in this book, for reasons outlined in the foreword, attention will here be limited to microsegregation#. The latter depends upon solute diffusion in the liquid and solid and is related to the dendrite shape and size. Understanding of this phenomenon is the key to interpreting the influence of solidification on the mechanical properties of cast products or welds. Furthermore, microsegregation reveals the original microstructure of any solidified alloy via the differences in etching tendency of regions of varying local composition.

In order to understand the segregation occurring at the scale of the dendrites, which is complicated due to the morphology of these crystals, it is useful to begin with a description of the solute distribution existing during directional solidification of a rod of constant cross-section with a planar solid/liquid interface (Fig. 1.4). In this special case, all of the changes occur in one dimension only and are therefore easier to analyse. Once this case is fully understood, it will be possible to use the results to study in a qualitative manner more complicated cases by imagining the changes occurring in small volume elements to be the same as those occurring during directional solidification (Fig. 1.5).

6.1 Mass-Balance in Directional Solidification

In chapter 3, the mass balance was treated only with respect to steady-state conditions. In order to understand microsegregation it is also essential to consider the initial and final transients. The first is required in order to establish the steady-state boundary layer, and the second arises from the interaction of the boundary layer at the

For an introduction to convection effects in solidification, the reader is referred to the bibliography at the end of the chapter.

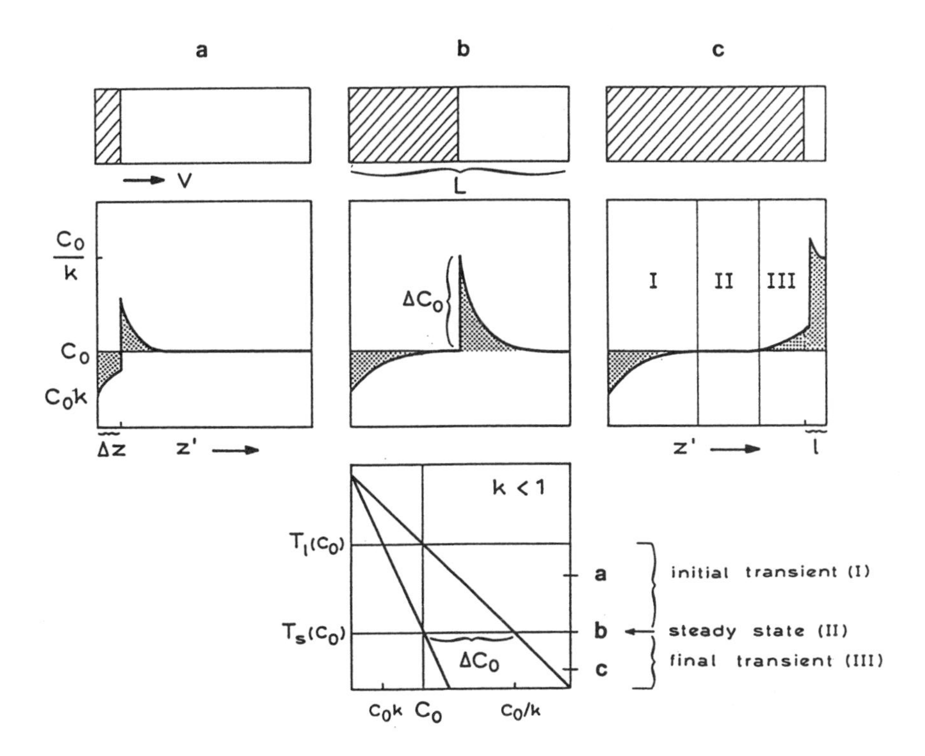

Figure 6.1 Initial and Terminal Transients

During directional solidification with a planar solid/liquid interface, in an apparatus such as the Bridgman furnace (Fig 1.4), the establishment of a steady-state boundary layer requires a distance of growth which corresponds to the length of the initial transient. This distance increases with decreasing growth rate. Within this transient, the concentration of the liquid at the interface increases from C_0 to C_0/k. With respect to the phase diagram (assuming the existence of local equilibrium), this means that the first solid to freeze has the composition, kC_0, and reaches the composition, C_0, and the interface temperature, T_s, corresponding to composition, C_0, at the steady-state. At the steady-state, the flux due to the interface advance (which arises from the difference in liquid and solid solubilities) is equal to the diffusional flux due to the concentration gradient at the solid/liquid boundary. In this case, the exponential decay described in chapter 3 is the exact solution. Finally, when the boundary layer becomes equal to the length of the remaining liquid region, diffusion into the liquid phase is hindered by the system boundary (the concentration gradient must be zero at the end of the crucible). Thus, the concentration in the liquid at the solid/liquid interface begins to increase to a value which is greater than C_0/k, and the solid concentration therefore becomes greater than C_0 and a terminal transient is created. In order to ensure mass conservation, the surface area of the grey region below C_0 must be equal to the grey area above C_0. Note that the lengths of the initial and terminal transients are unequal. These concentration variations can be a major problem in solidification processes and are known as segregation. The same phenomenon is usefully exploited in 'zone-refining', where the depleted (purer) part of the rod is used.

solid/liquid interface with the end of the specimen (Fig. 6.1). The diffusion boundary layer in the liquid ahead of a planar solid/liquid interface can be regarded as a limited region of the system which transports the solute missing from the initial transient in the solid where the concentration is below C_0, and maintains constant the overall composition of the system. This moving boundary layer disappears at the end of solidification by 'depositing' its solute content in the final transient. Thus, the *mean* composition of the solid is always the same as that of the liquid from which it is formed. Account is not taken here of reactions with the crucible or vaporisation of the alloy.

6.2 The Initial Transient

Figure 6.1 depicts in a schematic manner, the mechanism leading to the formation of the boundary layer and the initial transient. The process is shown for a bar of constant cross-section, A (appendix 11). When the first volume element, Adz', has solidified (where dz' is vanishingly small), the solute which has not been incorporated into the solid is equal to the incremental volume of the solid (Adz'), multiplied by the difference in composition between the liquid and the solid, $C_l^* - C_s^*$, which at $z' = 0$ is equal to $C_0 - kC_0$. This mass, divided by the time necessary for the advance of the interface by dz', leads to a solute flow at the interface where $V = dz'/dt$ (equation 3.4). Per unit interface area one gets

$$J_1 = VC_l^*(1 - k)$$

This flow leads to the creation of a pile-up at the interface and leads to the establishment of a concentration gradient, and diffusional flow into the liquid:

$$J_2 = -DG_c = V(C_l^* - C_0) \qquad [6.1]$$

Note that $G_c = -(C_l^* - C_0)V/D$ (appendix 2). The difference between the two flows, J_1, due to the creation of solid and J_2, due to diffusion in the liquid, gives the approximate variation in mean solute accumulation in a boundary layer of thickness, δ_c:

$$J_1 - J_2 = \left(\frac{d\overline{C}_l}{dt}\right)\delta_c \qquad [6.2]$$

This leads to a differential equation which, upon integration, gives (appendix 11):

$$C_l^* \cong \left(\frac{C_0}{k}\right)\left[1 - (1 - k)\exp\left(-\frac{kz'V}{D}\right)\right] \qquad [6.3]$$

As shown in figure 6.1, the initial transient exists until the increasing flow, J_1 (proportional to C_l^*) becomes equal to the initially smaller, but more rapidly increasing flow, J_2 (proportional to G_c). Figure 6.2 shows equation 6.3, for the solid composition, plotted as a function of the dimensionless distance. This permits an estimation of the length which must solidify before the steady state is reached. Note that in alloys with small values of k, solidifying at low rates, a large distance may be required. For example, if k is equal to 0.1, V is equal to 10^{-3}mm/s, and D is 5×10^{-3}mm^2/s, the characteristic distance required to establish a steady-state planar interface having 82% of the theoretical concentration is of the order of 100mm!

This is an important factor to bear in mind when applying steady-state theory to experiments involving a plane interface. Equation 6.3 is only an approximate, but useful, solution for the initial transient. The exact relationship can be found in Smith et al. (1955) and in appendix 11.

In the case of dendritic growth, the steady-state is reached much faster.

6.3 The Steady State

The steady state is established when $dC_l^*/dt = 0$. Then:

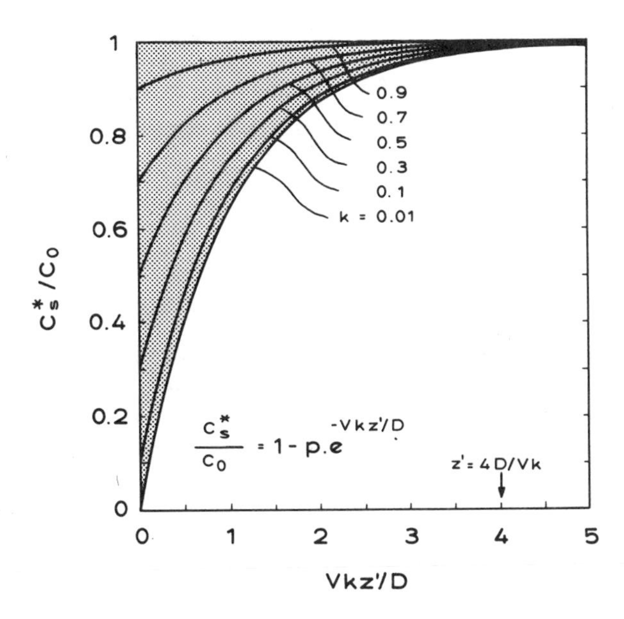

Figure 6.2 Length of the Initial Transient for a Planar Solid/Liquid Interface

In many solidification experiments, it is important to ensure that the steady-state has been reached. The diagram illustrates the length of the initial transient of figure 6.1 for typical distribution coefficient values of less than unity. A safe rule-of-thumb is that the distance of interface travel required for the establishment of the steady-state is $4D/Vk$.

$$J_1 = J_2 \qquad [6.4]$$

and:

$$\frac{C_l^* - C_0}{C_l^* (1 - k)} = 1 \qquad [6.5]$$

which is equivalent to $\Omega_c = 1$. In other words, the planar solid/liquid interface grows at the solidus temperature corresponding to the alloy composition. The concentration of the melt then decreases exponentially as a function of distance, z (equation 3.2):

$$C_l - C_0 = \Delta C_0 \exp\left(-\frac{Vz}{D}\right)$$

forming a diffusion boundary layer with the characteristic length, $\delta_c = 2D/V$ and a concentration gradient at the interface, $G_c = \Delta C_0 V/D$.

6.4 The Final Transient

When the boundary layer becomes comparable to the length of the remaining liquid zone, interaction with the boundary of the system begins to occur (Fig. 6.1). The end of the liquid zone can be regarded as being a perfectly impermeable wall which imposes a zero flux condition at that point:

$$\left(\frac{dC}{dz}\right)_{z'=L} = 0 \qquad [6.6]$$

This is equivalent to saying that the flow into the liquid decreases and therefore the concentration increases as shown in figure 6.1. If it is imagined that the end of the specimen is acting as a source with a strength equal to the flow which would leave this boundary if it were permeable, equation 6.6 is satisfied and the final concentration profile can be found by superposition of symmetric sources. This method leads to a series giving C_l^* as a function of the final liquid length, l (appendix 11).

6.5 Rapid Diffusion in the Liquid - Small Systems

The analysis of the solute distribution becomes much simpler when it is assumed that diffusion in the liquid is sufficiently rapid to avoid the establishment of any concentration gradient ahead of the solid/liquid interface. This is a reasonable assumption to make in the case of very high diffusion coefficients in the liquid, strong convection, and/or a very small system size, L, compared to the boundary layer thickness. Considering only the case of diffusion, no concentration gradient will exist in the liquid when the diffusion boundary layer is much greater than the system size, L,

$$\delta_c = \frac{2D_l}{V} \gg L \qquad [6.7]$$

This is so because the insulating effect of the end of the specimen will then smooth out any concentration gradient. Under these conditions, the most general treatment involves a combined approach (appendix 12). Figure 6.3 shows the corresponding concentration profile. Here, the boundary layer in the solid, δ_s, due to back diffusion, will take a value between zero and infinity depending upon the value of the diffusion coefficient in the solid. With the relative interface position, $s/L = f_s = (1 - f_l)$, the mass balance, $A_1 = A_2 + A_3$, can be written, to a first-order approximation:

$$(C_l - C_s^*)\, df_s = f_l\, dC_l + \frac{\delta_s}{2L}\, dC_s^* \qquad [6.8]$$

where $C_l = C_l^*$. Assuming that the interface position, s, is a parabolic function of time#, and integrating (appendix 12) gives:

$$\frac{C_l}{C_0} = (1 - uf_s)^{-p/u} \qquad [6.9]$$

where $u = 1 - 2\alpha' k$, $p = 1 - k$, and C_s^* can be obtained directly from $C_s^* = kC_l$. The parameter, α', can be calculated from:

$$\alpha' = \alpha\left[1 - \exp\left(-\frac{1}{\alpha}\right)\right] - 0.5 \exp\left(-\frac{1}{2\alpha}\right) \qquad [6.10]$$

where the dimensionless diffusion time (Fourier number) is:

$$\alpha = \frac{D_s t_f}{L^2} \qquad [6.11]$$

The behaviour of α' is such that, at small α-values (less than 0.1), $\alpha' = \alpha$, and at large α-values (greater than 50), $\alpha' = 0.5$ (Fig. A12.3).##

Substituting these limiting values of α' into equation 6.9 leads to two well-known equations. When α is equal to zero, α' is also equal to zero (no solid-state diffusion takes place) and:

This function gives a consistent solution to the problem. This is not the case when one assumes a linear relationship between s and t.

Due to the simplification made in deriving equation 6.9 it can only be used for k-values smaller than 1. For a more exact analysis of this problem see Kobayashi, 1988 (Appendix 12).

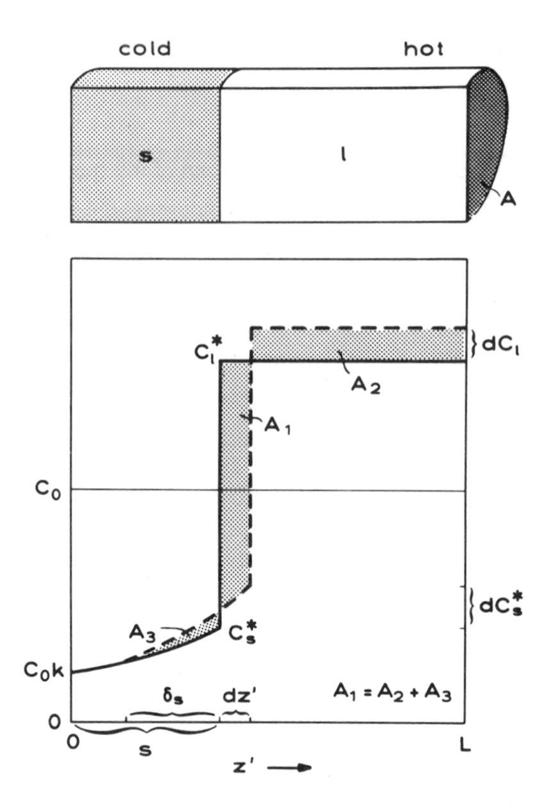

Figure 6.3 Segregation with Complete Liquid Mixing and Some Solid-State Diffusion

When mass transport in the liquid is very rapid, e.g. due to the effect of convection, the excess solute will be redistributed evenly over the entire volume of liquid. In this case, there will be an interaction with the far end of the crucible during the whole solidification process, and the entire solute distribution will essentially be a long terminal transient beginning at the solid concentration, C_0k. This behaviour is described by the 'Scheil' equation and predicts an infinite concentration at the end of solidification. In practice, eutectic solidification usually intervenes. In order to obtain a more realistic description of the concentration profile at the end of solidification, solid-state back diffusion must be taken into account. This can be done by using a simple mass balance. Thus, the solute rejected from the solid (represented by surface A_1) over the distance, dz', will partially increase the uniform liquid concentration by dC_l (surface A_2), and this in turn will increase the interface concentration and the associated concentration gradient in the solid, and therefore the flux into the solid (surface A_3). Mass conservation requires that the sum of the quantities described by the three surfaces must be zero. If diffusion in the solid is very rapid (e.g. C in δ-Fe), the boundary layer, δ_s, in the solid will be very large. In the limit, due to interaction with the initial boundary of the system ($z' = 0$), the concentration gradient in the solid will be decreased, leading, for the extreme case of infinitely rapid back-diffusion, to equilibrium solidification (lever-rule). In this case an homogeneous solid, with composition C_0, will result after the completion of solidification. The extent of back diffusion will depend on a dimensionless parameter, α, which can be regarded as describing the ratio of the diffusion boundary layer in the solid, δ_s, to the size of the system, L.

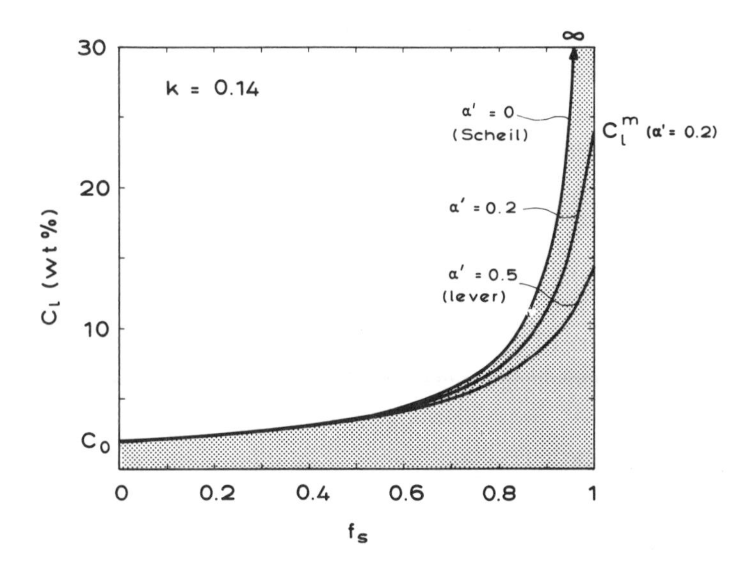

Figure 6.4 Segregation Curves in the Presence of Back Diffusion

The composition of the liquid (assumed to be homogeneous as in figure 6.3) increases at the end of the specimen. Under lever-rule conditions, the increase is from C_0 to C_0/k, while the Scheil equation predicts an increase from C_0 to infinity. All of the intermediate cases can be described by one relationship (equation 6.9) which contains a modified α-parameter, α', that can take values between 0 and 0.5. Note that the curve represents the path of the interface concentration, C_l^*, as a function of f_s. The final solute distribution profile in the solid cannot be determined in this way because it changes with time when $\alpha > 0$ due to back-diffusion. Therefore, only the end concentration ($f_s = 1$) represents a measurable value.

$$\frac{C_l}{C_0} = \frac{1}{(1-f_s)^p} = \frac{1}{f_l^p} \qquad [6.12]$$

This is known as the '*Scheil equation*'.

On the other hand, when α approches infinity, α' becomes equal to 0.5 and this implies that solidification will occur under equilibrium conditions. That is, solid-state back-diffusion is so rapid or the solid-state diffusion boundary layer, δ_s, is so much greater than L that, as in the case of the liquid, the insulating effect of the solid specimen end also smoothes out any concentration gradient in the solid. This case is described by the '*lever rule*':

$$\frac{C_l}{C_0} = \frac{1}{1-pf_s} \qquad [6.13]$$

Figure 6.4 illustrates the behaviour of equation 6.9 for an Al-2%Cu alloy, including the limiting cases described by equations 6.12 and 6.13. It is evident that the Scheil equation is a very poor approximation with regard to the final liquid composition, since the maximum liquid concentration, C_l^m is infinite. On the other hand, equilibrium solidification according to the lever-rule case, which leads to a final liquid concentration, $C_l^m = C_0/k$, is again unrealistic for most solutes because of their low solid-state diffusivity. However, there are very important exceptions such as those of interstitial solutes, especially in open crystal structures and small systems (e.g. interdendritic segregation of C in δ-Fe). The latter diffusion coefficient is so high that α-values greater than 100 are found. Knowledge of the value of C_l^m also permits the temperature of the remaining liquid to be calculated from the phase diagram (Fig 6.5).

6.6 Microsegregation

So far, solute distributions have only been considered for the case of relatively large

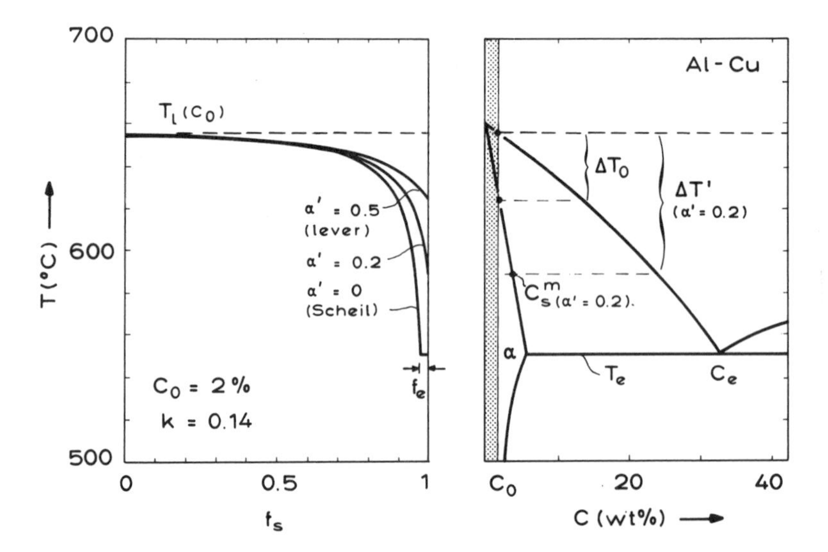

Figure 6.5 Relationship of the Segregate Freezing Point to the Phase Diagram

An increasing concentration (for distribution coefficients less than unity) is associated with a decreasing liquidus temperature since the slope, m, is then less than zero. Using the curves of figure 6.4, the temperature of the liquid as a function of volume fraction solidified can be derived. The use of realistic diffusion coefficients shows that, for small systems (such as interdendritic regions - figure 6.6) interstitial C in δ- and γ-Fe will behave according to the lever-rule. Hence, the last liquid of a binary Fe-C melt will solidify at a temperature close to the solidus while substitutional alloys, such as Al-Cu, which typically have much smaller solid-state diffusion coefficients will usually contain eutectic material in the last (interdendritic) regions to solidify; even when the overall composition is less than the solubility limit at T_e.

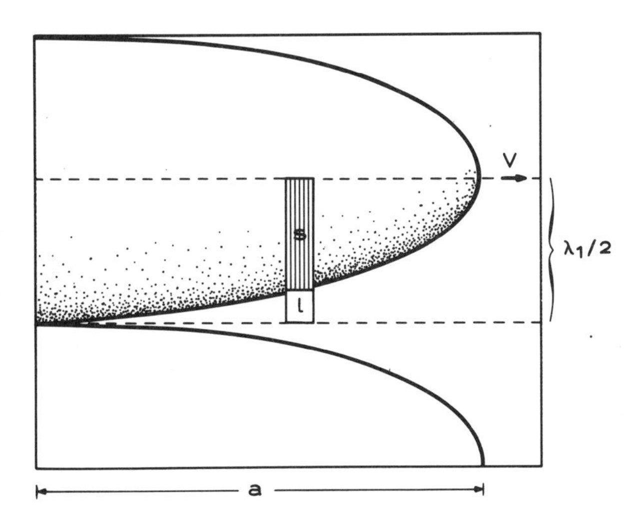

Figure 6.6 Characteristics of the Mushy Zone in Columnar Growth

The results for macroscopic directional solidification (figure 6.3) can be applied to the cellular or dendritic structure of a casting by considering a small volume element which is solidifying at right-angles to the imposed growth direction of the alloy. An increase in the concentration or a decrease in the liquidus temperature of the remaining liquid shown before can thus also be applied qualitatively to the mushy zone of a casting. Here, f_S is now the local volume fraction of solid in the two-phase region with $z' = 0$ at the centre of the cell trunk (corresponding to the position of the infinitely narrow volume element at the cell tip where $f_S = 0$) and $z' = L = \lambda_1/2$ at the last interdendritic liquid (corresponding to the position of the volume element at the cell root with $f_S = 1$). In the case of dendrite growth with secondary branching, the characteristic back-diffusion distance is not $\lambda_1/2$ as in cells, but instead, $\lambda_2/2$.

systems under directional solidification involving planar solid/liquid interfaces (Fig. 6.3). Because of the simplicity of such a unidirectional system and the approximations made, the equations derived are simple. The situation becomes extremely complex when the previously used approaches are applied to dendritic solidification. Further simplifications must therefore be made in order to make such problems tractable.

Firstly, the case of dendritic (or better cellular) solidification in the columnar zone of a casting is considered (Fig. 6.6). The mushy zone, of length a, is defined as the region within which liquid and solid coexist at various temperatures corresponding to the different concentrations due to solute redistribution. The zone length, a, is proportional to the non-equilibrium solidification range, $\Delta T'$, which is usually larger than the equilibrium melting range, ΔT_0 (Fig. 6.5).

In order to apply a mass balance, it is necessary to simplify the dendrite form in two respects. Firstly, it is assumed that there are no side-branches and, secondly, that the dendrite is plate-like rather than needle-like. It is now assumed that 'directional solidification' is occurring in an infinitesimally narrow volume element between, and perpendicular to, two cells or dendrites (Fig. 6.6). All of the relations which were developed before can now be applied in an approximate manner to the interdendritic region, where the solidification time, t_f, is:

$$t_f = -\frac{\Delta T'}{\dot{T}}$$

Here, the negative sign arises due to the negative value of $\dot{T}$, the cooling rate. The length of the volume element is then, $L = \lambda_1/2$, and $\Delta T' = m(C_l^* - C_l^m)$, where C_l^m corresponds to the composition of the last liquid. Substituting these values into equation 6.11 gives:

$$\alpha = -\frac{4D_s m\,(C_l^* - C_l^m)}{\lambda_1^2 \dot{T}} \qquad [6.14]$$

Under most conditions the dendrite tip concentration, C_l^* is very close to C_0 while C_l^m depends markedly on α, and can be obtained from equation 6.9 with $f_s = 1$. Thus:

$$C_l^m = C_0(2\alpha' k)^{-p/u} \qquad [6.15]$$

Considering now the dendrite form of figure 6.7, the back diffusion process, which is most marked at the end, will occur principally between the secondary arms and not between the primary trunks. Therefore, except in the case of cellular growth, where λ_1 is the characteristic spacing, λ_2 would be the appropriate dimension for most castings. This dimension is also important in equiaxed solidification, which is always dendritic in nature. It is known that (equation 4.21):

$$\lambda_2 = 5.5\,(M t_f)^{1/3}$$

so that the α-value for dendrites is deduced to be:

$$\alpha = 0.13\, D_s \Delta T'^{1/3} M^{-2/3} \,|\dot{T}|^{-1/3} \qquad [6.16]$$

where M is defined by equation 4.22. Because $\Delta T'$ depends on α, and this in turn depends on $\Delta T'$, the calculations have to be performed in an iterative manner, substituting for the initial $\Delta T' = \Delta T_0$. To a first approximation, one can also substitute for $\Delta T' = \Delta T_0$, leading to a constant value for the ripening parameter:

$$M \cong \frac{-2\,\Gamma D \ln(k)}{\Delta T_0\, p} \qquad [6.17]$$

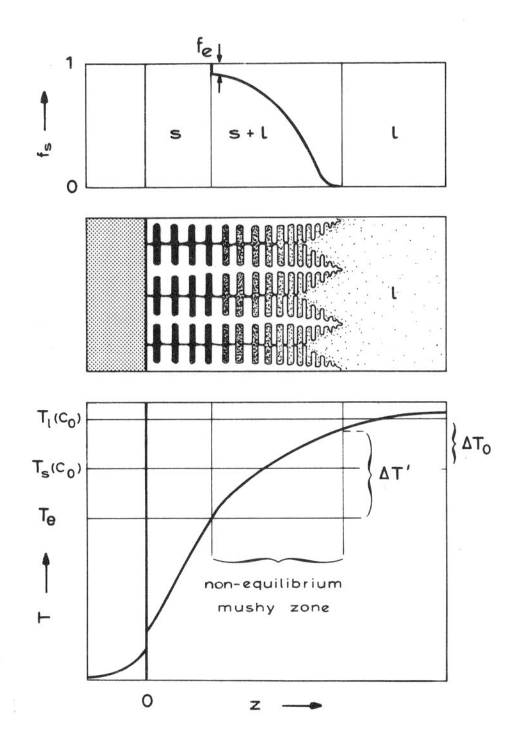

Figure 6.7 Microsegregation

The equilibrium melting range, ΔT_0, does not (except for the lever rule case) correspond to the range, $\Delta T'$, over which the mushy zone develops. The dendrite tips need a certain undercooling which is determined by the growth kinetics of the tip. The dendrite roots will usually have much higher concentrations than C_0/k, due to non-equilibrium solidification. This often leads to interdendritic precipitation of eutectic phases of volume fraction, f_e, even if the composition is not on the eutectic tie-line. In the columnar zone of a casting, as shown here, the volume fraction of solid, f_s, will follow an S-shaped curve like that in the uppermost diagram.

When C_l^m is greater than C_e, precipitation of the eutectic will generally occur and its volume fraction, f_e, can be calculated using equation 6.9, knowing that $f_e = 1 - f_s$ (for $C_l = C_e$). This shows that:

$$f_e = \left(\frac{1}{u}\right)\left[u - 1 + \left(\frac{C_0}{C_e}\right)^{u/p}\right] \qquad [6.18]$$

In multicomponent systems, the 'path of solidification' is more complicated due to the greater number of variables. In appendix 12, an example of such a situation is given for a ternary system. Futhermore, the practically important case of post-diffusion homogenisation of interdendritic segregation is treated in appendix 13. These equations permit an estimation of the degree of microsegregation existing after cooling a casting to room temperature and also of the time required to reach a certain degree of homogenisation during a given heat-treatment process.

Therefore, given the temperature gradient and the growth rate, the most important characteristics (C_l^m, f_e, $\Delta T'$, λ_1, and λ_2) of the solidified structure can be obtained approximately. That is, the solidification microstructure as well as the microscopic inhomogeneities in chemical composition can be determined.

Bibliography

Mass Balance

G.H.Gulliver, *Metallic Alloys*, Griffin, London, 1922.

E.Scheil, Zeitschrift für Metallkunde **34** (1942) 70.

W.G.Pfann, *Zone Melting*, 2nd Edition, Wiley, New York, 1966.

Transients

V.G.Smith, W.A.Tiller, J.W.Rutter, Canadian Journal of Physics **33** (1955) 723.

Microsegregation

H.D.Brody, M.C.Flemings, Transactions of the Metallurgical Society of AIME **236** (1966) 615.

T.W.Clyne, W.Kurz, Metallurgical Transactions **12A** (1981) 965.

D.H.Kirkwood, D.J.Evans, in *The Solidification of Metals*, Iron and Steel Institute, London, Publication 110, 1968.

T.F.Bower, H.D.Brody, M.C.Flemings, Transactions of the Metallurgical Society of AIME **236** (1966) 624.

K.Schwerdtfeger, Archiv für das Eisenhüttenwesen **41** (1970) 923.

E.A.Feest, R.D.Doherty, Metallurgical Transactions **4** (1973) 125.

S.N.Singh, B.P.Bardes, M.C.Flemings, Metallurgical Transactions **1** (1970) 1383.

Y.Ueshima, S.Mizoguchi, T.Masumiya, H.Kajioka, Metallurgical Transactions **17B** (1986) 845.

I.Ohnaka, Transactions of the Iron and Steel Institute of Japan **26** (1986) 1045.

S.Kobayashi, Transactions of the Iron and Steel Institute of Japan **28** (1988) 728.

Numerical Heat Flow Calculations with Segregation

P.N.Hansen, in *Solidification and Casting of Metals*, The Metals Society, London, 1979.

T.W.Caldwell et al., Metallurgical Transactions **8B** (1977) 261.

T.W.Clyne, Metal Science **16** (1982) 441.
M.Wolf, T.W.Clyne, W.Kurz, Archiv für das Eisenhüttenwesen **53** (1982) 91.
M.Rappaz, International Materials Reviews **34** (1989) 93.

Convection

G.S.Cole, in *Solidification*, American Society for Metals, Metals Park, Ohio, 1971, chapter 7.
G.H.Geiger, D.R.Poirier, *Transport Phenomena in Metallurgy*, Addison Wesley, 1973.

Macrosegregation

R.Mehrabian, M.C.Flemings, Metallurgical Transactions **1** (1970) 455.
M.C.Flemings, Scandinavian Journal of Metallurgy **5** (1976) 1.

Exercises

6.1 Write an equation for the $T_l^* - f_s$ relationship for the Lever rule and Scheil equation cases.

6.2 Determine α' values for Al-2wt%Cu and Fe-0.09wt%C alloys when t_f = 10s. Which system exhibits the greater tendency to segregate? First assume that λ_2 = 30 μm for both alloys and then check whether this assumption is reasonable.

6.3 It is desired to purify part of a cylindrical metal ingot by directional solidification. What interface morphology is required in order to accomplish this (planar, cellular, dendritic)? What conditions are most favourable: a short initial transient, or Scheil-type solidification? Give the maximum growth rate which can be used.

6.4 Devise a method for estimating part of the phase diagram (m, k values) of a transparent organic alloy by solidifying it under planar interface conditions at the limit of stability (while observing it with a microscope). Indicate how one might perform the experiment. It is assumed that the values of C_0 and D are known, and that G can be determined during the experiment.

6.5 What will happen if the rod in figure 6.1 is solidified under conditions of strong convection? Sketch the solute profile and indicate which equation applies to this situation.

6.6 In what respect is equation 6.3 an approximation to the initial transient? Examine the assumptions made concerning the boundary layer for the transient (see appendix 11 and figure A2.4)

6.7 Describe a method for the production of a control sample of a given composition for use in microprobe measurements. Such a standard should present a composition to the electron beam which is homogeneous at a scale of the order of 1 μm.

6.8 Write the equation, given in this chapter, which approximately describes the concentration variation along the curved interface of figure 6.6 under growth conditions where the tip concentration in the liquid, C_l^*, is roughly equal to C_0. In what region would one expect the concentration gradient in the liquid, perpendicular to the growth axis, to be (a) close to zero and (b) non-zero?

6.9 Indicate, with the aid of figure 4.12, the growth rates for which, in an alloy, no intercellular or interdendritic enrichment (segregation) will occur for (a) $G \leq 0$ and for (b) $G > 0$.

6.10 Why is equation 6.9 incorrect (under the original assumption that $u = 1 - 2\,\alpha k$) if $\alpha > 0.1$? What happens in this case? Sketch the solute profiles present in the solid and liquid under these conditions. Compare with figure A12.2.

6.11 Compare the fractions of eutectic predicted by the Scheil equation, by the Lever rule, and by equation 6.9, in the case of α-Fe-0.6wt%C and Al-2wt%Cu. Discuss the differences. When calculating α' use the same conditions as those assumed in exercise 6.2.

6.12 Zone melting is a process in which the tendency to segregation is exploited in order to produce a solid having a high purity. In practice, this is done by causing a molten zone with planar solid/liquid interfaces to pass many times through the material (see figure below). The equation for the impurity distribution after the first pass with strong mixing occurring in the liquid zone (C_l = constant) is: $C_l/C_0 = (1/k)[1 - (1-k)\exp(-kz/L)]$, where L is the zone length. Discuss this equation with regard to the equation for the initial transient (equation 6.3).

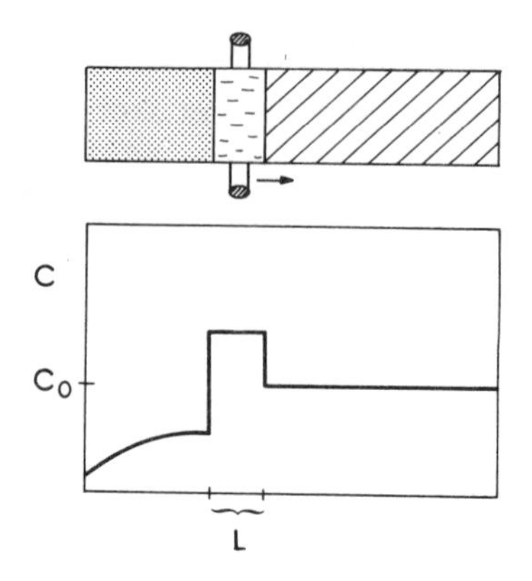

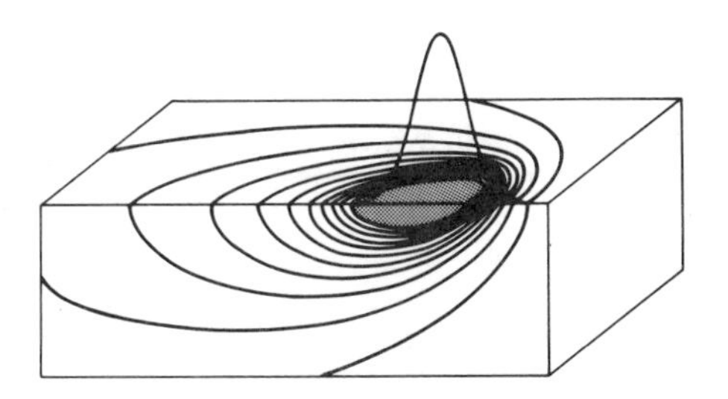

CHAPTER SEVEN

RAPID SOLIDIFICATION MICROSTRUCTURES

Rapid solidification processing (RSP) has become an important topic in solidification research and has shown itself to be useful in many potentially interesting applications. One generally understands the term, RSP, to mean the use of high cooling rates or large undercoolings to produce high rates of advance of the solidification front (typically $V > 1$cm/s). Under such conditions, the low Péclet number approximations which have been developed in the preceding chapters and which assume that the diffusion distance is larger than the scale of the microstructure are no longer valid and more general solutions are required. Such solutions will be discussed here for the benefit of those who are interested in RSP.

The equations to be developed are of use at both low *and* high growth rates. Only growth is treated here since the nucleation models presented are not affected by the rapid solidification conditions as long as the cluster populations change sufficiently rapidly. However, at very high cooling rates, the steady-state models over-estimate the nucleation rate (see Kelton & Greer, 1986). Depending upon the nucleation or growth temperature, a series of metastable phases can form. These phases include the quasicrystalline forms of the various microstructures (dendrites and eutectic). Therefore, a knowledge of the *metastable phase diagram* permits the prediction of the resultant microstructure.

Rapid growth can occur for one of two reasons:

1. *High undercooling of the melt*, which can be achieved by slow cooling in the absence of efficient heterogeneous nucleants (bulk undercooling), or by rapid quenching (powder fabrication).

2. *Rapidly moving temperature fields*, as observed during surface treatment or welding by means of high power density sources such as lasers or electron beams.

Another important type of process for rapid solidification is melt spinning or rotating chill block casting which, depending upon the nucleation behaviour of the melt, can belong to either of the above cases.

7.1 Departure from Local Equilibrium

In this section, it will be briefly shown what happens when the solid/liquid interface is no longer at equilibrium. As shown in table 7.1, the concept of local equilibrium at the solid/liquid interface (m and k corresponding to equilibrium conditions) is no longer valid when the growth rate becomes large. In an alloy, this condition will be fulfilled when the growth rate, i.e. the rate of interface displacement, V, is comparable to, or larger than, the rate of diffusion over an interatomic distance. If V is of the same order of magnitude as the diffusion rate, D_i/δ_i, where D_i is the interface diffusion coefficient (which is smaller than the bulk liquid diffusion coefficient) and δ_i is a length of interatomic dimensions which characterises compositional rearrangement at the solid/liquid interface, the crystal will not have time to change its composition so as to reach the same chemical potential as the melt. When $D_i \sim 10^{-10}$m^2/s and $\delta_i \sim 10^{-9}$m, the critical growth rate will be of the order of 0.1m/s.

Under these conditions, and following Baker and Cahn (1971) and Boettinger and Coriell (1984, 1986), one must express the interface temperature, T^*, and the composition of the solid at the interface, C_s^*, as:

$$T^* = T(V,C_l^*) - \Gamma K \qquad [7.1]$$

$$C_s^* = C_l^* k(V, C_l^*) \quad [7.2]$$

where $T(V,C_l^*)$ and $k(V,C_l^*)$ reflect the changes in phase equilibrium which are produced by the high growth rate. At equilibrium, $V = 0$ and the two functions, equations 7.1 and 7.2, are simply related to the phase diagram. That is, $T(0,C_l^*)$ represents the equilibrium liquidus line and $k(0,C_l^*)$ is the equilibrium distribution coefficient. The effect of curvature upon k will be neglected here. At high growth rates, the growth rate dependent distribution coefficient will be written as k_V, instead of the equilibrium distribution coefficient symbol, k, which has been used in the preceding chapters.

Table 7.1 (after Boettinger and Coriell, 1986)

Increasing solidification rate/undercooling ↓

I *Full Diffusional Equilibrium*
- No chemical potential gradients (phase compositions are uniform)
- No temperature gradients
- Lever rule applies

II *Local Interfacial Equilibrium*
- Phase diagram gives compositions and temperatures only at liquid/solid interface
- Corrections made for interface curvature (Gibbs-Thomson effect)

III *Metastable Local Interfacial Equilibrium*
- Stable phase cannot nucleate or grow sufficiently fast
- Metastable phase diagram (true thermodynamic phase diagram) gives the interface conditions

IV *Interfacial Non-Equilibrium*
- Phase diagram fails at the interface
- Chemical potentials are not equal at the interface
- Free energy functions of phases still lead to criteria which predict impossible reactions

In view of the above, imagine what happens to these parameters when a solid/liquid interface sweeps through a melt at an extremely high rate ($V >> D_i/\delta_i$). In this situation, the atoms have no time to rearrange themselves in the interface so as to equalise the chemical potential of both phases and complete solute trapping will result. The composition of the solid will be the same as that of the liquid and k_V, the velocity-dependent distribution coefficient, will be equal to unity. Figure 7.1 illustrates this situation, and the related chemical potential gradients.

However, a crystal can be formed from its melt, without a change in composition, only under specific thermodynamic conditions (Baker and Cahn, 1971). It will occur only when the system can decrease its free enthalpy. For this to be true, the interface temperature must be below the T_0 line of the corresponding phase diagram (Fig. 7.2).

In order to be thermodynamically consistent there are now two conditions to fulfil - one permits the liquid and solid compositions at the interface to converge at high rates

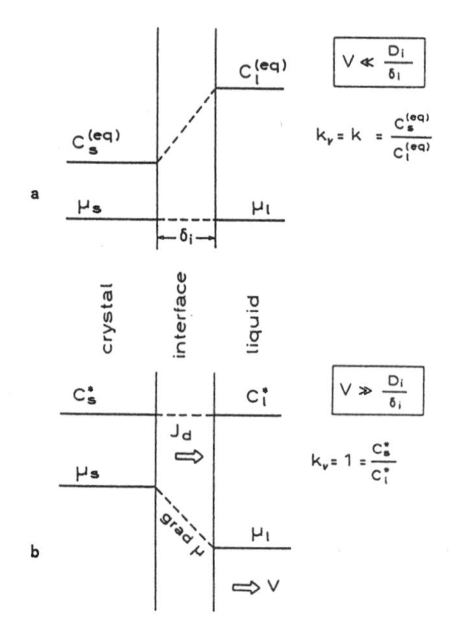

Figure 7.1 Loss of Local Equilibrium at the Solid/liquid Interface

When the interface moves at a low rate, the atomic movements will be rapid enough to permit at least local equilibrium to be established due to local changes in composition (equality of chemical potentials). When V is much smaller than the diffusion rate, D_i/δ_i, k_v tends towards k; the equilibrium distribution coefficient (a). At the other extreme (b), the growth rate is so high that the solute atoms are frozen into solid of the same composition as they arrive at the interface (an effect known as solute trapping). In this case, where V is much larger than D_i/δ_i, k_v tends towards unity and the chemical potential of the solid will be higher than that of the liquid. The chemical potential gradient is the cause of the back-flux of solute atoms which, at very high rates, will be much slower than the rate of incorporation of atoms.

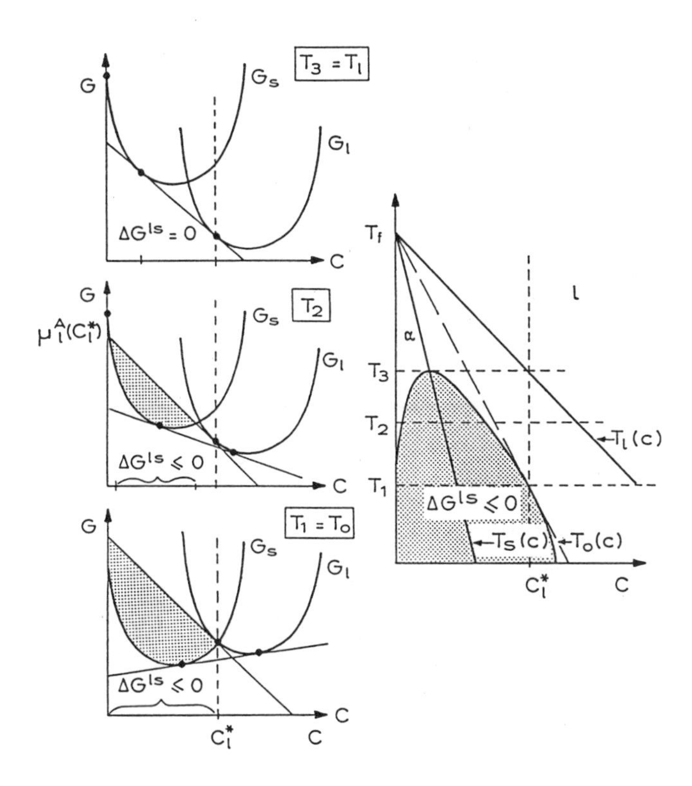

Figure 7.2 The Importance of T_0 for Diffusionless Transformation

According to the principles of thermodynamics, complete solute trapping or diffusionless solidification ($k_v = 1$) can occur only if, in the transformation process, the free energy of the system is reduced, i.e. if the interface temperature is below T_0. Therefore T_0 is the highest temperature at which α can form with the same composition as the melt. It is the temperature at which both phases have the same free energy. T_0 is a function of composition and lies between the liquidus and the solidus. The hatched region represents the range of thermodynamically allowed compositions of α-crystals that can form when a liquid of composition C_l^* solidifies.

(k_v tends to unity), and the other demands that this can happen only if the interface is below the T_0-temperature of the corresponding composition.

A simple relationship for the first phenomenon has been given by Aziz (1982) (see appendix 6):

$$k_v = \frac{k + (\delta_i V/D_i)}{1 + (\delta_i V/D_i)} \tag{7.3}$$

where the parameter, $\delta_i V/D_i$, can be regarded as being an interface Péclet number (P_i), where δ_i is a characteristic interface width (Fig. 7.1), and D_i is the interface diffusion

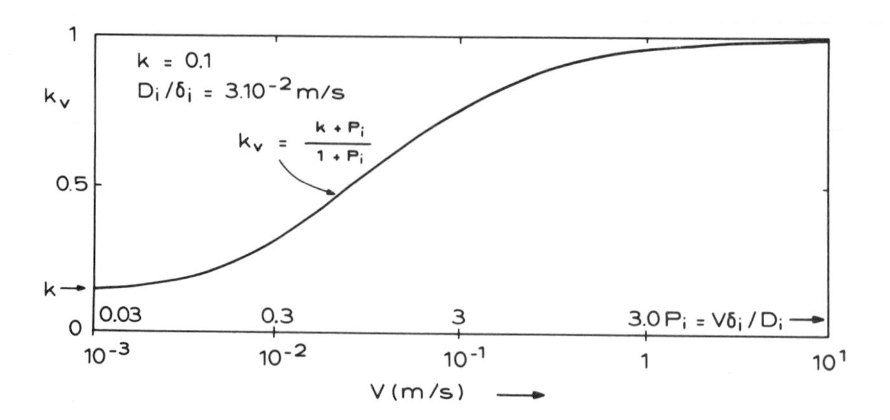

Figure 7.3 The Variation of the Distribution Coefficient with Growth Rate

The non-equilibrium distribution coefficient, k_v, changes, over a critical velocity range, from the equilibrium value, k, to unity. Thus, at high growth rates, diffusionless solidification occurs and the interface temperature must be below T_0. The variation in k_v has a marked effect upon the microstructures formed under rapid solidification conditions, since the degree of solute rejection will depend upon its magnitude.

coefficient. Figure 7.3 shows the behaviour of this expression as a function of V.

The second condition to be satisfied ensures that $T^* \leq T_0$ when $k_v \to 1$. If the attachment kinetics of the solvent are also included in the treatment, but the curvature undercooling is neglected for the moment, one obtains for equation 7.1 (appendix 6):

$$T^* = T_f + m'C_l^* - (R_g T_f/\Delta S_f)\, V/V_0 \qquad [7.4]$$

with the apparent liquidus slope at high growth rates, m':

$$m' = m[1 + f(k)]$$

and

$$f(k) = \frac{k - k_v[1 - \ln(k_v/k)]}{(1-k)} \qquad [7.5]$$

Here, V_0 corresponds to the limit at which crystallisation, i.e. atom attachment to the crystal, can still occur. The upper limit to V_0 is the velocity of sound; which is of the order of some 1000m/s for metals. It can be easily seen that, at low growth rates ($k_v \to k$ and $V << V_0$), $f(k)$ becomes equal to zero, i.e. $m' = m$ and the third term on the right-hand side of equation 7.4 approaches zero; leading to equation 3.17 when K^* (the curvature of the interface) = 0.

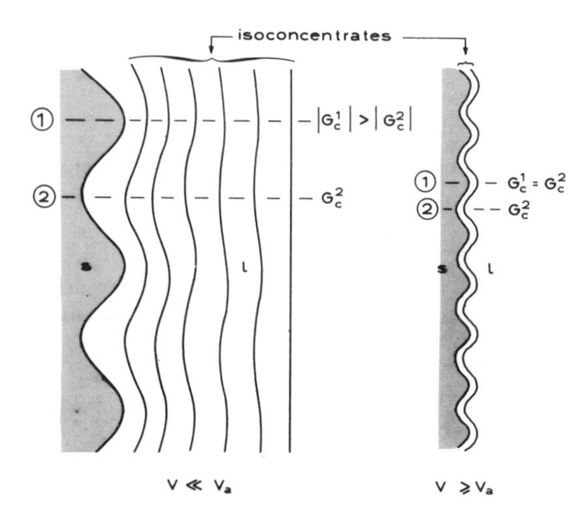

Figure 7.4 Diffusion Field Near to Absolute Stability

At high growth rates, the solute diffusion distance (D/V) becomes small and restricted to the interface itself while capillarity phenomena become dominant due to the fine structures formed. These phenomena again stabilise a planar interface at high growth rates. The critical growth rate at which this occurs is called the rate of absolute stability, V_a. The stabilisation of a planar interface can be expected from the fact that, at $V = V_a$, the concentration gradients ahead of the points and depressions of a perturbation are the same. Therefore, the difference in growth rate between the points and depressions, which is the reason for microstructure formation, vanishes.

7.2 Absolute Stability

As has already been shown in figure 4.12, a planar interface will also become stable at high growth rates. This happens when the solute diffusion distance, δ_c ($\sim D/V$), approaches the solute capillarity length, s_c ($\sim \Gamma/\Delta T_0$). The critical growth rate for absolute stability, V_a (appendix 7) is:

$$(V_a)_c = \Delta T_0 D / k_v \Gamma \qquad [7.6]$$

The suffix, c, indicates that this limiting velocity is controlled by solute diffusion#. The main physical reason for this, at first sight strange, phenomenon is that the diffusion distance narrows with increasing growth rate and diffusion becomes more and more localised (Fig. 7.4). On the other hand, at high rates capillarity becomes a dominant

Note that equation 7.6 is valid only for temperature gradients such that $G \ll k_v \Gamma V^2/D^2$. That is, under most conditions of laser surface treatment (G ~ 10^6K/m) at $V > 1$ cm/s.

feature of the process and does not allow the microstructure to become even finer. In other words, the diffusion distance decreases with V^{-1}, while the microstructure (e.g. the wavelength, λ_i, of equation 3.25) decreases as $V^{-1/2}$. Therefore, at some critical velocity, the microstructure becomes too coarse for lateral diffusional processes; thus leading to stabilization of a flat interface.

From the above, one can now predict that, for a given alloy and positive temperature gradient, planar growth will always occur when the growth rate is sufficiently high. With increasing growth rate, for a given alloy and not too high a temperature gradient, there will be a transition from plane front to cells, to dendrites, to cells again, and back to plane front (Fig. 7.5).

Pure undercooled substances ($G < 0$) exhibit an analogous behaviour, but the controlling diffusion process in this case is thermal in nature. Because, for metals, the thermal diffusion coefficient, a, is much greater than D, the stabilisation of a flat interface occurs at much higher growth rates than that given by equation 7.6. It is shown in appendix 7 that crystallisation from undercooled alloy melts will produce morphologically stable growth fronts if the growth rate due to the undercooling is given by:

$$V_a = (V_a)_c + (V_a)_t \qquad [7.7]$$

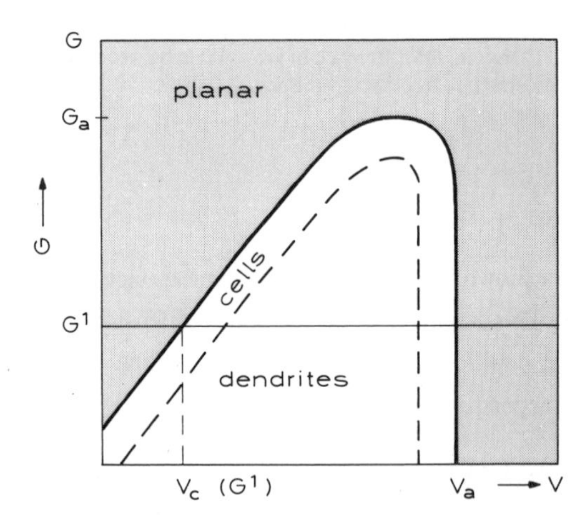

Figure 7.5 Range of Stable (Planar) Interface Morphology

When a positive temperature gradient, G, is imposed upon an alloy of given composition there is, at low growth rates ($V = V_c$), a transition from a planar to a cellular morphology due to constitutional undercooling. A reverse transition from cells to a planar front is observed at high rates ($V > V_a$). The latter is essentially independent of the temperature gradient. Above a certain temperature gradient, G_a, the planar form is always stable.

where $(V_a)_c$ is determined by equation 7.6 and $(V_a)_t = a\theta_t/\Gamma$. Here, $\theta_t = \Delta h_f/c$ is the unit thermal undercooling.

In order to model dendrite growth, it is important not only to know the limit of absolute stability but also the wavelength of the marginally stable state, λ_i. This value can be obtained from the generalised stability equation (compare with the large bracket on the right-hand side of equation 3.23):

$$-\Gamma\omega^2 - G^* + mG_c^* = 0 \qquad [7.8]$$

where $\omega = 2\pi/\lambda$, the effective temperature gradient, $G^* = \bar{\kappa}_l G_l \xi_l + \bar{\kappa}_s G_s \xi_s$, $\bar{\kappa}_l = \kappa_l/(\kappa_s+\kappa_l)$, $\bar{\kappa}_s = \kappa_s/(\kappa_s+\kappa_l)$, G_l and G_s are the temperature gradients in the liquid and solid, respectively, at the interface and G_c^* $(= G_c\xi_c)$ is the effective concentration gradient in the liquid at the solid/liquid interface. The stability parameters, ξ_l, ξ_s, and ξ_c, are functions of the corresponding Péclet numbers and are given in appendix 7 (equation A7.23).

From equation 7.8, one obtains the critical wavelength for zero amplification of the instabilities:

$$\lambda_i = \left\{(1/\sigma^*)\left[\frac{\Gamma}{(mG_c^* - G^*)}\right]\right\}^{1/2} \qquad [7.9]$$

where σ^* is a stability constant which is of the order of $1/4\,\pi^2$.

Equation 7.9 has the same form as that of equation 3.22, except that ϕ is here the difference in the effective gradients, which in turn depends upon the stability parameters, ξ_l, ξ_s, and ξ_c. The latter three parameters approach unity at small Péclet numbers, and equation 7.9 approaches equation 3.22. Stability is again obtained when the denominator of equation 7.9 becomes equal to zero. This happens

at *low growth rates*:

- when $mG_c^* = G^*$,

and at *high growth rates*:

- under conditions of directional solidification ($G^* > 0$) when the effective concentration gradient, G_c^*, approaches zero ($\xi_c \to 0$ when $P_c >> 1$),
- under conditions of solidification in an undercooled melt ($G^* < 0$) when both G_c^* and G^* approach zero.

7.3 Rapid Dendritic/Cellular Growth

As shown in chapter 4 and appendix 9, dendritic growth can be modelled by solving the relevant heat and mass transfer equations and using stability arguments. Here, general equations for equiaxed and columnar dendrites will be developed. For simplicity, the kinetic undercooling for atom attachment will be neglected at first but will be re-introduced into the equations later.

The capillarity-corrected transport equation can then be written in terms of undercooling as:

$$\Delta T = \Delta T_t + \Delta T_c + \Delta T_r \qquad [7.10]$$

where the various undercoolings are defined by equations A9.17 - A9.19.

As shown by figure A8.1, ΔT_t is the thermal undercooling due to latent heat release at the tip, ΔT_c is the solutal undercooling due to solute rejection by the dendrite, and ΔT_r is the curvature undercooling.

As was done in chapter 4, the second equation required is obtained by applying stability arguments. This defines the tip radius as a function of the local effective gradients of composition, G_c^*, and temperature, G^*, via equation 7.9; assuming that $R = \lambda_i$ (note that G_c^* and G^* have to be evaluated at the dendrite tip). Substituting equation 7.9 (with R replacing λ_i) into 7.10 gives different relationships for the two cases of columnar and equiaxed growth.

Columnar growth ($G>0$)

In this case, there is an imposed macroscopic heat flux, from the superheated liquid to the solid, which produces a positive temperature gradient at the dendrite tips. If the thermal conductivities and diffusivities of the liquid and solid are assumed to be equal ($\kappa_l = \kappa_s = \kappa$ and $a_l = a_s = a$) and if, for simplicity, the effect of latent heat is neglected, the temperature gradients in both phases at the tip are also equal, i.e. $G_l = G_s = G$. In this case, equation 7.9 with $R = \lambda_i$ reduces to:

$$R = \left\{(1/\sigma^*)\left[\frac{\Gamma}{(mG_c\xi_c - G)}\right]\right\}^{1/2}, \qquad (G > 0) \qquad [7.11]$$

Note, that in this case, the increase in ξ_s cancels out the decrease in ξ_l with changing Péclet number, and the effective temperature gradient, G^* (equation 7.8), becomes equal to the imposed temperature gradient, G (see appendix 7).

Furthermore, noting that the imposed temperature field does not create any thermal undercooling (since the temperature field is imposed from the exterior, i.e. depends upon the heating/cooling system and not upon the dendrite tip), equation 7.10 reduces to :

$$\Delta T = \Delta T_c + \Delta T_r \qquad [7.12]$$

By substituting ΔT_c and ΔT_r from equations A9.18 and A9.19 into equation 7.12, one obtains the final result for columnar growth:

$$V^2 A' + VB' + G = 0 \qquad [7.13]$$

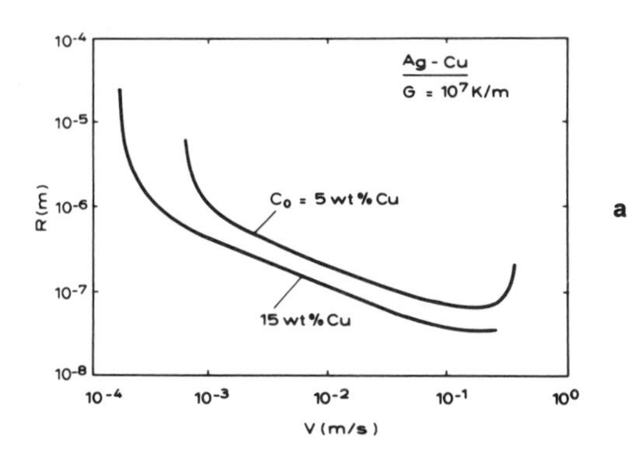

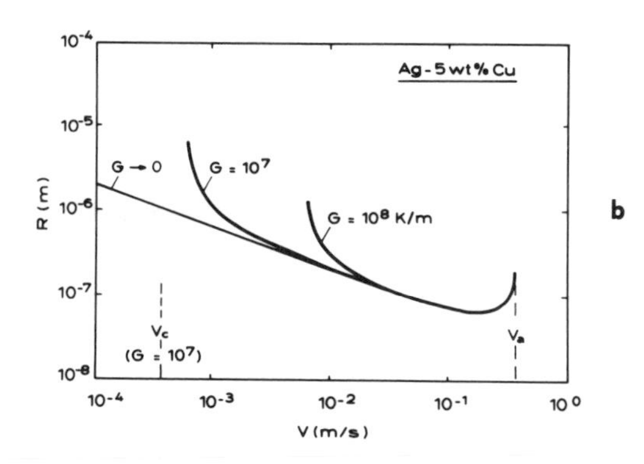

Figure 7.6 Columnar Dendritic Growth

This growth phenomenon is typical of laser remelting (welding, surface treatment, etc.). Within the cellular/dendritic regime of figure 7.5, a typical variation of the tip radius as a function of growth rate has the form shown. Close to the limit of constitutional undercooling, V_c (large negative slope of the curves) cells will form, then dendrites, and finally (close to the limit of absolute stability, V_a) cells will again be seen. There, R increases with increasing V. A change in composition will modify the whole range of microstructures formed (a) while a change in G will modify the cellular regime at low rates (b).

where

$$A' = \frac{\pi^2 \Gamma}{P_c^2 D^2}; \qquad B' = \frac{\theta_c \xi_c}{D}$$

Here, the unit solutal undercooling, θ_c, is equal to $(\Delta T_0\, k_v\, A(P_c))$, where $A(P_c) = C_l^* / C_0$ (equation A9.18a), and the stability function, ξ_c, is given by equation A7.23c. From equation 7.13, the growth rate, V, and the tip radius, R, can be deduced for various Péclet numbers. This leads to the relationship between R and V which is shown in figure 7.6.

Figure 7.6b is similar to figure 4.11; the important difference being that, due to the decrease in ξ_c at high Péclet numbers, R increases slightly with increasing V just before reaching the limit of absolute stability. The increase in R around V_a arises from an increasing localisation of the diffusion field and indicates a tendency to more planar growth. At high growth rates ($V < V_a$) the interface has a cellular morphology, which changes to a planar one beyond V_a.

In order to calculate the tip concentration, one uses the definition of the supersaturation (equation A8.1), leading to:

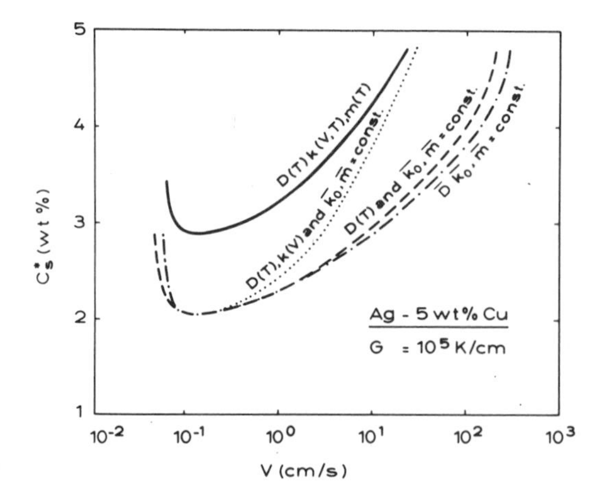

Figure 7.7 Growth Rate Dependent Composition of Columnar Dendrites

The composition of the dendrite tips which corresponds approximately to the minimum in concentration of the microstructure depends very much upon the completeness of the physical modelling. At high growth rates (and high tip undercoolings) the parameters which depend upon V and T [that is, k_v, $D(T)$, $k(T)$, $m(T)$] will have a marked effect upon the results of the calculations. The highest curve takes account of all of the possible variations in the above parameters for the binary system, Ag-5%Cu.

$$C_l^* = C_0 A(P_c)$$
$$C_s^* = k_v C_l^* \qquad [7.14]$$

Finally, the tip temperature can be calculated by using equation 7.4; to which the curvature undercooling has been added, giving:

$$T^* = T_f - 2\Gamma/R + C_l^* m[1 + f(k)] - (R T_f/\Delta S_f) V/V_0 \qquad [7.15]$$

Equation 7.15 also contains the thermodynamic correction for the interface temperature (for dilute solutions), as expressed by $f(k)$ in equation 7.5, and the kinetic undercooling (last term on right hand side).

It becomes evident that, at high rates, C_l^* and the undercooling $(T_f - T^*)$ can become large and several of the parameters will change drastically. These parameters are k_v, which is a function of V (equation 7.3) and of the temperature via the equilibrium distribution coefficient, and $D(T)$. Figure 7.7 shows how these variations will affect the concentration versus growth rate curve. Figure 7.8 shows, superposed upon the phase diagram, the manner in which the liquid and solid compositions and the temperature of the tip will change as a function of the growth rate.

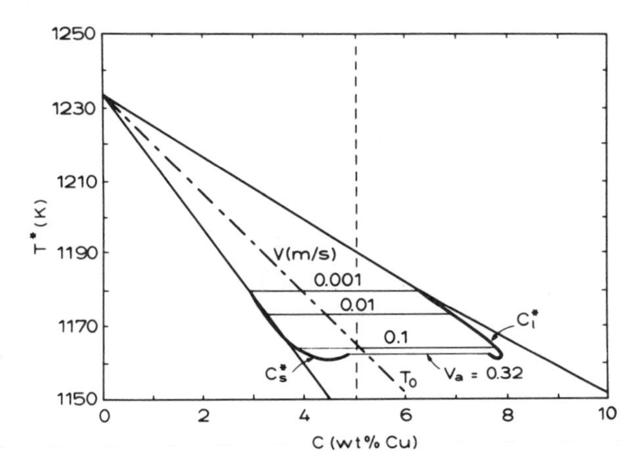

Figure 7.8 Modelling Laser Resolidification of an Ag-5wt%Cu Alloy

The liquid and solid compositions, and the temperature of the dendrite tip, as a function of growth rate are as indicated (the phase diagram is also plotted for reference). With increasing solidification rate, the tip composition / temperature locus remains slightly below the phase diagram liquidus and solidus, due to curvature undercooling. However, above a certain rate (where k_v deviates from k), the solid composition increases more rapidly and approaches C_0 at V_a.

Equiaxed growth in undercooled melts ($G_l < 0$)

In this case, an isothermal dendrite tip is assumed to form in an undercooled melt. Therefore, the temperature gradient in the solid, G_s, is equal to zero, making the term, $\bar{\kappa}_s G_s \xi_s$, in G^* of equation 7.8 equal to zero. Thus, one obtains for the dendrite tip radius, again assuming equal conductivities in liquid and solid:

$$R = \{(1/\sigma^*)[\frac{\Gamma}{(mG_c\xi_c - 0.5G_l\xi_l)}]\}^{1/2} \quad , \quad (G < 0) \qquad [7.16]$$

Now the full transport equation (7.10) has to be solved since, in undercooled alloy melts, the tip is a source of both heat and solute. By determining the gradients with the aid of the usual flux balance equations (A9.24-A9.26), one obtains:

$$R = \frac{(\Gamma/\sigma^*)}{(2P_c\theta_c\xi_c + P_t\theta_t\xi_t)} \qquad [7.17]$$

where $\xi_t \equiv \xi_l$ (equation A7.23a).and ξ_c is defined by equation A7.23c. Finally, one finds the undercooling, tip radius, and growth rate, respectively, as shown in appendix 9.

The results of applying this model are shown in figure 7.9 for a pure substance and an alloy. Due to the presence of the thermal field around the dendrite tip, the behaviour is more complicated in undercooled alloys than it is in those solidified under constrained growth conditions.

When the growth rate dependent distribution coefficient (equation 7.3) is also taken into account (assuming, for simplicity, constant D and negligible ΔT_k) the growth rate versus undercooling relationship will change drastically, as shown by figure 7.9. Above a critical undercooling, ΔT^+, the distribution coefficient approaches unity and the dendrite becomes a purely thermal one (Fig. 7.10). That is, the curve for $V(\Delta T)$ joins the curve for $C_0 = 0$. This abrupt transition can also be seen in figure 7.11.

7.4 Rapid Eutectic Growth

As with dendrite growth, eutectic growth exhibits a different behaviour at high Péclet numbers. This arises from the fact that, in order to simplify the equations as, for example, in the Jackson-Hunt analysis, it is assumed that equation A10.3 becomes b = $2\pi/\lambda$. This is the case only when $2\pi/\lambda >> V/2D$ or if $P = V\lambda/2D << 2\pi$. Furthermore, it is again assumed for reasons of simplicity that, in the flux balance (equation A10.5), the interface composition in the liquid, C_l^*, is approximately equal to C_e. For small undercoolings, this is certainly a reasonable assumption but is not so at high undercoolings.

Relaxing these two low-Péclet number approximations leads to a more general solution which applies at both low and high Péclet numbers. For example, equation 5.10 (see also A10.31) indicates that:

$$\lambda^2 V = \frac{2D}{C'P'}\left[f\,\frac{\Gamma_\beta \sin\theta_\beta}{m_\beta} - (1-f)\frac{\Gamma_\alpha \sin\theta_\alpha}{m_\alpha}\right], \qquad P < 1 \qquad [7.18]$$

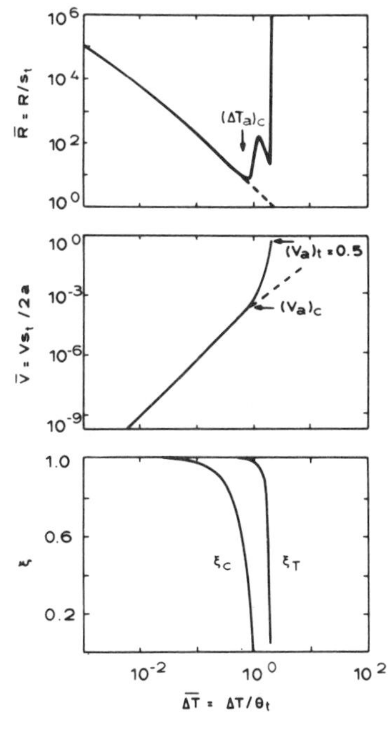

Figure 7.9 Dendritic Growth in Undercooled Melts

This situation, which produces many equiaxed grains in a casting under normal conditions, will not necessarily do so during rapid solidification. For example, in highly undercooled powder particles one particle is usually just one grain. However, growth will be controlled by the undercooling, and the thermal diffusion fields at the dendrite tips will modify the overall growth behaviour as compared with columnar growth. Here, it is shown how, for an alloy, the tip radius (a) and the growth rate (b) change with undercooling (k = constant). For the chosen composition and at low undercoolings, the dendrites are essentially solute diffusion controlled. When these dendrites reach an undercooling which corresponds to the limit of absolute stability for solute diffusion, $(V_a)_c$, a transition from a mostly solutal to a purely thermal dendrite is observed. As the thermal diffusion coefficient of metals is much larger than D, the tip radius suddenly increases. After reaching the limit of absolute stability for thermal diffusion, $(V_a)_t$, the tip radius increases again. Beyond this undercooling ($\Delta T_{max} = \theta_t + \Delta T_0$), it is predicted that plane front solidification will occur. However, note that an absence of microsegregation is predicted above ΔT_a (corresponding to $(V_a)_c$). At low ΔT, the slope of the growth rate variation indicates a relationship of the form, $V = \Delta T^n$, where n is between 2.5 and 3; depending upon the composition. This exponent increases sharply as $(V_a)_t$ is approached. The corresponding changes in the stability parameters, ξ_c and ξ_t, are shown (c). In this figure $\bar{R}$, $\bar{V}$ and $\overline{\Delta T}$ are dimensionless quantities (Lipton, Kurz, Trivedi 1987).

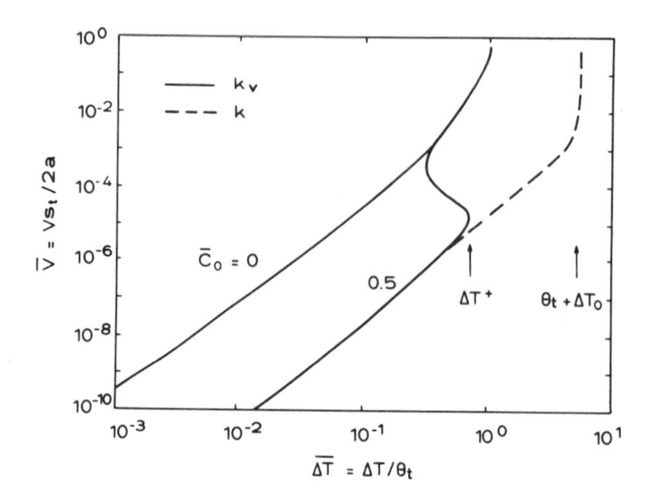

Figure 7.10 Effect of Solute Trapping Upon Undercooled Dendrite Growth

The high growth rates which can be obtained by large melt undercoolings will make k_v tend towards unity. If one takes this effect into account, the $V(\Delta T)$ relationship will drastically change. Above a certain undercooling, ΔT^+, where k_v tends towards unity, the essentially solutal (alloy) dendrites will become pure thermal dendrites, i.e. behave as if their composition, C_0, was equal to zero. ($\bar{V}$ and $\overline{\Delta T}$ are dimensionless quantities as in figure 7.9).

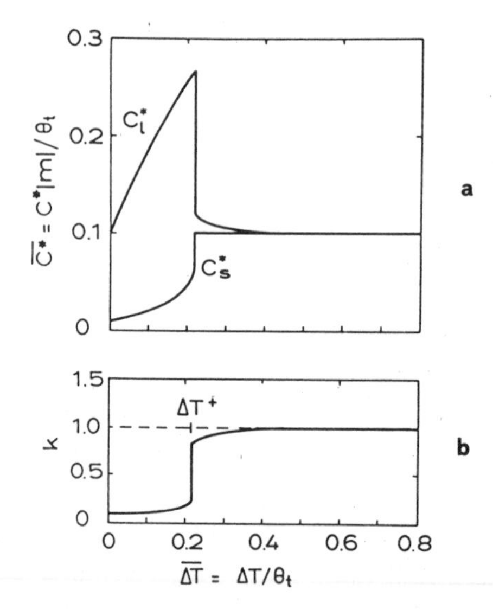

Figure 7.11 Dendrite Tip Composition as a Function of Melt Undercooling

The change in interface composition as a function of the total undercooling is shown (a) for the case of figure 7.10. Here, k_v changes abruptly at the critical undercooling, ΔT^+, since V increases sharply there (b). Above ΔT^+, an homogeneous solid will be formed.

A more general analytical solution, for any Péclet number has been obtained for the special case where $k_\alpha = k_\beta = k$, (Trivedi et al. 1987) in which the product, $C'P'$, in equation 7.18 is replaced by $(1-k)\psi'$. One sees that the essential difference at high Péclet numbers is the replacement of the series, P', which is a function only of the volume fraction of β, f, (see equation A10.31) by the series, ψ', which is a function of f, k, and P (see appendix 10).

Therefore, as in the case of dendritic growth, $\lambda^2 V$ is constant only at small Péclet numbers and increases steeply when $P >> 1$ (Fig. 7.12).

This behaviour again leads to a sort of absolute stability for eutectic growth - a limiting rate above which coupled two-phase growth is less favourable than planar single-phase growth. This occurs when the interface temperature reaches the solidus temperature of the α- or β-phase; whichever is the higher. At that limit, λ increases strongly as shown in figure 7.13 for distribution coefficients close to unity.

For smaller distribution coefficients (small solid solubility) the eutectic undercooling becomes very high (Fig. 7.14) and the temperature-dependent diffusion coefficient, $D(T)$, will slow down the transformation. This is why the curves for small k in figure 7.13 bend back.

If, for eutectic with a small solid solubility, the interface temperature crosses the glass transition temperature of the system, two-phase growth will stop and a single amorphous phase will form. Such a behaviour is shown in figure 7.15. It can now be

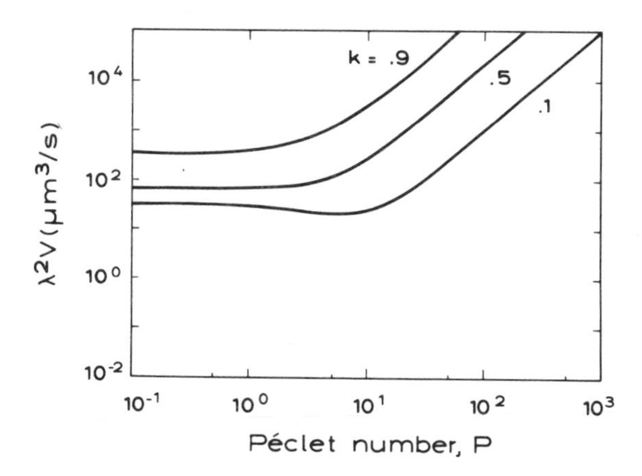

Figure 7.12 Eutectic Growth at High Péclet Numbers

As in the case of dendrites, eutectics also exhibit an increase in the scale of their microstructures at high growth rates (high Péclet numbers). At low growth rates, $\lambda^2 V$ = constant (as shown in chapter 5), but its value increases markedly when $P = \lambda V/2D$ becomes large with respect to unity. Therefore, it is important to remember that the rule-of-thumb, "finer structures at higher growth rates" is no longer valid at very high growth rates (of the order of 1m/s).

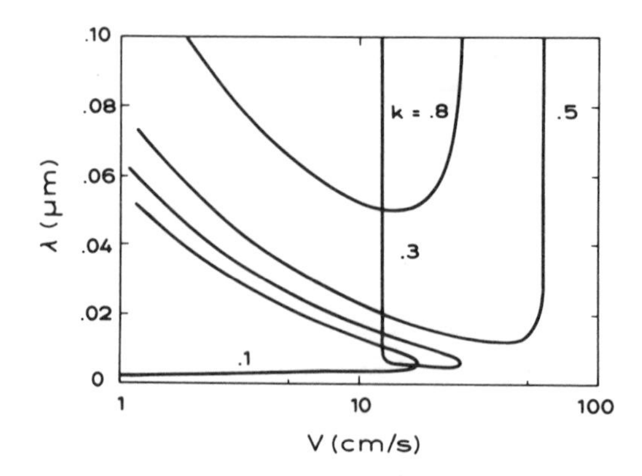

Figure 7.13 Eutectic Spacing - Growth Rate Relationship

When this figure, for an eutectic phase distribution coefficient, k, of 0.8 is compared with figure 7.6, the analogy between $\lambda(V)$ for eutectics and $R(V)$ for dendrites becomes obvious. That is, both λ and R decrease with V before again increasing upon approaching the limit of absolute stability. For small k-values, the behaviour is more complicated due to the increased eutectic growth undercooling, which decreases $D(T)$ and bends the curve back before it increases again.

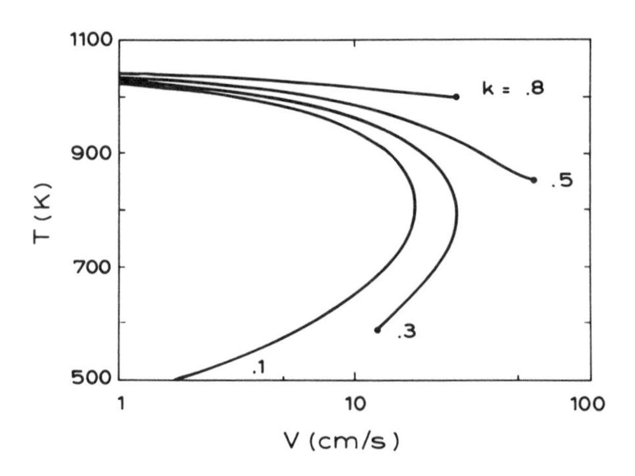

Figure 7.14 Eutectic Interface Temperature - Growth Rate Relationship

The growth undercooling increases, with increasing V, at a rate which increases as the k-value of the eutectic phases decreases. However, since the diffusion coefficient decreases with decreasing temperature, growth will ultimately slow down. Therefore, fine eutectic structures can be obtained at low temperatures and low growth rates, if k is sufficiently small, as observed during crystallisation from the amorphous state.

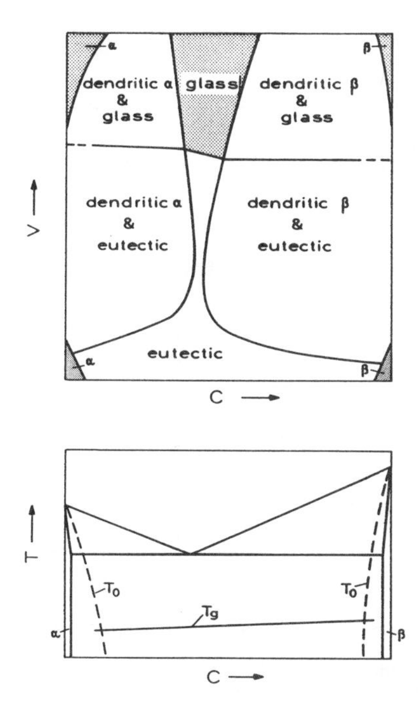

Figure 7.15 Glass Formation in Eutectic Systems

In the case of low k-values (low solubilities of the crystalline phases), the T_0 lines drop steeply and interface undercooling has to become high before either solute trapping (below T_0) or absolute stability (at T_s) can appear. If the glass transition temperature is high relative to the eutectic temperature, rapid solidification of the eutectic will ultimately produce a glass. This behaviour is shown here with the aid of a microstructure selection diagram for $G>0$ and various growth rates. The hatched regions at high V indicate homogeneous solid, crystalline α or β and glass at around the eutectic composition. (after W.J.Boettinger, 1982).

seen why so-called "deep" eutectics are associated with such a transition. It is because they usually have very low k-values and a high glass transition temperature with respect to the low melting point of the eutectic.

7.5 Intercellular Solute Redistribution

Another phenomenon whose theoretical treatment has to be modified at high growth rates is interdendritic solute redistribution. In chapter 6, the case of microsegregation where solidification starts at the liquidus temperature, i.e. where the tip undercooling is zero, has been discussed. By further assuming complete mixing in the interdendritic liquid, various segregation equations have been developed (lever rule, Scheil, Brody-Flemings). However, at the high tip undercoolings which are observed under rapid solidification conditions, these equations can no longer be applied (Fig. 7.16).

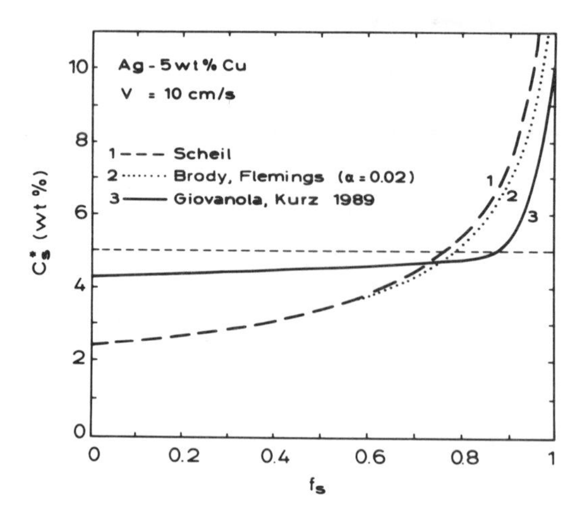

Figure 7.16 Microsegregation Profile

Under rapid solidification conditions, the interdendritic solute distribution deviates markedly from the Scheil and Brody-Flemings equations. The reason for this is the large solute build-up at the dendrite tip. Therefore, the composition of the dendrite trunk is much higher and makes the microsegregation less pronounced. The solute redistribution model for rapid solidification (3) is compared with the Scheil and Brody-Flemings analyses. It can be seen that, at high rates, an increased homogenization is obtained. This is one of the advantages of the powder atomisation process. The alloy is broken up into many small homogeneous grains which are then compacted to give a larger piece.

An exact solution of this problem is extremely difficult to obtain as one has to take account of the transient regime of the solute fields around the dendrite or cell tips (Fig. 7.17). However, the problem can be solved approximately in the following way:

Assume that the initial transient of the composition at the solid/liquid interface before the interdendritic liquid becomes homogeneous can be described by a second-order polynomial up to a certain volume fraction, f_X. Above f_X, the liquid composition is approximately homogeneous and the Scheil equation can be applied. A detailed description of this model can be found in Giovanola and Kurz (1989). The result of such a calculation is shown in figure 7.16, and compares favourably with experimental measurements.

To sum up this chapter, one now has a complete theory which has proved to be useful in interpreting the manifold phenomena and microstructures which are observed in rapid solidification processing. Evidently, in order to make the picture really complete precise knowledge is required concerning the various metastable phases which might appear at high undercoolings

Note that all of the equations which have been developed in this chapter, for the rapid solidification case, are general in nature and can equally well describe steady-state solidification under normal growth conditions. However, the simplified equations which have been developed in chapters 3 to 6 are easier to use when P is less than unity.

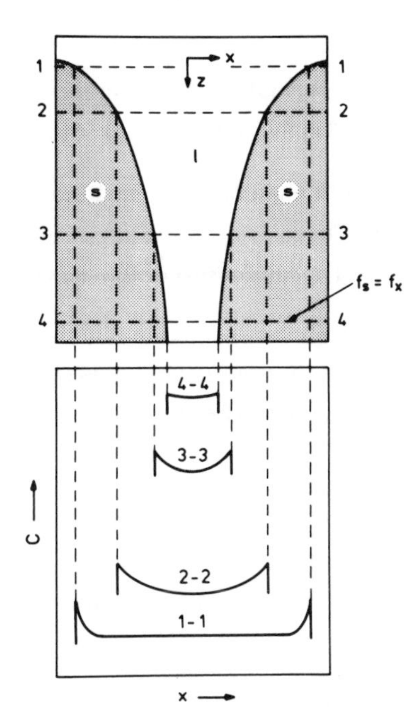

Figure 7.17 Intercellular Solute Concentration Profile

In the transient region of a dendrite or cell (1,2), the concentration gradients at the interface are large - especially under rapid solidification conditions. Therefore, the assumption of an homogeneous liquid is valid only at a high volume fraction of solid (3,4).

Bibliography

Nucleation in RSP

K.F.Kelton, A.L.Greer, J. Non-Crystalline Solids **79** (1986) 295.

Thermodynamics of RSP

J.C.Baker, J.W.Cahn, in *Solidification*, ASM, Metals Park, OH, 1971, p. 23.

M.J.Aziz, J. Applied Physics **53** (1982) 1158.

W.J.Boettinger, S.R.Corriell, R.F.Sekerka, Mater. Science Engineering **65** (1984) 27.

J.H.Perepezko, W.J.Boettinger, Mater. Res. Soc. Symp. Proc. **19** (1983) 223.

W.J.Boettinger, S.R.Corriell, in *Science and Technology of the Undercooled Melt*, (P.R.Sahm, H.Jones, C.M.Adam, Eds.) Martinus Nijhoff Publication, Dordrecht, 1986, p. 81.

Interface Stability (large thermal Péclet numbers)

R.Trivedi and W.Kurz, Acta Metallurgica **34** (1986) 1663.

Directional Dendritic Growth (Laser and electron beam processing)

W.Kurz, B.Giovanola and R.Trivedi, Acta Metallurgica **34** (1986) 823.

W.Kurz, B.Giovanola and R.Trivedi, Journal of Crystal Growth **91** (1988) 123.

Equiaxed Dendritic Growth (Atomisation process, bulk undercooling)

J.Lipton, W.Kurz and R.Trivedi, Acta Metallurgica **35** (1987) 957.

R.Trivedi, J.Lipton and W.Kurz, Acta Metallurgica **35** (1987) 965.

Eutectic Growth

R.Trivedi, P.Magnin and W.Kurz, Acta Metallurgica **35** (1987) 971.

W.J.Boettinger in *Rapidly Solidified Amorphous & Crystalline Alloys* (B.H.Kear, B.C.Giessen, Eds.) Elsevier North Holland, New York, 1982.

Solute Distribution

B.Giovanola, W.Kurz, in *State of the Art of Computer Simulation of Casting and Solidification Processes*, E-MRS Conference Proceedings, Les Éditions de Physique, 1986.

B.Giovanola, W.Kurz, Metallurgical Transactions **21A** (1990) 260.

Exercises

7.1 Describe two different mechanisms which lead to a microsegregation-free solid.

7.2 Develop a simple relationship for the estimation of the concentration range of various alloys at which either solute trapping or plane-front growth with local equilibrium beyond V_a will be the major supersaturation mechanism. How will the form of the phase diagram ($k \rightarrow 1$ or $k \rightarrow 0$) influence the critical composition?

7.3 Develop a relationship for the T_0-curve. (Hint: use equation 7.4).

7.4 Show under what conditions the characteristic diffusion distance will be equal to the relevant microstructural length for interface instabilities, for dendrites/cells and for eutectic. Calculate V_a approximately for dendrites by using the low Péclet number approximation. Sketch the structures and the diffusion fields for all three cases, for $P << 1$ as well as for $P >> 1$.

7.5 Why does the tip radius, R, of the dendrite model which was developed in this chapter for $G > 0$ not reach infinity at the limit of absolute stability?

7.6 Show that the limit of absolute stability, V_a, at low alloy compositions increases with the square root of the alloy concentration, C_0.

7.7 What is the main effect of the temperature dependence of the diffusion coefficient? Can you relate this effect to the form of TTT diagrams?

7.8 In which systems will the effect of the temperature-dependent diffusion coefficient become important at high rates; in alloys with low or high solubility? Does your reasoning hold for dendrites, for eutectic or for both?

7.9 What is the maximum value of the undercooling in columnar dendritic growth? Which growth rate corresponds to that undercooling?

7.10 Which assumptions in the original Jackson-Hunt treatment restrict their solution to low Péclet numbers?

7.11 In an eutectic system, which eutectic phase will reach absolute stability first?

7.12 Under what conditions would you predict glass formation during laser resolidification? Note that, at the bottom of the trace, crystals will always grow first as the growth rate is zero there and increases to a maximum at the surface.

SUMMARY

As outlined at the beginning of this book, the study of solidification microstructures is of ever-increasing importance in the field of engineering since the properties of most alloys, whether cast, forged, welded or rapidly solidified, will depend to a large extent upon the degree of control which can be exercised during solidification. Furthermore, various new solidification processes show signs of great future promise. These include the enlargement of the equiaxed zone by means of the electromagnetic stirring of castings, surface remelting and alloying by high-energy beams such as lasers, directional monocrystalline castings, melt atomization, chill block casting, and other techniques. Therefore, there is a need for a better understanding of the common mechanisms which are associated with this important phase transformation.

It has been the aim of the authors to describe, from an engineering point of view, and via the use of simple physical models, some of the most important solidification phenomena which occur. The latter range from atomistic to macroscopic in scale. However, one phenomenon which is not discussed in this book (for reasons which have been explained in the foreword) is convection.

In the introduction, the importance of *heat flow* phenomena has been pointed out. Currently, the cooling rates employed range from 10^{-5} to 10^{10}K/s and correspond to 'castings' ranging in size from several metres to a fraction of a micrometre. These various cooling rates produce differing microstructures which can be divided essentially into four categories, i.e: columnar or equiaxed grains of dendrites or eutectic. In each of these cases, capillarity effects and the diffusion of solute and/or heat determine the microstructure which results from the solidification conditions. A fifth possibility arises when the cooling rate is too high for nucleation to occur. This leads to the formation of an amorphous solid.

As far as solidification at the *atomic level* is concerned, two phenomena: nucleation and interface structure, have been described. Nucleation is a predominantly interface-controlled process involving an activation energy which is strongly affected by the presence of a catalyst. The nucleation rate, and therefore the equiaxed grain size, is very sensitive to the existence of heterogeneous particles. From the well-known laws of nucleation, the time-temperature-transformation (TTT) diagram for the beginning of solidification of a substance can be derived. This permits, among other things, an explanation for the appearance of the amorphous phases mentioned above.

The interfacial atomic attachment kinetics largely determine the crystal habit. Substances such as metals, which have a low entropy of fusion, grow in a non-faceted manner. Some transparent organic substances which behave like metals in this respect are especially useful for the *in situ* observation of solidification using an optical microscope. On the basis of such observations, the likely behaviour of a metallic system can be deduced without the difficulties posed by the opacity and high melting point of metals. In substances having a high entropy of melting, the crystal growth behaviour is largely determined by growth defects. These play a decisive role in the growth of irregular eutectics, such as the most important casting alloys (Al-Si, cast iron).

On a larger scale, the first phenomenon which becomes apparent is the *morphological instability* of solid/liquid interfaces, that normally occurs on a scale which is of the order of micrometres. This is only a transient stage in the development of the steady-state growth forms of cells, dendrites or eutectics.

Equiaxed grains of pure or alloyed materials are inherently unstable when their diameter exceeds a critical value (generally of the order of some micrometres). On the other hand, columnar grains are always stable (plane front growth) in pure samples and can be stable in the case of alloys if the temperature gradient is sufficiently high. At very high growth rates, the onset of so-called absolute stability will lead to the occurrence of solidification involving a structureless solid/liquid interface leading to a micro segregation free solid.

The most frequently observed solidification microstructure is the *dendrite*. This growth morphology is characterised by its paraboloid-like tip and by the formation of branches along simple low-index orientations. The growth of these crystals is determined by the following diffusion phenomena: thermal (equiaxed growth of pure substances), solutal (columnar growth of alloys), and thermal/solutal (equiaxed growth in alloys). The solution of the diffusion equations, combined with a stability criterion, leads to a unique description of dendrite tip growth.

In the case of columnar growth, the imposed positive temperature gradient constrains the dendrites to grow in the form of arrays having a characteristic primary spacing, λ_1. During growth, the length of contact time which the initially fine branches have with their melt causes ripening to occur to such a degree that the final branch spacing, λ_2, is very much greater than the initial one at the tip.

As far as *eutectics* are concerned, there are two types of microstructure: regular and irregular, whose occurrence depends upon the entropy of melting of one of the phases. Regular structures are typically exhibited by phases having a low entropy of melting, and grow in a steady-state manner, whereas irregular structures (in which one phase is faceted) oscillate between two operating points. As in the case of dendrites, the combination of the diffusion solution with a criterion which decides the operating point of the system leads to the deduction of the growth law. Eutectic microstructures store an appreciable fraction of their transformation energy in the form of α/β interfaces (which are usually those having the minimum energy).

The competitive growth of eutectic and dendrites will lead to mixed microstructures which can be summarised with the aid of the so-called coupled zone. Due to the very high undercoolings necessary for rapid eutectic growth of systems with small solubilities a transition to a glass may be observed.

Peritectic reactions develop around dendrites mostly in the form of interdendritic precipitates.

Any local equilibrium between two phases will result in *concentration differences* which are necessary in order to satisfy thermodynamic requirements. These differences create local concentration variations which are known as segregation. In the case of a dendritic interface, excess solute is built into the interdendritic regions. Back-diffusion into the solid dendrite plays an important role in governing the magnitude of the composition of the last liquid to solidify.

The use of very *high growth rates* (high Péclet numbers) leads to increasingly large departures from equilibrium. In particular, the distribution coefficient becomes a funcion of the growth rate. Also, the extreme narrowing of heat and mass diffusion fields leads to phenomena such as the reappearance of a stable planar interface in single-phase alloys at high growth rates and a minimum possible spacing in the case of two-phase (eutectic) alloys. The solute segregation behaviour is also affected.

The coupling of microstructural models with segregation models, and their incorporation into heat-flux calculations, today permits the modelling of and a better understanding of diverse solidification processes occurring over a wide range of conditions. In the future, it will be possible to relate thermodynamic information on multicomponent systems and the microstructural models presented here, together with others yet to be developed, to multidimensional, macroscopic heat flow calculations. In this way, the much-needed improved control of solidification processes will be made feasible.

For Further Reading - Books and Articles of General Interest

H.I.Aaronson (Ed.), *Lectures on the Theory of Phase Transformations*, Metallurgical Society of AIME, New York, 1975.

H.D.Brody, D.Apelian (Eds.), *Modelling of Casting and Welding Processes*, Metallurgical Society of AIME, New York, 1981.

J.W.Christian, *The Theory of Transformations in Metals and Alloys*, 2nd Edition, Pergamon, Oxford, 1975.

F.Durand (Ed.), *Solidification des Alliages*, Les Editions de Physique, F-91944 Les Ulis, France, 1988.

R.Elliott, *Eutectic Solidification Processing - Crystalline and Glassy Alloys*, Butterworths, London, 1983.

M.C.Flemings, *Solidification Processing*, McGraw-Hill, New York, 1974.

H.Fredriksson (Ed.), *State of the Art of Computer Simulation of Casting and Solidification Processes*, Les Editions de Physique, 1986.

H.Fredriksson, M.Hillert (Eds.), *The Physical Metallurgy of Cast Iron*, North Holland, New York, 1985.

M.Glicksman, R.F.Sekerka, Solidification Kinetics, in *Problems in Materials Science*, Volume 1, (Ed. H.D.Merchant), Gordon & Breach, New York, 1972.

A Guide to the Solidification of Steels, Jernkontorets, Stockholm, 1977.

H.Hermann (Ed.), *Ultrarapid Quenching of Liquid Alloys*, Academic Press, New York, 1981 (Treatise on Materials Science and Technology, Volume 20).

H.Jones, *Rapid Solidification of Metals and Alloys*, The Institution of Metallurgists, London, 1982.

H.Jones, W.Kurz (Eds.), *Solidification Microstructures, 30 years after Constitutional Supercooling*, Materials Science and Engineering, Special Issue, vol. **65**, 1984.

B.H.Kear et al. (Eds.), *Rapidly Solidified Amorphous and Crystalline Alloys*, Elsevier North-Holland, 1982.

W.Kurz, P.R.Sahm, *Gerichtet erstarrte eutektische Werkstoffe*, Springer, Berlin, 1975.

B.Lux, I.Minkoff, F.Mollard (Eds.), *The Metallurgy of Cast Iron*, Georgi, St Saphorin, Switzerland, 1975.

M.McLean, *Directionally Solidified Materials for High Temperature Service*, The Metals Society, London, 1983.

I.Minkoff, *The Physical Metallurgy of Cast Iron*, Wiley, New York, 1983.

I.Minkoff, *Solidification and Cast Structure*, Wiley, New York, 1986.

H.Nieswaag, J.W.Schut (Eds.), *Quality Control of Engineering Alloys and the Role of Metal Science*, Delft, 1978.

Nucleation Phenomena, American Chemical Society, Washington DC, 1966.

B.R.Pamplin, *Crystal Growth*, Pergamon, Oxford, 1975.

R.L.Parker, Crystal Growth Mechanisms, in *Solid State Physics* (Eds. H.Ehrenreich, F.Seitz, D.Turnbull), **25** (1970) 151.

W.G.Pfann, *Zone Melting*, 2nd Edition, Wiley, New York, 1966.

P.R.Sahm, P.N.Hansen, *Numerical Simulation and Modelling of Casting and Solidification Processes for Foundry and Cast-House*, International Commercial Foundry Technical Association, New York, 1984.

P.R.Sahm, H.Jones, C.M.Adam, *Science and Technology of the Undercooled Melt*, Martinus Nijhoff, Dordrecht, 1986.

Solidification, American Society for Metals, Metals Park, Ohio, 1971.

The Solidification of Metals, ISI Publication 110, London, 1968.

Solidification and Casting of Metals, The Metals Society, London, 1979.

Solidification Technology in the Foundry and Casthouse, The Metals Society, London, 1983.

Solidification Processing 1987, The Institute of Metals, London, 1988.

D.M.Stefanescu, G.J.Abbaschian, R.J.Bayuzick (Eds.), *Solidification Processing of Eutectic Alloys*, The Metallurgical Society, 1988.

APPENDIX 1

MATHEMATICAL MODELLING OF THE MACROSCOPIC HEAT FLUX

Heat Diffusion Equation

Solidification of a material with a melting point which is above the ambient temperature will occur spontaneously once some of the solid has formed. The heat flux from the hot melt to the surroundings allows the liquid to cool, to transform to the solid and the solid to cool further to the temperature of the surrounding medium. Transformation to the solid can mean the formation of crystals, as in most metals (crystallisation), or the creation of an amorphous state (glass formation). In the case of crystallisation, latent heat is evolved at the solid/liquid interface. It is this heat source, together with an often highly complicated interface morphology, which makes difficult the solution of the heat diffusion equation.

In order to deal with heat flow problems in solidification, one can use two basically different approaches.

Front-Tracking Method (Two-Domain Method, Stefan Problem)

In these methods, the overall volume within which the heat-flow equation is to be solved is sub-divided into 2 domains (liquid and solid) with a solid/liquid interface (see Fig. A1.1). Accordingly, one has 2 diffusion equations to solve; each for a single-phase region. For a two-dimensional problem, this is written as:

$$\frac{\partial}{\partial y}(\kappa_l \frac{\partial T}{\partial y}) + \frac{\partial}{\partial z}(\kappa_l \frac{\partial T}{\partial z}) = c_l \frac{\partial T}{\partial t} \qquad [A1.1]$$

$$\frac{\partial}{\partial y}(\kappa_s \frac{\partial T}{\partial y}) + \frac{\partial}{\partial z}(\kappa_s \frac{\partial T}{\partial z}) = c_s \frac{\partial T}{\partial t} \qquad [A1.2]$$

where c_l and c_s are the volumic specific heats of liquid and solid, respectively.

Appropriate boundary conditions are applied at the fixed boundaries (e.g. equality of heat flux). At the free moving solid/liquid interface, 2 boundary conditions have to be satisfied:

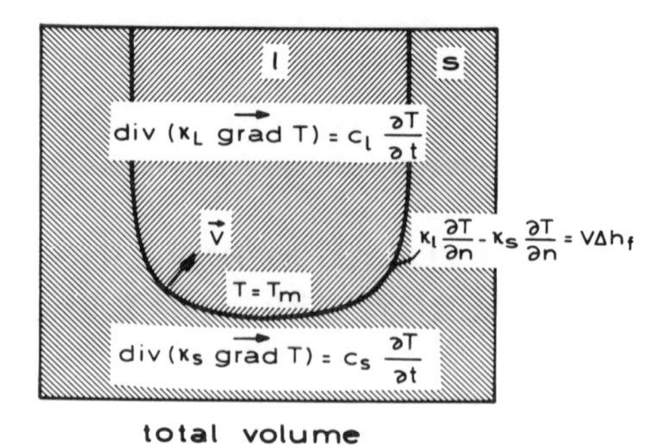

Figure A1.1

- continuity of temperature (for negligible atom attachment kinetics)

$$T_l^* = T_s^* = T_f - \Gamma K \qquad [A1.3]$$

- continuity of heat flow

$$\kappa_s \frac{\partial T}{\partial n}\bigg|_s - \kappa_l \frac{\partial T}{\partial n}\bigg|_l = \Delta h_f V \qquad [A1.4]$$

where the temperature gradients are those normal to the solid/liquid interface. This problem (without the curvature term ΓK, in A.1.3) is known as the Stefan problem. In some cases, analytical solutions are available. In most problems, one has to use numerical methods involving a front-tracking procedure, i.e. a method which can follow the solid/liquid interface with time. This approach, which is quite complicated, is primarily used:

i) in non-stationary calculations of the solidification of pure metals [1,2]

ii) in calculations of steady-state microstructures [3,4]. In such cases, solute diffusion within the 2 media (solid/liquid) also have to be considered, as well as appropriate interface boundary conditions for both the concentration and the solute flow.

Averaging Method (One-Domain Method)

Since front-tracking methods are very difficult to apply to general problems of solidification, especially when dealing with alloys which have a mushy zone, a simpler approach is commonly used (see Fig. A1.2). That is, one defines a local solid fraction,

f_s, at a given "point". Or, to be more precise, within a volume element which is large with respect to the microstructure and small compared to the temperature inhomogeneities.

With the enthalpy, H, defined as:

$$H(T) = \int_0^T c(\theta)d\theta + \Delta h_f(1-f_s) \qquad [A1.5]$$

where Δh_f is assumed to be independent of temperature, one has:

$$\mathrm{div}(\kappa\,\overrightarrow{\mathrm{grad}}T) = \frac{dH}{dt} \qquad [A1.6]$$

The change in enthalpy during heating or cooling of the material is, according to equation A1.5:

$$\frac{dH}{dt} = c(T)\frac{\partial T}{\partial t} - \Delta h_f\frac{\partial f_s}{\partial t} \qquad [A1.7]$$

Equation A1.6 can be written in terms of a single spatial dimension (z or r) as:

$$\frac{\partial}{\partial z}\left(\kappa\frac{\partial T}{\partial z}\right) = \frac{dH}{dt} \qquad \text{(plate)} \qquad [A1.8]$$

$$\frac{1}{r}\cdot\frac{\partial}{\partial r}\left(\kappa r\frac{\partial T}{\partial r}\right) = \frac{dH}{dt} \qquad \text{(cylinder)} \qquad [A1.9]$$

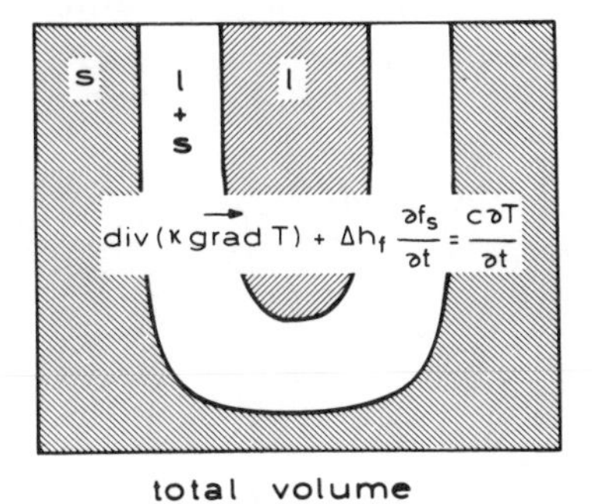

Figure A1.2

$$\frac{1}{r^2} \cdot \frac{\partial}{\partial r}\left(\kappa r^2 \frac{\partial T}{\partial r}\right) = \frac{dH}{dt} \qquad \text{(sphere)} \qquad [A1.10]$$

For pure metals, as well as for alloys, equation A1.7 is highly non-linear and very few analytical solutions exist. Therefore, numerical solutions are usually required; especially in the case of multidimensional problems. It is beyond the scope of this textbook to provide a full introduction to the subject. However, interested readers will here find some basic notions and some useful references to the field.

Analytical Solutions for Semi-Infinite Single-Phase Systems

Neglecting heat generation for the moment, and keeping a (= κ/c) constant, equations A1.1 or A1.6/A1.7 can be written for a unidirectional heat flux as [5-7]:

$$a\frac{\partial^2 T}{\partial z^2} = \frac{\partial T}{\partial t} \qquad 0 \le z \le \infty \qquad [A1.11]$$

One possible general solution of equation A1.11 is:

$$T(z, t) = A + B\,\text{erf}(Z) \qquad [A1.12]$$

where A and B are constants and

$$Z = \frac{z}{2(at)^{1/2}} \qquad [A1.13]$$

The error function, erf, (and its complement, erfc, which can be a better choice for problems involving heating), have the properties (Fig. A1.3)#:
erf(0) = 0; erf(∞) = 1; erf(–Z) = – erf(Z); erfc(Z) = 1 – erf(Z).
Noting that:

$$\text{erf}(Z) = \frac{2}{\pi^{1/2}} \int_0^Z \exp(-Z'^2)\,dZ' \qquad [A1.14]$$

the first derivative is:

$$\frac{d(\text{erf}[Z])}{dZ} = \frac{2}{\pi^{1/2}} \exp[-Z^2] \qquad [A1.15]$$

With the aid of a programable calculator, the error function, for $0 \le Z < \infty$, can easily be determined by using the approximation [8]:

The exponential integral, E_1, which is also plotted in figure A1.3 will be used in appendix 8 (equation A8.17)

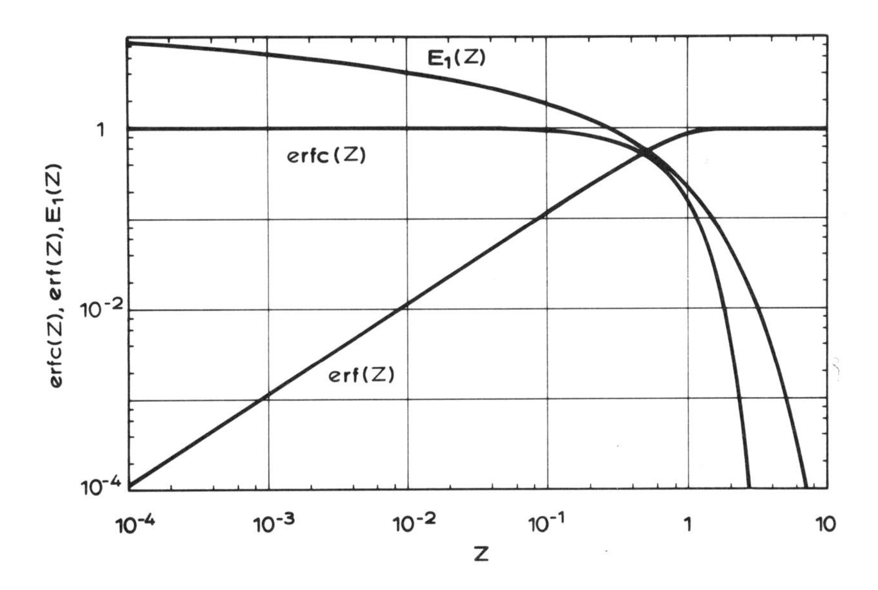

Figure A1.3

$$\mathrm{erf}(Z) = 1 - (a_1 t + a_2 t^2 + a_3 t^3 + a_4 t^4 + a_5 t^5)\exp(-Z^2) \qquad [A1.16]$$

$$t = \frac{1}{1 + a_0 Z}$$

and

$a_0 = 0.3275911$ $\quad a_1 = 0.254829592$

$a_2 = -0.284496736$ $\quad a_3 = 1.421413741$

$a_4 = -1.453152027$ $\quad a_5 = 1.061405429$

Exact values of erf(Z) are given in table A1.1. Some useful approximations are:

$$\mathrm{erf}(Z) \cong \left[1 - \exp\left(-\frac{4Z^2}{\pi}\right)\right]^{1/2}$$

$$\mathrm{erf}(Z) \cong 1 \qquad z > 2$$

$$\mathrm{erf}(Z) \cong \frac{2Z}{\pi^{1/2}} \qquad z < 0.2$$

A good approximation, at least over the range of Z values which is covered by table A1.1, is provided by the simple Padé-type expression:

$$\mathrm{erf}(Z) \cong \frac{2Z}{\pi^{1/2}} \cdot \frac{3}{(Z^2+3)} \qquad [A1.17]$$

In order to demonstrate that equation A1.12 is a general solution of equation A1.11, one can use equation A1.15, so that:

$$\frac{\partial T}{\partial z} = \frac{B\exp[-Z^2]}{(\pi a t)^{1/2}} \qquad [A1.18]$$

and

$$\frac{\partial^2 T}{\partial z^2} = -\frac{Bz\exp[-Z^2]}{2(\pi a^3 t^3)^{1/2}} \qquad [A1.19]$$

The time derivative is:

$$\frac{\partial T}{\partial t} = -\frac{Bz\exp[-Z^2]}{2(\pi a t^3)^{1/2}} \qquad [A1.20]$$

Substituting the latter two expressions into the differential equation (A1.11) shows that the error function is indeed a solution. Other solutions might be $T = A + B\mathrm{erfc}(Z)$ or $T = (1/t^{1/2})\exp(-Z^2)$. The choice of the solution (assuming that one exists) will depend upon the boundary conditions.

The Moving Boundary Problem

As figure A1.4 indicates, several regions exist in a casting; each of which exhibits a specific thermal behaviour. These include the mould, the air-gap, the solid, the mushy (solid-plus-liquid) zone, and the liquid. An analytical solution of the general problem is not possible. For this reason, and to permit the reader to make a quick rule-of-thumb estimate, attention will here be restricted to just one simple case.

Simple Analytical Solution

Under the assumptions of one-dimensional heat flow, semi-infinite system, highly-cooled mould at T_m, no air-gap, planar solid-liquid interface, and zero superheat of the melt, the boundary conditions (Fig. A1.5) are:

$$T_m = T_0 = \text{constant} \qquad 0 > z > -\infty$$

$$T_l = T_f = \text{constant} \qquad z > s$$

Table A1.1 Values of the Error Function [8]

Z	erf(Z)	Z	erf(Z)	Z	erf(Z)	Z	erf(Z)
0.00	0.0000	0.26	0.2869	0.52	0.5379	0.78	0.7300
0.01	0.0113	0.27	0.2974	0.53	0.5465	0.79	0.7361
0.02	0.0226	0.28	0.3079	0.54	0.5549	0.80	0.7421
0.03	0.0338	0.29	0.3183	0.55	0.5633	0.81	0.7480
0.04	0.0451	0.30	0.3286	0.56	0.5716	0.82	0.7538
0.05	0.0564	0.31	0.3389	0.57	0.5798	0.83	0.7595
0.06	0.0676	0.32	0.3491	0.58	0.5879	0.84	0.7651
0.07	0.0789	0.33	0.3593	0.59	0.5959	0.85	0.7707
0.08	0.0901	0.34	0.3694	0.60	0.6039	0.86	0.7761
0.09	0.1013	0.35	0.3794	0.61	0.6117	0.87	0.7814
0.10	0.1125	0.36	0.3893	0.62	0.6194	0.88	0.7867
0.11	0.1236	0.37	0.3992	0.63	0.6270	0.89	0.7918
0.12	0.1348	0.38	0.4090	0.64	0.6346	0.90	0.7969
0.13	0.1459	0.39	0.4187	0.65	0.6420	0.91	0.8019
0.14	0.1569	0.40	0.4284	0.66	0.6494	0.92	0.8068
0.15	0.1680	0.41	0.4380	0.67	0.6566	0.93	0.8116
0.16	0.1790	0.42	0.4475	0.68	0.6638	0.94	0.8163
0.17	0.1900	0.43	0.4569	0.69	0.6708	0.95	0.8209
0.18	0.2009	0.44	0.4662	0.70	0.6778	0.96	0.8254
0.19	0.2118	0.45	0.4755	0.71	0.6847	0.97	0.8299
0.20	0.2227	0.46	0.4847	0.72	0.6914	0.98	0.8342
0.21	0.2335	0.47	0.4937	0.73	0.6981	0.99	0.8385
0.22	0.2443	0.48	0.5027	0.74	0.7047	1.00	0.8427
0.23	0.2550	0.49	0.5117	0.75	0.7112	1.30	0.9340
0.24	0.2657	0.50	0.5205	0.76	0.7175	1.80	0.9891
0.25	0.2763	0.51	0.5292	0.77	0.7238	2.00	0.9953

where the interface position on the z-axis is given by the function, $s(t)$. The temperature in the solid can be determined with the aid of equations A1.12 and A1.13, giving:

$$T_s = A_s + B_s \operatorname{erf}\left[\frac{z}{2(a_s t)^{1/2}}\right] \qquad \text{[A1.21]}$$

and, applying the boundary conditions:

$$z = 0 \qquad A_s = T_0$$

$$z = s \qquad B_s = \frac{T_f - T_0}{\mathrm{erf}[\Phi]}$$

$$\Phi = \frac{s}{2(a_s t)^{1/2}}$$

Since, under the above assumptions, T_f and T_0 are constant (Fig. A1.5), Φ is constant and

$$s = 2\Phi(a_s t)^{1/2} \qquad \text{or} \qquad t = \frac{s^2}{4a_s \Phi^2} \qquad [A1.22]$$

In order to determine the value of Φ, it is necessary to consider the (Neumann) boundary condition which reflects the heat flow at the solid/liquid interface (see also appendix 2). The latent heat generated during interface advance (defined in this book to be positive for solidification) must be conducted away through the solid, giving the flux balance (for zero superheat in the liquid):

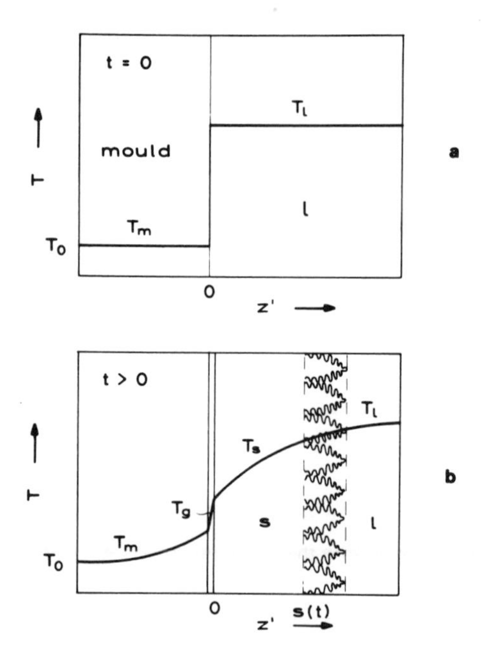

Figure A1.4

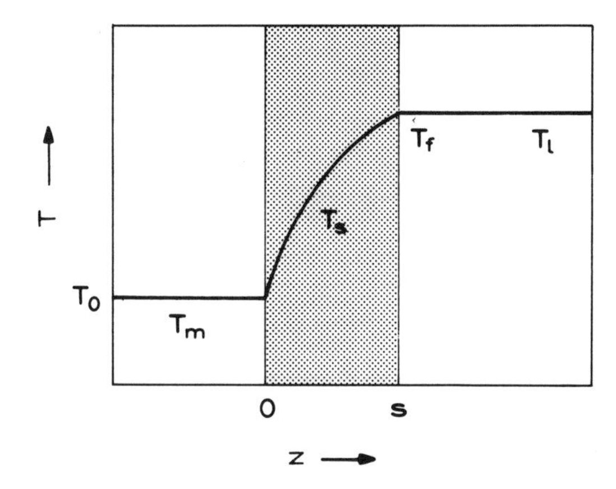

Figure A1.5

$$\Delta h_f \frac{ds}{dt} = \kappa_s \left(\frac{\partial T_s}{\partial z}\right)_{z=s} \qquad [A1.23]$$

From equation A1.22:

$$V = \frac{ds}{dt} = \left(\frac{a_s}{t}\right)^{1/2} \Phi \qquad [A1.24a]$$

while, from equation A1.18 at $z = s$,

$$G_s = \left(\frac{\partial T_s}{\partial z}\right)_{z=s} = \frac{B_s \exp[-\Phi^2]}{(\pi a_s t)^{1/2}} \qquad [A1.24b]$$

Substituting V and G_s from equation A1.24 into equation A1.23, with B_s given by equation A1.21, gives (with $a_s = \kappa_s/c_s$):

$$\Delta T_s^\circ = \pi^{1/2} \Phi \operatorname{erf}(\Phi) \exp(\Phi^2) \qquad [A1.25]$$

where the dimensionless temperature, ΔT_s°, is equal to $(T_f - T_0)c_s/\Delta h_f$. Evaluation of ΔT_s° permits the calculation of Φ by iteration and, using equation A1.22, the position of the interface as a function of time can be found to obey a square-root law. For typical values of ΔT_s°, ranging from 0 to 4, equation A1.25 can be approximated quite well by (Fig. A1.6):

$$\Delta T_s^\circ \cong 2\Phi^2(\Phi^2 + 1) \qquad [A1.26]$$

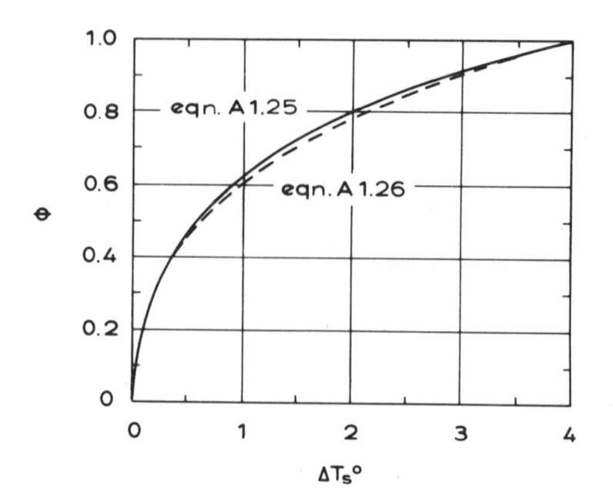

Figure A1.6

A more general solution, which permits a better description of the real situation shown in figure A1.4b, has been given by Garcia et al. [9,10]. An analytical treatment which takes account of the mushy zone has been developed by Lipton et al. [11].

Numerical Solutions

It is beyond the scope of the present appendix to describe in detail the various numerical techniques which are applied to solidification problems. The interested reader is referred to other sources [12,13]. The resolution of equations A1.6/1.7 (Average Method) will be illustrated by using a finite difference method (FDM), assuming for simplicity that $\kappa_l = \kappa_s$, that $c_l = c_s$ and that both are independent of temperature. A regular mesh, Δx by Δy, in two dimensions (Fig. A1.7) will be used. Considering one point, C, of the enmeshment, and its 4 neighbours, (E, N, W, S), and writing equation A1.6/1.7 in finite difference form, gives simply:

$$\kappa\left(\frac{T_E - T_C}{\Delta x} + \frac{T_W - T_C}{\Delta x}\right)\Delta y$$

$$+ \kappa\left(\frac{T_N - T_C}{\Delta y} + \frac{T_S - T_C}{\Delta y}\right)\Delta x =$$

$$\Delta x \Delta y\left(c\frac{\Delta T_C}{\Delta t} - \Delta h_f \frac{\Delta f_{s,c}}{\Delta t}\right) \qquad \text{[A1.27]}$$

The first term on the LHS of equation A1.27 represents the heat leaving or entering the vertical sides of the volume, V_c, while the second term corresponds to the horizontal

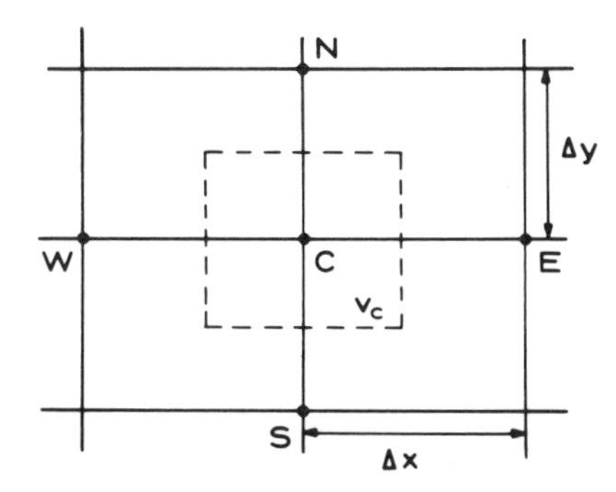

Figure A1.7

sides at north and south. The sum of these two contributions will be labelled Q_C in the following equations. The RHS term is just the variation of the enthalpy within the volume of the mesh V_C.

The solution of equation A1.27 at all of the nodes of the enmeshment can only be found if a *solidification path* is defined, i.e. if a relationship between f_S and T exists. Two basic methods can be used.

Macroscopic Modelling (Unique Solidification Path)

In the macroscopic modelling of solidification, one neglects any nucleation or growth undercooling and, accordingly, the relationship between f_S and T does not depend upon additional parameters such as the velocity of the interface. The behaviour of $f_S(T)$, which will then appear in equations A1.27, can be deduced from a segregation model as described in chapter 6 (e.g. equation 6.9). Two methods can be used to do this.

Effective specific heat method: $\Delta f_S(T)/\Delta t$ in equation A1.27 is replaced by $\Delta f_S/\Delta T \cdot \Delta T/\Delta t$ so as to obtain:

$$Q_c = \Delta x \Delta y \cdot c^*(T) \frac{\Delta T_c}{\Delta t} \qquad [A1.28]$$

where

$$c^*(T) = c - \Delta h_f \frac{\Delta f_s(T)}{\Delta T} \qquad [A1.29]$$

The effective specific heat per unit volume, $c^*(T)$, can be easily introduced into standard commercial computer codes as a tabulated thermophysical property. This is the reason for its popularity. However, because it exhibits discontinuities or strong variations when close to the liquidus or to the eutectic temperature, numerical methods which are based upon an effective specific heat have some difficulty in ensuring energy conservation. For more details of this method, see references 14 and 15.

Enthalpy method: equation A1.27 is combined with equation A1.5 to give

$$Q_c = \Delta x \Delta y \cdot \frac{\Delta H}{\Delta t} \qquad [A1.30]$$

Instead of treating the temperature as a variable, one adjusts the enthalpy, H. Therefore, equation A1.30 has to be combined with the function, $T(H)$, obtained from equation A1.5 using a segregation model. This method has the advantage of ensuring energy conservation during the phase transformation [16,17].

Micro-Macroscopic Modelling (Solidification Path Non-Unique)

In the micro-macroscopic modelling of solidification, transformation kinetics effects and associated nucleation and/or growth undercoolings are taken into account. For *columnar growth*, expressions for $f_S\,(T,V)$ have been derived, where V is the growth rate of the dendrite tips [18,19]. In *equiaxed growth* situations, the fraction of solid is written as [20,21]:

$$f_s = n(t)\frac{4}{3}\pi R^3(t) f_i\,(t) \qquad [A1.31]$$

where n is the grain density, R is the average grain radius and f_i is the fraction of solid within the grains (assumed to be spherical). The grain density can be derived from a nucleation model (see chapter 2 and reference 24), while the radius, $R(t)$, is associated with the growth kinetics (of the dendrite tip or eutectic front), which relate $V = dR/dt$ to the undercooling, ΔT (as explained in chapters 4 and 5). Finally, f_i has to be deduced from a solute balance at the scale of the grains [22,23]. For a more detailed discussion of the macroscopic modelling of heat flow and of its combination with microstructural models, the reader is referred to the review by Rappaz [24].

References

[1] R.Viskanta, C.Beckermann, *Mathematical Modelling of Solidification*, Symp. Interdisciplinary Issues in Materials Processing and Manufacturing, Amer. Soc. Mech. Eng. Annual Meeting, Boston, USA, 1987.

[2] J.Crank, *Free and Moving Boundary Problems*, Clarendon Press, Oxford, 1984.

[3] S.R.Coriell, G.B.McFadden, R.F.Sekerka, Ann. Rev. Mater. Sci. **15** (1985) 119.

[4] Y.Saito, G.Goldbeck-Wood, H.Müller-Krumbhaar, Physical Review **A38** (1988) 2148.

[5] H.S.Carslaw, J.C.Jaeger, *Conduction of Heat in Solids*, 2nd Edition, Oxford University Press, London, 1959.

[6] G.H.Geiger, D.R.Poirier, *Transport Phenomena in Metallurgy*, Addison Wesley, Reading, 1973.

[7] J.Szekely, N.J.Themelis, *Rate Phenomena in Process Metallurgy*, Wiley Interscience, New York, 1971.

[8] M.Abramowitz, I.A.Stegun (Eds.), *Handbook of Mathematical Functions*, Dover, New York, 1965.

[9] A.Garcia, T.W.Clyne, M.Prates, Metallurgical Transactions **10B** (1979) 85.

[10] T.W.Clyne, A.Garcia, International Journal of Heat & Mass Transfer **23** (1980) 773.

[11] J.Lipton, A.Garcia, W.Heinemann, Archiv für das Eisenhüttenwesen **53** (1982) 469.

[12] D.R.Croft, D.G.Lilley, *Heat Transfer Calculations using Finite Difference Equations*, Applied Science Publ., 1977.

[13] Suhas V.Patankar, *Numerical Heat Transfer and Fluid Flow*, Hemisphere Publ. Corp., 1980.

[14] J.A.Dantzig, J.W.Wiese, Metallurgical Transactions **16B** (1985) 195 and 203.

[15] T.W.Clyne, Metallurgical Transactions **13B** (1982) 471.

[16] Q.T.Pham, International Journal of Heat & Mass Transfer **29** (1986) 285.

[17] J.L.Desbiolles, J.J.Droux, J.Rappaz, M.Rappaz, Computer Physics Reports **6** (1987) 371.

[18] B.Giovanola, *Solidification Rapide de Surfaces Métalliques par Laser*, Sc. D. Thesis (1986), Ecole Polytechnique Fédérale de Lausanne.

[19] S.C.Flood, J.D.Hunt, Journal of Crystal Growth **82** (1987) 543 and 552.

[20] M.Rappaz, D.M.Stefanescu, in *Solidification Processing of Eutectic Alloys*, (edited by D.M.Stefanescu, G.J.Abbaschian and R.J.Bayuzick) TMS, 1988.

[21] M.Rappaz, J.L.Desbiolles, Ph.Thévoz, W.Kurz, in *Erstarrung metallischer Werkstoffe*, DGM Informationsgesellschaft-Verlag, 1988.

[22] M.Rappaz, Ph.Thévoz, Acta Metallurgica **35** (1987) 1478.

[23] M.Rappaz, Ph.Thévoz, Acta Metallurgica **35** (1987) 2929.

[24] M.Rappaz, International Materials Reviews **34** (1989) 93.

APPENDIX 2

SOLUTE AND HEAT FLUX CALCULATIONS RELATED TO MICROSTRUCTURE FORMATION

The literature of solidification microstructure theory, especially in recent times, has become highly complex and different mathematical methods have been used in developing the various models. However, it is the point of view of this book that a reasonably accurate result, which reveals the principles involved and the influence of the physical variables, can be obtained for any problem by using a coherent approach. In fact, exactly the same general principles apply to all solidification problems and the various published analyses differ only with respect to the approximations which are made, and to the weight which is given to various aspects of the problem in question.

The most common approximation which is made is that solidification is occurring under steady-state conditions and that, therefore, the concentrations and solid/liquid interface morphology are independent of time. The principal disadvantage of this assumption is that no evolution of the interface shape can occur. The result of this constraint is that the solution to the basic diffusion problem is indeterminate and a whole range of morphologies is permissible from the mathematical point of view. In order to distinguish the solution which is the most likely to correspond to reality, it is necessary to find some additional criterion. Examination of the stability of a slightly perturbed growth form is probably the most reasonable manner in which to treat this situation.

In the present appendix, one aim is to supply the reader with mathematical techniques which are sufficient to attack the problems of microscopic heat and mass transfer which are treated elsewhere in the book. Another aim is to provide the reader with a general and systematic method for approaching steady-state solidification problems.

The general features of a solidification problem can be described as follows: a solid/liquid interface whose form is defined by a given mathematical function containing one or more variable parameters, is assumed to be advancing without change into the melt. As it advances, heat and/or solute are evolved at each point of the interface and diffuse into the solid and the melt. The diffusing solute will build up ahead of the interface when $k<1$ and form a boundary layer, while a uniform level, C_0, of the solute is supposed to exist at a sufficiently large distance from the interface. The boundary layer can be characterised by the ratio of the diffusion coefficient to the growth rate. Typical orders of magnitude of the equivalent boundary layers for a planar interface (characteristic diffusion distance) are shown in table A2.1. From this, it is seen that the boundary layer thickness for mass transfer of substitutional elements in the solid is so small as to be negligible#, while the boundary layer thicknesses for heat transfer (at low growth rates) in the solid or liquid are much larger than the scale of most castings. Convection must also be considered since the presence of a hydrodynamic boundary layer will reduce the thickness of the thermal boundary layer.

Table A2.1 Equivalent Boundary Layers for a Planar Interface

Type of Diffusion	Diffusing Species	Matrix at T_f	Layer Thickness (mm) at $V = 0.01$mm/s
solutal: $\delta_c = 2D/V$	substitutional atom	crystal	10^{-4}
	interstitial atom	crystal	10^{-1}
	either	liquid	10^{0}
thermal: $\delta_t = 2a/V$	heat	crystal	$>10^{2}$
		liquid	$>10^{2}$

Differential Equation for Diffusion

In view of the previous arguments, attention will be restricted here to solute diffusion occurring in the liquid. For simplicity, the suffix, l, will be dropped from the concentrations since these always refer to the liquid in the present appendix. That is, $C_l \equiv C$. The equation governing any diffusion process is:

$$\frac{\partial^2 C}{\partial x^2} + \frac{\partial^2 C}{\partial y^2} + \frac{\partial^2 C}{\partial z^2} = \frac{1}{D} \cdot \frac{\partial C}{\partial t} \qquad \text{[A2.1]}$$

In physics texts this is usually written:

$$\nabla^2 C = \frac{1}{D} \cdot \frac{\partial C}{\partial t}$$

while, in mathematical texts, it is often written in the suffix notation:

$$C_{xx} + C_{yy} + C_{zz} = \frac{1}{D} C_t$$

An exception is the case of microsegregation where, even for substitutional solutes, solid-state diffusion plays an important role due to the very high concentration gradients existing in the last stages of solidification (appendix 12).

Equation A2.1, which is analogous to equation A1.1, applies to three-dimensional space. However, the problems which are treated in the present book usually require the consideration of no more than two spatial dimensions:

$$\frac{\partial^2 C}{\partial y^2} + \frac{\partial^2 C}{\partial z^2} = \frac{1}{D} \cdot \frac{\partial C}{\partial t} \qquad [A2.2]$$

Solutions can often be simplified by using a suitable coordinate system. For example, in spherical polar coordinates# equation A2.1 becomes [1 - 3]:

$$C_{rr} + \frac{2}{r}C_r + \frac{1}{r^2}C_{\theta\theta} + \frac{\cot\theta}{r^2}C_\theta + \frac{1}{r^2\sin^2\theta}C_{\psi\psi} = \frac{1}{D}C_t$$

where θ, ψ, and r are the equatorial and azimuthal angles, and radial distance, respectively, and the suffix notation has been used for greater clarity. When the diffusional behaviour is independent of the angular orientations, θ and ψ, this equation becomes:

$$\frac{\partial C}{\partial t} = D\left(\frac{\partial^2 C}{\partial r^2} + \frac{2}{r}\frac{\partial C}{\partial r}\right)$$

More generally, one can write:

$$\frac{\partial C}{\partial t} = D\left(\frac{\partial^2 C}{\partial r^2} + \frac{n}{r}\frac{\partial C}{\partial r}\right)$$

where $n = 2$. Further reductions to the equation for a cylindrical or plate geometry can be obtained by replacing n in the above expression by unity or zero, respectively.Thus, the steady-state growth of a sphere is governed by:

$$\frac{d^2 C}{dr^2} + \frac{2}{r}\frac{dC}{dr} = 0$$

The principal characteristic of the diffusion equation is its conservative nature; i.e. it acts so as to even out any irregularities. This can be seen firstly by considering the one-dimensional equation for rectangular coordinates:

$$\frac{\partial^2 C}{\partial z^2} = \frac{1}{D} \cdot \frac{\partial C}{\partial t} \qquad [A2.3]$$

For other coordinate systems, see reference 3. In principle, any interface shape can be given a suitable system of coordinates. However, this would be pointless unless the resultant transformed differential equation could be separated. This can only be achieved in some eleven systems, including the cylindrical, spherical, and parabolic. The latter system is very useful for treating dendrite tip problems.

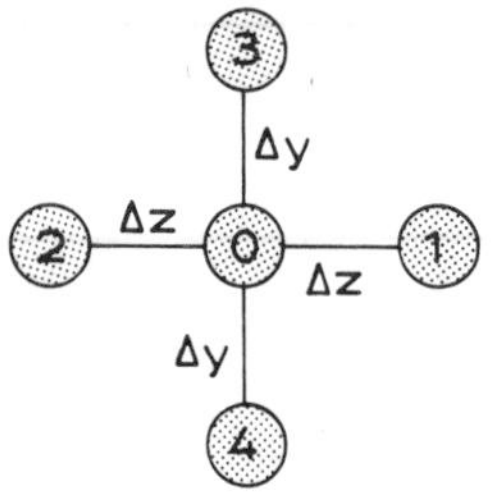

$$\frac{\partial^2 c}{\partial y^2} + \frac{\partial^2 c}{\partial z^2} = 0$$

$$\frac{(C_3 - C_0) - (C_0 - C_4)}{\Delta y^2} + \frac{(C_1 - C_0) - (C_0 - C_2)}{\Delta z^2} = 0$$

$$\boxed{C_0 = \frac{C_1 + C_2 + C_3 + C_4}{4}} \qquad \Delta y = \Delta z$$

Figure A2.1

Note that the left-hand-side is the expression which defines the sense of the curvature of a function. Thus, when the second derivative is positive it denotes a concave-upwards part of a function. In the present case, it would correspond to a local minimum in the concentration distribution. However, from equation A2.3, the local change in concentration with time is also positive and therefore the depression in the concentration distribution will tend to be removed. The reverse is true for negative values of the left-hand-side. The conservative nature of the two-dimensional, time-independent (Laplace) equation can also be seen clearly from the finite-difference scheme (Fig. A2.1) where each cell must take a value which is the average of the values of the four surrounding cells. A little reflection will show that the only 'bumps' which can persist, in this steady-state situation, are those imposed by the boundary conditions. Finally, it can be proved that the direction of flow of heat or mass at any point is at right-angles to the isotherms or isoconcentrates. It is less well known that both the isotherms and flux lines can be drawn by inspection, without ever solving the diffusion equation. Unfortunately, this 'flux-plotting' technique [4] does not work well for solidification problems because of the complicated boundary conditions which are usually imposed.

Directional Growth Equation

Equation A2.3, which is known as Fick's second law, would be of little use in treating moving boundary problems because the interface movement would have to be accounted for. Instead, an equation which is expressed in a coordinate system that is moving with the interface can be used. In figure A2.2, the coordinates of the point, P, with respect to axes moving with the interface are (y,z). Its coordinates with respect to a stationary observer are (y',z'). From the diagram, it is evident that:

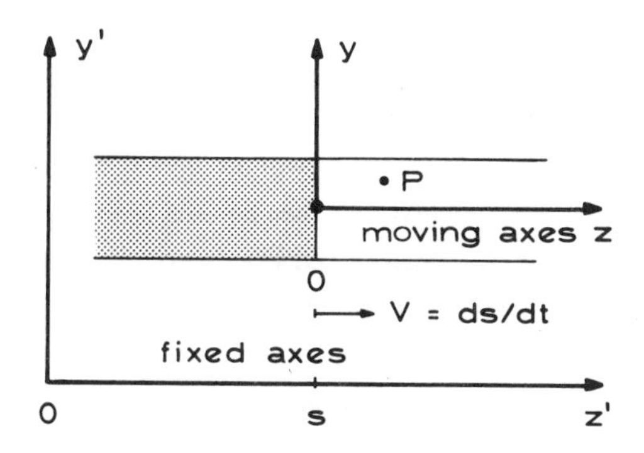

Figure A2.2

$$z = z' - Vt \tag{A2.4}$$

Since $\partial z/\partial z' = 1$, $\partial C/\partial z' = (\partial C/\partial z)(\partial z/\partial z') = \partial C/\partial z$, and similarly, $\partial^2 C/\partial z'^2 = \partial^2 C/\partial z^2$, the left-hand-side of equation A2.3 is unchanged.

The concentration is a function of z' and t, but must be transformed so as to become a function of $z(t)$ and t. In the moving reference frame, the local variation in concentration as a function of time, $\partial C/\partial t$, becomes:

$$\frac{\partial C}{\partial t} \rightarrow \frac{\partial C}{\partial t} - V\frac{\partial C}{\partial z}$$

As the frame is moving in the z-direction, one therefore has:

$$D\frac{\partial^2 C}{\partial z^2} = -V\frac{\partial C}{\partial z} + \frac{\partial C}{\partial t} \tag{A2.5}$$

Re-introducing a second spatial coordinate (which is unaffected by the above transformation of coordinates) in order to describe lateral diffusion, and rearranging gives:

$$\frac{\partial^2 C}{\partial y^2} + \frac{\partial^2 C}{\partial z^2} + \frac{V}{D}\cdot\frac{\partial C}{\partial z} = \frac{1}{D}\cdot\frac{\partial C}{\partial t}$$

or, in its time-independent form (steady-state growth):

$$\frac{\partial^2 C}{\partial y^2} + \frac{\partial^2 C}{\partial z^2} + \frac{V}{D}\cdot\frac{\partial C}{\partial z} = 0 \tag{A2.6}$$

This equation, which is known as the directional growth equation, will be used to solve most of the problems in this book.

Solutions of the Directional Growth Equation

The first step in solving a directional growth problem is to discover what functions satisfy equation A2.6. These functions can then be used as the starting point in solving any problem (in rectangular coordinates), and an exact solution would be obtained if all of the boundary conditions could be satisfied everywhere.

It is assumed firstly that the solution of equation A2.6 can be expressed as the product of separate functions of y and z alone [1 - 3]:

$$C(y,z) = Y(y)\,Z(z) \qquad [A2.7]$$

Inserting the relevant derivatives of this expression into equation A2.6 gives:

$$\frac{d^2Y}{dy^2}Z + \frac{d^2Z}{dz^2}Y + \frac{V}{D}\cdot\frac{dZ}{dz}Y = 0$$

and dividing throughout by $C(y,z)$ gives:

$$\frac{1}{Y}\cdot\frac{d^2Y}{dy^2} + \frac{1}{Z}\cdot\frac{d^2Z}{dz^2} + \frac{V}{D}\cdot\frac{1}{Z}\cdot\frac{dZ}{dz} = 0 \qquad [A2.8]$$

Each term which involves y or z alone must be equal to a constant known as the separation constant. This can be seen by considering the term in Z for instance: either it is a constant or it is a function of z. In the latter case, the other terms in equation A2.8 must be functions of z in order to satisfy the equation. However, this contradicts the assumption that the functions each depend upon only one variable. Hence, each term is equal to a constant, a, and the sum of the constants must be zero (from equation A2.8). The sign of the separation constant is determined by inspection after considering the properties which the solution must have in order to reflect the characteristics of the physical situation. Thus, one can write:

$$\frac{1}{Z}\cdot\frac{d^2Z}{dz^2} + \frac{V}{D}\cdot\frac{1}{Z}\cdot\frac{dZ}{dz} = a$$

$$\frac{d^2Z}{dz^2} + \frac{V}{D}\cdot\frac{dZ}{dz} - aZ = 0 \qquad [A2.9]$$

In the overall direction of advance of the interface (z axis), one expects the existence of a boundary layer which is theoretically of infinite extent. This fact, together with the form of equation A2.9 (i.e. a weighted sum of successive differentials) makes the exponential function a likely candidate. Therefore, setting:

$$Z = \exp[bz] \qquad [A2.10]$$

and performing the differentiations indicated by equation A2.9 gives:

$$b^2\exp[bz] + \frac{Vb}{D}\exp[bz] - a\exp[bz] = 0$$

The factor, exp[bz], cancels to leave:

$$b^2 + \frac{Vb}{D} - a = 0$$

This quadratic algebraic equation is solved by elementary means to give:

$$b = -\frac{V}{2D} - \left[\left(\frac{V}{2D}\right)^2 + a\right]^{1/2}$$

The positive root is not considered because the solution to the problem might then predict infinite values of concentration in the liquid far from the interface. The general solution to equation A2.9 is thus:

$$Z = \exp\left\{\left(-\frac{V}{2D} - \left[\left(\frac{V}{2D}\right)^2 + a\right]^{1/2}\right)z\right\} \qquad [A2.11]$$

Considering now the term in Y (equation A2.8), one can set:

$$\frac{1}{Y}\cdot\frac{d^2Y}{dy^2} = -a$$

or:

$$\frac{d^2Y}{dy^2} + aY = 0 \qquad [A2.12]$$

The form of this equation suggests that the function to be substituted should be such that its second derivative is of the same form as the original function, but of opposite sign. As the second derivative of an exponential function has the same sign as the original function, a circular function is a more likely candidate. Thus:

$$Y = \cos[cy] \quad \text{or} \quad Y = \sin[cy] \qquad [A2.13]$$

Substitution of either expression into equation A2.12 gives:

$$c = a^{1/2}$$

and

$$Y = \cos[a^{1/2}y] \quad \text{or} \quad Y = \sin[a^{1/2}y] \qquad [A2.14]$$

Finally, substituting equations A2.14 and A2.11 into equation A2.7 gives:

$$C = \cos[a^{1/2}y]\exp\left\{\left(-\frac{V}{2D} - \left[\left(\frac{V}{2D}\right)^2 + a\right]^{1/2}\right)z\right\} \qquad [A2.15a]$$

or

$$C = \sin[a^{1/2}y]\exp\left\{\left(-\frac{V}{2D} - \left[\left(\frac{V}{2D}\right)^2 + a\right]^{1/2}\right)z\right\} \qquad [A2.15b]$$

Examination of the form of these equations shows that, overall, the lateral distribution will be cyclic in form, and that the concentration will decrease exponentially away from the interface in the growth direction (Fig. A2.3). Using the theory of Fourier series, any number of cosine (or sine) functions can be added together in order to satisfy the boundary conditions. Note that when the constant, a, is very large it will dominate the exponential decrease of the boundary layer. When it is very small, the boundary layer will be the same as that for a uniform plane interface. These two cases correspond to a frequently varying lateral concentration, and to a slowly varying lateral concentration respectively, since the value of a is inversely proportional to the wavelength of the interface morphology. This is seen in the case of eutectic growth at normal speeds where the rate of exponential decrease is dominated by the wavelength, and the thickness of the boundary layer becomes proportional to the eutectic spacing (appendix 10). When the scale of the interface morphology can vary over a wide range or the boundary layer is reduced, the full solution (equation A2.15) must be used. These situations arise when carrying out stability analyses (appendix 7) or studies of eutectic growth at high rates (appendix 10).

Boundary Conditions

Three types of mathematical boundary condition are generally imposed on the solution of a differential equation. These are the Dirichlet condition, which defines the absolute value of the solution at a boundary point, the Neumann condition, which defines the normal gradient of the solution at the boundary, and the Robin (mixed) condition which establishes a relationship between the absolute value and the gradient of the solution at the boundary. The latter condition is the main source of difficulty in solidification problems because the interface concentrations and their gradients are usually not given explicitly but must be found as part of the solution. This condition, since it arises from the balance between solute rejection and diffusion at the interface, will be called the flux condition. The other conditions will also be given names which reflect their significance in solidification problems (Fig. A2.3).

Flux Condition

As a solid/liquid interface advances at the local normal growth rate, V_n, with an interface concentration in the liquid, C^*, and solid concentration, kC^*, the quantity of solute rejected per unit time will be $V_n(1-k)C^*$. This must be balanced by the creation of a concentration gradient in the liquid, normal to the isoconcentration contours, which permits solute removal at the same rate via diffusion. Thus:

$$V_n(1-k)C^* = -D\frac{\partial C}{\partial m} \qquad \text{[A2.16]}$$

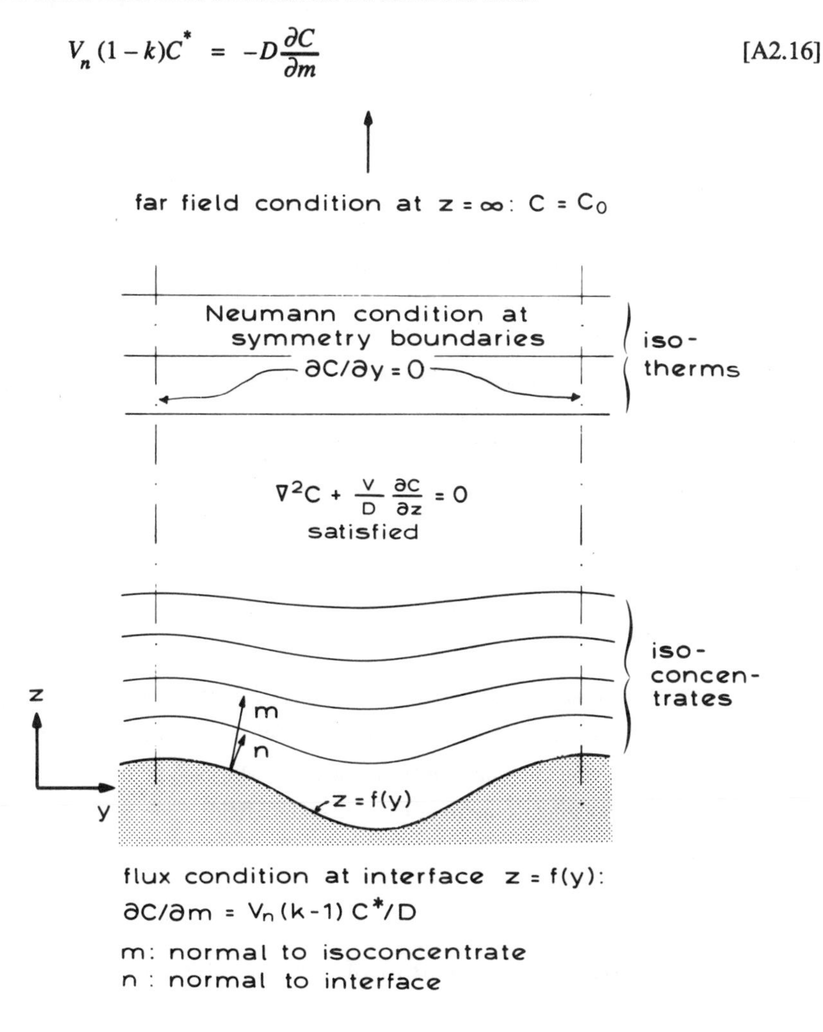

Figure A2.3

One complication is that the normal to the interface, n, along which the interface advances locally, is not generally the same as the normal, m, to the isoconcentration contours in the liquid. Thus, in order to simplify the problem, the interface should be assumed to be of uniform concentration (or isothermal). Otherwise, the flux condition can be easily applied only on an axis of symmetry, where the two normals are bound to be aligned. Such a point would be the tip (or trough) of the perturbation shown in figure A2.3.

Far-Field Condition

A Dirichlet condition can be imposed far ahead of the interface because here the original composition is expected to be unaffected by the advance of the interface, i.e.:

$$C = C_0, \qquad z = \infty \qquad [A2.17]$$

Symmetry Condition

Most interface morphologies consist of arrays of similar shapes. Advantage can be taken of this fact by studying just one half-period of the shape along the y-axis. If the shapes are presumed to be identical, there can be no mass transfer between them. Hence, zero concentration gradient (Neumann) conditions can be imposed at the boundaries of a typical interface 'motif', i.e.:

$$\frac{\partial C}{\partial y} = 0, \qquad y = \frac{n\lambda}{2} \qquad [A2.18]$$

where $n = 0, 1, 2, ...$

Coupling Condition

Under normal solidification conditions for metals (V<100mm/s), each location along the interface will have a local freezing point which is a function of the local concentration and the local curvature. In steady-state growth, each point of the interface must lie on the corresponding isotherm of the temperature field:

$$T^* = T_f + m\Delta C - \Gamma K \qquad [A2.19]$$

Satisfaction of Boundary Conditions

In a previous section, a general solution to the Laplace equation was obtained in terms of elementary functions. It would be overly optimistic to expect any real situation to involve boundary conditions which permitted such a solution to be used without any further effort. Unfortunately, the greater part of the problem still lies ahead. Indeed, the requirement that the basic solution (or its derivatives) should have certain values at the boundaries of the region studied accounts for much of the effort expended by applied mathematicians. Their researches over the past two hundred years have produced an enormous range of methods for attacking the problem [2]. A number of methods, which are used elsewhere in the text, will be described below.

Firstly, two cases will be considered in which the boundary conditions *can* be satisfied exactly.

Steady-State Diffusion Field Ahead of a Moving Planar Interface

As was shown earlier in this appendix, the diffusion equation in a coordinate system with its origin fixed at the solid/liquid interface takes the steady-state form for unidirectional diffusion (Fig. A2.4a), obtained from equation A2.5, of:

$$\frac{\partial^2 C}{\partial z^2} + \frac{V}{D} \cdot \frac{\partial C}{\partial z} = 0$$

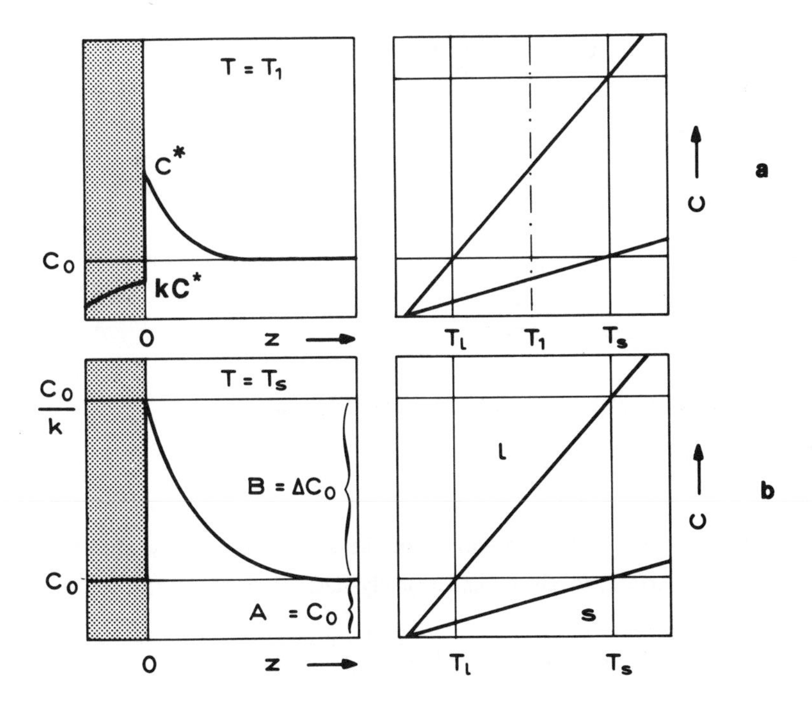

Figure A2.4

Noting the similarity of the above expression to equation A2.9, setting the separation constant, a, equal to zero, and repeating the steps after equation A2.10, gives:

$$Db^2 + Vb = 0$$

The solutions of this so-called 'auxiliary' equation are $b = 0$ and $b = -V/D$. Therefore, the general solution is:

$$C = A + B\exp\left(-\frac{Vz}{D}\right) \qquad [A2.20]$$

Again following the principles described above, a far-field condition is applied. That is, far from the interface the concentration must be equal to the original composition, C_0. Letting $C = C_0$ when z tends to infinity shows that $A = C_0$. Therefore:

$$C = C_0 + B\exp\left(-\frac{Vz}{D}\right)$$

Next, one can apply the flux (Robin) condition at the solid/liquid interface. Here, the rate of solute rejection must be equal to the diffusional flux in the liquid at the interface:

$$C^*(1-k)V = -D\left(\frac{dC}{dz}\right)_{z=0}$$

Therefore, when $z = 0$:

$$C^* = C_0 + B$$

and

$$\left(\frac{dC}{dz}\right)_{z=0} = -\frac{VB}{D}$$

Substituting these expressions into the above shows that:

$$B = C_0\frac{1-k}{k} = \Delta C_0$$

Therefore, the complete solution for the solute distribution ahead of a planar solid/liquid interface, advancing under steady-state conditions, is [5]:

$$C = C_0 + \left(\frac{C_0}{k} - C_0\right)\exp\left[-\frac{Vz}{D}\right] \qquad [A2.21]$$

The boundary layer shown in figure A2.5 is of infinite extent. In order to obtain a convenient practical estimate of its thickness, an equivalent boundary layer, δ_c, is often

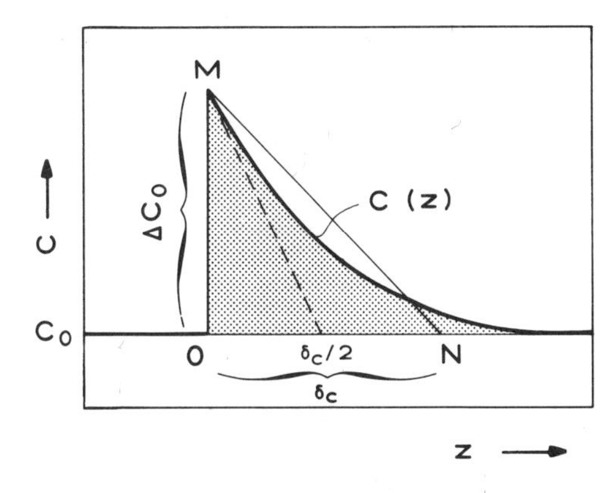

Figure A2.5

defined. This equivalent layer is chosen so as to contain the same total solute content as the infinite layer, and has a constant concentration gradient across its thickness. Thus, the area of the triangle, OMN, must be equal to the area of the shaded surface. That is:

$$\frac{\Delta C_0 \delta_c}{2} = \Delta C_0 \int_0^\infty \exp\left[-\frac{Vz}{D}\right] dz$$

giving:

$$\delta_c = \frac{2D}{V}$$

Differentiating equation A2.21 at $z = 0$ gives:

$$G_c = \left(\frac{dC}{dz}\right)_{z=0} = -\frac{\Delta C_0 V}{D} \qquad \text{[A2.22]}$$

From this, it can be seen that the absolute value of the concentration gradient at the interface is equal to twice the absolute mean concentration gradient of the equivalent boundary layer. Such relationships are very useful in understanding the constitutional undercooling criterion (chapter 3.3).

Diffusion Field Around a Growing Sphere

When no tangential diffusion is occurring, the diffusion equation can be written in terms of the radial coordinate alone [6,7]:

$$\frac{\partial C}{\partial t} = D\left(\frac{\partial^2 C}{\partial r^2} + \frac{2}{r}\cdot\frac{\partial C}{\partial r}\right) \qquad [A2.23]$$

where r is the radius. Under steady-state conditions,

$$\frac{d}{dr}\left(r^2\frac{dC}{dr}\right) = 0$$

the general solution of this equation is:

$$C = A + \frac{B}{r} \qquad [A2.24]$$

For the situation described in figure A2.6, the boundary conditions are:

$$C = C_0 \qquad r = \infty$$

$$C = C^* \qquad r = R$$

Satisfaction of these conditions shows that $A = C_0$ and $B = R(C^* - C_0)$. Thus, equation A2.24 becomes:

$$C = C_0 + \frac{R}{r}(C^* - C_0) \qquad [A2.25]$$

The concentration gradient in the liquid becomes:

$$\frac{dC}{dr} = -\frac{R}{r^2}(C^* - C_0)$$

and, at the interface

$$\left(\frac{dC}{dr}\right)_R = -\frac{C^* - C_0}{R} \qquad [A2.26]$$

This shows that, to a first approximation, the thickness of the boundary layer around a growing sphere is equal to the radius and increases with increasing size of the sphere. For other solutions see references 8 and 9.

Some methods which are available for the treatment of more difficult cases can now be considered:

Classical Method

When using this method, advantage is taken of the linearity of equation A2.6. That is, the sum of any series of terms having the same form as the basic solution (such as the sine and cosine functions found previously) will also be a solution. This means that,

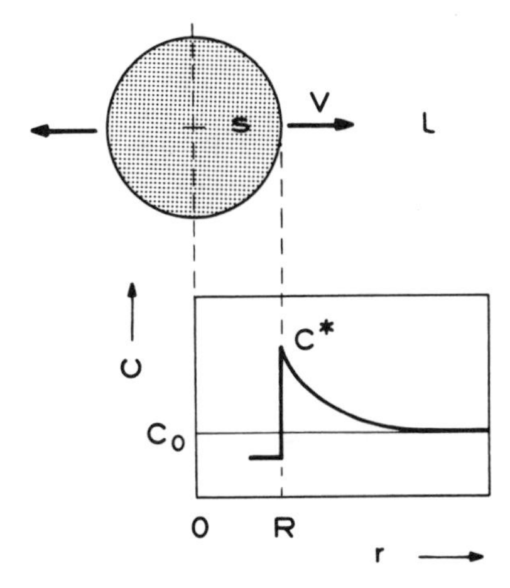

Figure A2.6

although the basic solution (often called an eigenfunction) is unlikely to satisfy the boundary conditions, 'adjustment' using terms of the same form will allow the conditions to be approximated more and more closely. For example, if the basic solution is of the form, $\cos[2\pi y/\lambda]$, one can consider adding terms such as $\cos[4\pi y/\lambda]$ etc., e.g.:

$$C(y) = A_1 \cos\left[\frac{2\pi y}{\lambda}\right] + A_2 \cos\left[\frac{4\pi y}{\lambda}\right] + \ldots + A_i \cos\left[\frac{2\pi i y}{\lambda}\right] \qquad \text{[A2.27]}$$

where the A_i are constants. Because the oscillation of the cosine functions increases in frequency with increasing value of i, finer and finer adjustments can be made to the basic solution. In order to carry this out in practice, the constants before each term in a series such as the one above have to be suitably chosen. Many methods have been devised in order to find the required values of these constants. The best method, but one which is rarely feasible, is to take advantage of the orthogonality of functions such as the circular ones (sine, cosine). In this method, each term in the above series would be multiplied by $\cos[2\pi jy/\lambda]$:

$$C(y)\cos\left[\frac{2\pi j y}{\lambda}\right] = A_1 \cos\left[\frac{2\pi y}{\lambda}\right]\cos\left[\frac{2\pi j y}{\lambda}\right] + A_2 \cos\left[\frac{4\pi y}{\lambda}\right]\cos\left[\frac{2\pi j y}{\lambda}\right]$$
$$+ A_i \cos\left[\frac{2\pi i y}{\lambda}\right]\cos\left[\frac{2\pi j y}{\lambda}\right]$$

When both sides are integrated over one wavelength:

$$\int_0^\lambda C(y)\cos\left[\frac{2\pi jy}{\lambda}\right]dy = A_1\int_0^\lambda \cos\left[\frac{2\pi y}{\lambda}\right]\cos\left[\frac{2\pi jy}{\lambda}\right]dy$$

$$+ A_2\int_0^\lambda \cos\left[\frac{4\pi y}{\lambda}\right]\cos\left[\frac{2\pi jy}{\lambda}\right]dy + \ldots$$

$$+ A_i\int_0^\lambda \cos\left[\frac{2\pi iy}{\lambda}\right]\cos\left[\frac{2\pi jy}{\lambda}\right]dy \quad \text{[A2.28]}$$

all of the terms on the right-hand side will disappear unless $j = i$. Therefore, by setting j equal successively to 1, 2, 3, etc, the value of any A_i can be 'singled out'. A general expression is usually obtained for all of the adjustable constants of the series. The reader should satisfy himself that if this 'trick' were not available, the A-values (Fourier coefficients) could only be determined exactly by solving an infinite set of simultaneous algebraic equations. Unfortunately, this is usually the case since the above method can only be employed when the boundary coincides with a coordinate line over which the functions are also orthogonal.

In response to this common difficulty, approximate methods have been developed and will be described in the next section. Meanwhile, a relatively little-known technique for obtaining the Fourier coefficients without integration will be described. However, this technique is only applicable when the boundary conditions are discontinuous in some way. For example, see figure A13.1 where the first and higher derivatives of the concentration distribution are discontinuous at regular intervals.

In the case of a cosine series, the Fourier coefficients are given directly by:

$$A_i = \frac{1}{i\pi}\left[-\sum_{s=1}^{m} J_s \sin(iy_s) - \frac{1}{i}\sum_{s=1}^{m} J'_s \cos(iy_s) + \ldots\right.$$

$$\left.\ldots + \frac{1}{i^2}\sum_{s=1}^{m} J''_s \sin(iy_s) + \ldots\right] \quad \text{[A2.29]}$$

where the J, J', J'', etc are 'jumps' in the function, first derivative, second derivative, etc. The definition of such a jump will be given in later appendices where the technique is applied to various problems. The derivation of equation A2.29 is quite simple [10] but is beyond the scope of the present book.

Method of Weighted Residuals

In effect, this technique [11] sets out to satisfy the boundary conditions of the problem in the same way as does the classical method. Thus, a series consisting of functions having adjustable multiplying constants is used. However, in this case, the functions are rarely orthogonal and the constants can only be found by solving a set of simultaneous algebraic equations. The number of terms in the series is chosen so as to be equal to the number of adjustable constants. The method is approximate in nature and the higher-order approximations can only be handled by using numerical analysis techniques and a computer. Nevertheless, surprisingly accurate analytical results can often be obtained by using just a few terms; and sometimes only one [12]. Such an analytical solution has the advantage that it will reveal the influence of the various experimental and physical parameters directly whereas a more accurate but numerical method will not. The method is extremely flexible and can be applied to any problem. As an example, it will be used to find a solution to the dendritic growth problem in two dimensions

Simple dendrite analysis

A function will be chosen which satisfies the boundary conditions. This is known as the external method. The trial solution can also be such that, alone, it does not satisfy either the differential equation, *or* the boundary conditions. The reader will doubtless appreciate the flexibility of the method, but on the other hand he must also have a firm grasp of the physics of the real situation in order to be able to use it properly.

In the present case, a Dirichlet condition will be applied at the interface, assuming that the concentration is constant over the parabolic surface of the dendrite. Thus:

$$C = C^* \qquad z = -\beta y^2$$

Also, applying the far-field condition,

$$C = C_0 \qquad z = \infty$$

one can easily construct an expression which satisfies these conditions. Firstly, take the basic solution for a planar interface which was derived above:

$$C = C_0 + (C^* - C_0)\exp[-bz] \qquad \text{[A2.30]}$$

This describes a solute distribution in which the value at infinity is C_0. Thus, the far-field condition is satisfied. It is also required that the concentration be equal to C^* when the value of the exponential term is unity. In the case of the planar interface, this only occurs when z is equal to zero. Therefore, it is necessary to replace z by an expression which is equal to zero whenever a coordinate pair, (y,z), corresponds to the surface of the parabolic plate dendrite. This is true when $z = -\beta y^2$, so that the required expression is:

$$C = C_0 + (C^* - C_0)\exp[-b(\beta y^2 + z)] \quad \text{[A2.31]}$$

The reader should prove for himself that this satisfies the boundary conditions and that the concentration decreases exponentially with y^2 and z. The value of b can be determined by means of the flux condition; applied at the tip:

$$V(k-1)C^* = D\left(\frac{\partial C}{\partial z}\right)_{y=z=0} \quad \text{[A2.32]}$$

Substituting the z-derivative of equation A2.31 into equation A2.32 gives:

$$b = \frac{V}{D} \cdot \frac{(k-1)C^*}{C_0 - C^*}$$

so that:

$$C = C_0 + (C^* - C_0)\exp\left[-\frac{V(k-1)C^*(\beta y^2 + z)}{D(C_0 - C^*)}\right] \quad \text{[A2.33]}$$

The differential equation (equation A2.6) is:

$$\frac{\partial^2 C}{\partial y^2} + \frac{\partial^2 C}{\partial z^2} + \frac{V}{D} \cdot \frac{\partial C}{\partial z} = 0$$

Substituting equation A2.33 into A2.6 and simplifying gives:

$$\frac{V}{D} \cdot \frac{4y^2\beta^2(k-1)C^*}{(C_0 - C^*)} - 2\beta + \frac{V}{D} \cdot \frac{(k-1)C^*}{(C_0 - C^*)} - \frac{V}{D} = 0$$

The LHS of the above equation is the 'residual' which gives the method its name. The equation must be satisfied in order to solve the original problem. The first term can be eliminated by considering only the point of the dendrite. It will be equal to zero when $y = 0$, so that:

$$\frac{V}{D} \cdot \frac{(k-1)C^*}{C_0 - C^*} = \frac{V}{D} + 2\beta \quad \text{[A2.34]}$$

The curvature at the tip of the parabola, $z = -\beta y^2$, is given by the second derivative of z with respect to y. Thus, the curvature is 2β and the radius of curvature is:

$$R = \frac{1}{2\beta} \qquad [A2.35]$$

Substituting this value into equation A2.34 gives:

$$\frac{V}{D} \cdot \frac{(k-1)C^*}{C_0 - C^*} = \frac{V}{D} + \frac{1}{R}$$

and:

$$\frac{(k-1)C^*}{C_0 - C^*} = 1 + \frac{D}{VR}$$

but $VR/2D$ is the Péclet number, so that:

$$\frac{(k-1)C^*}{C_0 - C^*} = 1 + \frac{1}{2P}$$

Now, $(C^* - C_0)/(C^* - kC^*)$ is the dimensionless supersaturation, Ω, so that:

$$\Omega = \frac{2P}{2P+1} \qquad [A2.36]$$

which is the same as the Zener-Hillert [9] solution (appendix 8).

Perturbation Method

If the geometry of the problem is such that the interface form corresponds closely to some simple shape, any exact solution which is available for the simple shape can be assumed to be similar to that for the slightly different morphology. This principle will be illustrated by treating an almost planar solid/liquid interface growing under steady-state conditions. A related problem will be studied in appendix 7.

Slightly-Perturbed Interface

The form of the planar solid/liquid interface, described by the equation, $z = 0$, is assumed to be changed so that it is then represented by the expression:

$$z = \varepsilon \sin[\omega y] \qquad [A2.37]$$

where ε is assumed to be a very small amplitude, and ω is the wave number ($= 2\pi/\lambda$)

of the perturbation.

Recall that the exact solution for a planar solid/liquid interface under steady-state conditions (equation A2.21) is:

$$C = C_0 + \left(\frac{C_0}{k} - C_0\right)\exp\left[-\frac{Vz}{D}\right]$$

where C_0/k is the concentration in the liquid at the interface and C_0 is the original composition. Now, using the perturbation technique, a term having the same form as the perturbation (equation A2.37) is added to the exact solution for the unperturbed interface, i.e.:

$$C = C_0 + \left(\frac{C_0}{k} - C_0\right)\exp\left[-\frac{Vz}{D}\right] + A\varepsilon\sin[\omega y]\exp[-bz] \qquad \text{[A2.38]}$$

where b has to be equal to $(V/2D) + [(V/2D)^2 + \omega^2)]^{1/2}$ in order that the added term should satisfy equation A2.6 (see equation A2.11), and A is a constant whose value is to be determined by forcing equation A2.38 to satisfy the boundary conditions. These are, for $z = \varepsilon\sin[\omega y]$:

$$C = C^* \qquad \text{[A2.39]}$$

$$V(1-k)C^* = -D\frac{\partial C}{\partial z} \qquad \text{[A2.40]}$$

Note that equation A2.38 already satisfies the far-field condition, $C = C_0$ when $z = \infty$, since the unknown constant, A, disappears. Rather more work is required in order to make it satisfy the other boundary conditions (e.g. equation A2.39). The first step is to substitute C^* for C and $\varepsilon\sin[\omega y]$ for z:

$$C^* = C_0 + \left(\frac{C_0}{k} - C_0\right)\exp\left[-\frac{VS}{D}\right] + AS\exp[-bS] \qquad \text{[A2.41]}$$

where, for clarity, S has been used to represent $\varepsilon\sin[\omega y]$. The above expression can only be evaluated because of the assumption that ε (and S) are small. In this case, an exponential function, $\exp[-x]$, can be approximated by $1 - x$. Again, because ε is small, terms involving ε^2 (and S^2) can be neglected. This leads to:

$$C^* = C_0 + \left(\frac{C_0}{k} - C_0\right)\left(1 - \frac{VS}{D}\right) + AS\,(1 - bS)$$

By differentiating equation A2.21 with respect to z and then setting z equal to zero, the concentration gradient, G_c, at the plane interface is found to be (equation A2.22):

$$G_c = -\frac{V}{D}\left(\frac{C_0}{k} - C_0\right)$$

Substituting this into the previous equation and dropping terms in S^2 gives:

$$C^* = \frac{C_0}{k} + G_c S + AS \qquad [A2.42]$$

Thus far, no real progress has been made since the value of C^* is also unknown. It is necessary to find another equation which links C^* and A, i.e. the flux condition (equation A2.40). Thus, differentiating equation A2.38 gives:

$$\frac{dC}{dz} = -\frac{V}{D}\left(\frac{C_0}{k} - C_0\right)\exp\left[-\frac{Vz}{D}\right] - bAS\exp[-bz]$$

At the interface this becomes:

$$\left(\frac{dC}{dz}\right)_{z=S} = -\frac{V}{D}\left(\frac{C_0}{k} - C_0\right)\left(1 - \frac{VS}{D}\right) - bAS(1 - bS)$$

Substituting G_c for $-(V/D)(C_0/k - C_0)$ gives:

$$\left(\frac{dC}{dz}\right)_{z=S} = G_c\left(1 - \frac{VS}{D}\right) - bAS \qquad [A2.43]$$

Substituting equations A2.42 and A2.43 into equation A2.40 and cancelling S throughout leads to:

$$A = \frac{kVG_c}{Vp - Db}$$

where $p = 1 - k$. Thus, the original expression becomes:

$$C = C_0 + \left(\frac{C_0}{k} - C_0\right)\exp\left[-\frac{Vz}{D}\right] + \left(\frac{kVG_c\,\varepsilon\sin[\omega y]}{Vp - Db}\right)\exp[-bz] \qquad [A2.44]$$

This expression describes the solute distribution ahead of the slightly perturbed solid/liquid interface. Use is made of perturbation analysis in appendix 7, where the stability of a planar solid/liquid interface is considered.

References

[1] H.S.Carslaw, J.C.Jaeger, *Conduction of Heat in Solids*, 2nd Edition, Oxford University Press, London, 1959.

[2] J.Crank, *The Mathematics of Diffusion*, Oxford University Press, London, 1956.

[3] P.Moon, D.E.Spencer, *Field Theory Handbook*, Springer-Verlag, Heidelberg, 1961.

[4] L.F.Richardson, Philosophical Magazine **15** (1908) 237.

[5] W.A.Tiller, K.A.Jackson, J.W.Rutter, B.Chalmers, Acta Metallurgica **1** (1953) 428.

[6] C.Zener, Journal of Applied Physics **20** (1949) 950.

[7] R.L.Parker, Solid State Physics **25** (1970) 151.

[8] F.C.Frank, Proceedings of the Royal Society of London **A201** (1950) 586.

[9] G.Engberg, M.Hillert, A.Oden, Scandinavian Journal of Metallurgy **4** (1975) 93.

[10] E.Kreyszig, *Advanced Engineering Mathematics*, Wiley, New York, 1968.

[11] B.A.Finlayson, *The Method of Weighted Residuals and Variational Principles*, Academic Press, New York, 1972.

[12] R.Aris, *Mathematical Modelling Techniques*, Pitman, London, 1978.

APPENDIX 3

LOCAL EQUILIBRIUM AT THE SOLID/LIQUID INTERFACE

The Phase Diagram

In most analyses of alloy solidification, it is assumed that the solid/liquid interface behaves locally as if it were in a state of equilibrium. This means that the reaction rates, in the small volume which makes up the very thin but finite interface layer, are expected to be rapid in comparison with the rate of interface advance. As a result, the transfer of atoms and changes in their arrangement, which are required in order to maintain the constancy of the chemical potentials in both phases, are relatively rapid and can therefore be neglected. Such a simplification is permissible in the case of metals solidifying at the rates encountered in normal casting and welding operations#. The assumption of local equilibrium means that, if the interface temperature is known, then one can obtain the liquid and solid compositions at the interface by reference to the ***equilibrium*** phase diagram. This does not mean that the system as a whole is at equilibrium, as gradients of temperature and composition are present.

In order to avoid non-essential variables which would complicate the analysis without revealing any new principles, it is usually assumed that both the liquidus and the solidus lines of the relevant part of the phase diagram are straight. During solidification, the liquidus line is the more important and represents the point where the liquid 'first' transforms to solid, i.e., *at the liquidus an alloy starts to freeze.* Thus, the slope, m, of the liquidus which is defined as:

$$T_l[C] = T_f + mC \qquad [A3.1]$$

is used in calculations.

The solidus composition can be determined at any time from the definition of the distribution coefficient:

At very high growth rates (V>100mm/s), such as those which occur during rapid solidification processing, conditions of local equilibrium no longer exist. Therefore, the value of k changes [1]. See appendix 6.

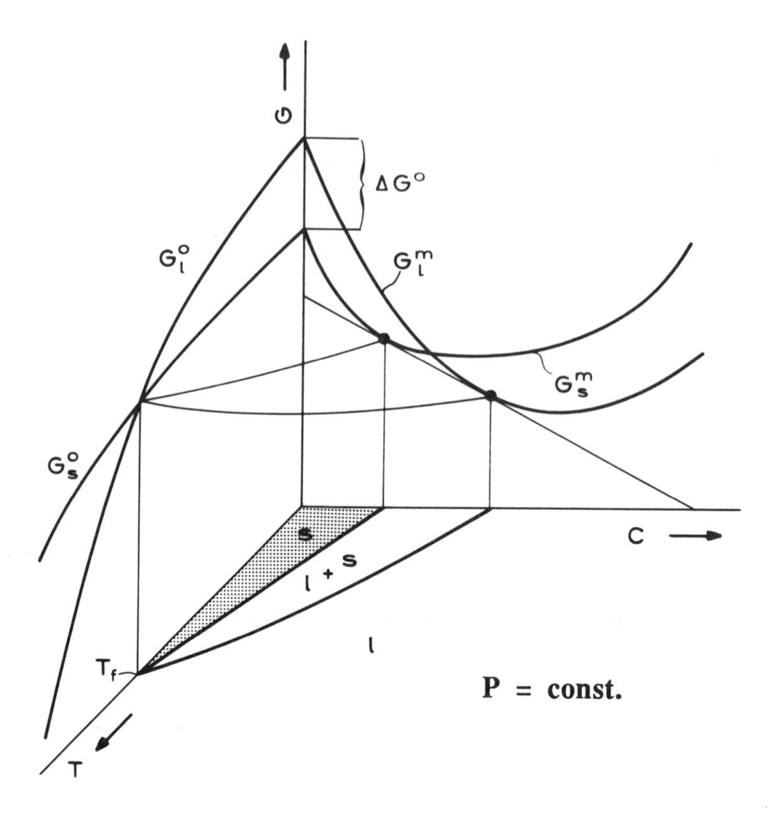

Figure A3.1

$$k = \left(\frac{C_s}{C_l}\right)_{T,P} \quad [A3.2]$$

The constants, m and k, are always defined in the present book in such a way that the product, $m(k-1)$ is positive. In general, m can be positive or negative and k can be greater or less than unity, respectively.

Two important properties characterise the range of coexistence of solid and liquid for a given alloy:

$$\Delta T_0 = -m\Delta C_0 \quad [A3.3]$$

$$\Delta C_0 = \frac{C_0(1-k)}{k} \quad [A3.4]$$

All of these properties depend upon the Gibbs free energy (free enthalpy) of the alloy system, as shown by figure A3.1. The latter relates a free-enthalpy versus concentration diagram and a free-enthalpy versus temperature diagram to a temperature versus concentration (phase) diagram. The form of the curves in the ΔG–C diagram can be described by using the regular solution model [2 - 4] which shows that for, X, the mole fraction of solute B:

$$G_m = (1-X)G_A^{\bullet} + XG_B^{\bullet} + \Omega X(1-X) + RT[X\ln(X) + (1-X)\ln(1-X)] \quad [A3.5]$$

where Ω is the interaction parameter.

The curves in the ΔG–T diagram depend upon the standard values for the pure components:

$$G^{\bullet} = H^{\bullet} - TS^{\bullet}$$

where the dots indicate values for the pure component.

The free enthalpy (Gibbs free energy) difference which exists between the pure liquid and pure solid is:

$$\Delta G^{\bullet} = \Delta H^{\bullet} - T\Delta S^{\bullet} \quad [A3.6]$$

where

$$\Delta H^{\bullet} = \Delta H_f^{\bullet} - \int_T^{T_f} \Delta c^{\bullet} dT$$

$$\Delta S^{\bullet} = \Delta S_f^{\bullet} - \int_T^{T_f} \left(\frac{\Delta c^{\bullet}}{T}\right) dT$$

and $\Delta c^{\bullet} = c^l - c^s$.

The quantity, $\Delta G^{\bullet}$, in equation A3.6 is very important with regard to nucleation and growth processes and can be evaluated when the temperature dependence of $\Delta c^{\bullet}$ is known. At high temperatures, the difference in specific heat of the liquid and solid can be described by:

$$\Delta c^{\bullet} = K_1 T + K_2 \quad [A3.7]$$

For an undercooled melt, $\Delta T \equiv T_f - T$ and [5]:

$$|\Delta G^{\bullet}| = \frac{\Delta H_f^{\bullet} \Delta T}{T_f} - \Delta T^2 \left[\frac{K_1}{2} + \frac{K_2}{T_f + T}\right] \quad [A3.8]$$

If the value of $\Delta c^{\bullet}$ is unknown, the simplest assumption (and one which is quite reasonable for metals) is that $\Delta c^{\bullet} = 0$. This leads to:

$$|\Delta G^{\bullet}| = \frac{\Delta H_f^{\bullet} \Delta T}{T_f} = \Delta S_f^{\bullet} \Delta T \qquad [A3.9]$$

The quantity, $\Delta S_f^{\bullet}$, is the difference in slope of the G–T function of two phases (Fig. A3.1). For a more detailed discussion of this problem the reader is referred to reference 6.

From equations A3.5 and A3.9, one can now see which parameters influence the magnitudes of m and k. These are the interaction parameter, Ω, for the atoms of both species in the alloy, which determines the form of the G_m–X curves, and the melting entropy, $\Delta S_f^{\bullet}$, which separates the origins of the ΔG_m–X curves. A description of further relationships between k, m, and the thermodynamic properties is given by Flemings [7].

For small concentrations of an alloying element, and assuming ideal solution behaviour, a useful relationship between the melting entropy of the solvent, (A) and the liquidus slope and distribution coefficient of the solute (B) can be obtained. Under these assumptions, the chemical potential of the solvent is:

$$\mu_A^s = \mu_A^{s\bullet} + RT\ln(1 - X_s)$$

$$\mu_A^l = \mu_A^{l\bullet} + RT\ln(1 - X_l) \qquad [A3.10]$$

where X_s and X_l are the mole fractions of B atoms in the solid and liquid, respectively. At equilibrium ($\mu_A^s = \mu_A^l$):

$$\Delta G_A^{\bullet} = \mu_A^{l\bullet} - \mu_A^{s\bullet} = RT\ln\left[\frac{1 - X_s}{1 - X_l}\right] \qquad [A3.11]$$

From equation A3.9, one obtains:

$$\frac{\Delta S_f^{\bullet} \Delta T}{RT} = \ln\left[\frac{1 - X_s}{1 - X_l}\right] \qquad [A3.12]$$

For small solute concentrations and noting that, for $z \rightarrow 1$, $\ln(z)$ is approximately equal to $z - 1$, it can be seen that the RHS of equation A3.12 becomes approximately equal to $X_l - X_s = X_l(1 - k)$ when X approaches unity. Therefore:

$$\frac{\Delta S_f^{\bullet}}{RT} \cdot \frac{\Delta T}{X_l} = 1 - k$$

and substituting for $\Delta T/X_l = m_a$; the liquidus slope expressed as degrees per atomic fraction, one obtains:

$$\frac{1-k}{m_a} = \frac{\Delta S_f^{\bullet}}{RT} \qquad \text{[A3.13]}$$

Or, as X_l tends to zero and T tends to T_f:

$$\frac{1-k}{m_a} = \frac{\Delta S_f^{\bullet}}{RT_f} \qquad \text{[A3.14]}$$

Here, $\Delta S_f^{\bullet}$ is the melting entropy of the pure solvent of melting point, T_f, and both m and k have to be defined in terms of mole (atom) fractions.

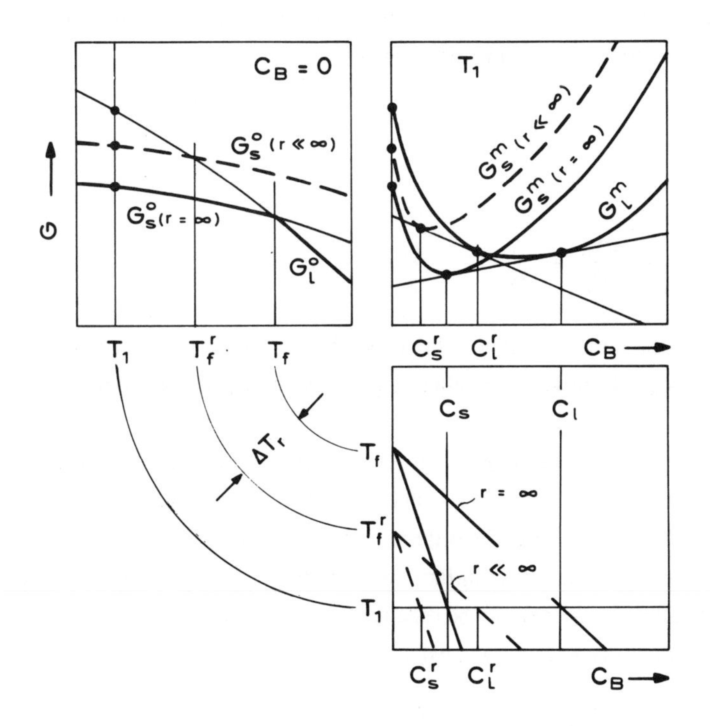

Figure A3.2

Capillarity Effects [8,9]

The total Gibbs free energy of a small solid particle in a melt is inversely proportional to its size. Because the free enthalpy of the solid increases with decreasing diameter, while the free enthalpy of the liquid remains constant (if the amount of liquid is much greater than the amount of solid), the melting point decreases for the pure solid as well as for the alloy (Fig. A3.2). The increase in the free enthalpy of the particle, due to its curved surface (of radius, r), can be regarded as being an increase in internal pressure (the dot suffix will be dropped for the remainder of this appendix):

$$\Delta G_r = v_m \Delta P \qquad [A3.15]$$

where v_m is the molar volume (assumed to be constant), and ΔP is given by the specific interface energy and the curvature:

$$\Delta P = \sigma K \qquad [A3.16]$$

Combining equations A3.9, A3.15, and A3.16 leads to a relationship between the equilibrium temperature drop and the curvature:

$$T_f - T_f^r = \Delta T_r = \Gamma K \qquad [A3.17]$$

and

$$\Gamma = \frac{\sigma v_m}{\Delta S_f} = \frac{\sigma}{\Delta s_f} \qquad [A3.18]$$

where ΔS_f is the absolute value of the molar freezing entropy and Δs_f is the volumic freezing entropy. In order to use these equations, the parameters σ and K have to be defined:

Specific Interface Energy, σ: it is assumed that the solid/liquid interface (in fact a volume) is a surface across which the properties change discontinuously. The specific surface energy is defined there as being the reversible work, dw, which is required in order to create new surface area, dA. In the case of a solid/liquid interface, the specific interface energy can be set equal to the interfacial tension, γ. According to figure A3.3, the work necessary to extend the surface by dz is:

$$dw = f dz = \sigma l dz = \sigma dA$$

from which one can obtain the definition:

$$\sigma \equiv \left(\frac{dw}{dA}\right)_{T,\,V,\,\mu} \qquad [A3.19]$$

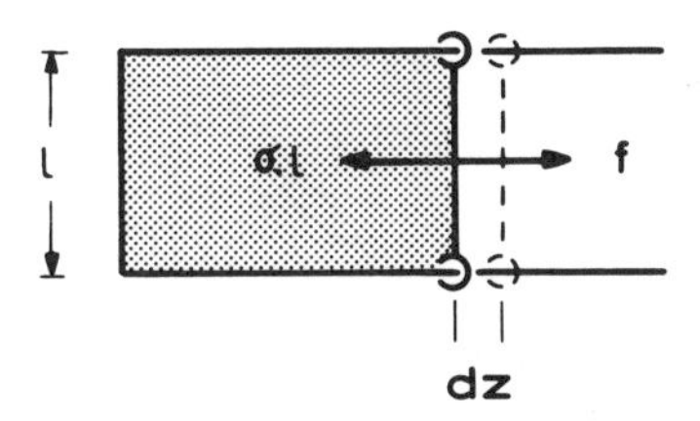

Figure A3.3

Curvature, K: in two dimensions, the curvature of a function is defined as the change in the slope, $\delta\theta$, of that function over a length of arc, δl (Fig. A3.4a):

$$K = \frac{\delta\theta}{\delta l}$$

and, since $\delta l = r\delta\theta$,

$$K = \frac{1}{r}$$

More generally, it can be shown that the curvature at a point ($\delta l \to 0$) is given, in Cartesian coordinates, by:

$$K = \frac{z''}{[1 + z'^2]^{3/2}} \qquad [A3.20]$$

where z' and z'' are the first and second derivatives, respectively, of the function, $z(y)$.

The average curvature of an arbitrary line segment depends only upon the gradients of the curve at its end-points, and upon the distance between the latter (Fig. A3.5). In the case of surfaces in three dimensions, the curvature can be defined as the variation in surface area divided by the corresponding variation in volume:

$$K = \frac{dA}{dv} \qquad [A3.21]$$

For a general surface with constant principal radii (minimum and maximum at 90° to each other - Fig. A3.4b) one can define (for small angles):

$$l_1 = r_1\theta$$
$$l_2 = r_2\theta$$

Increasing the radii by dr gives:

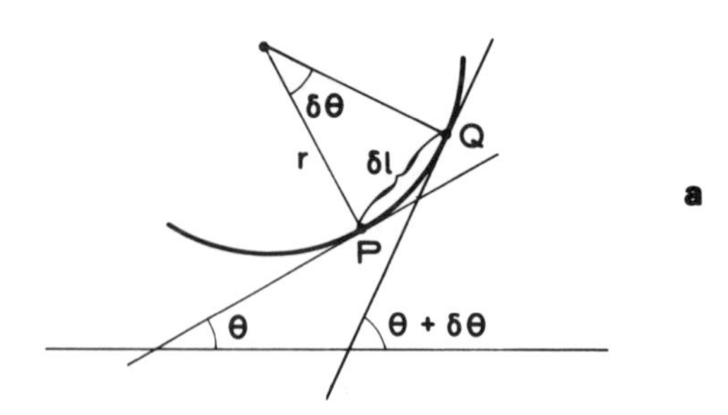

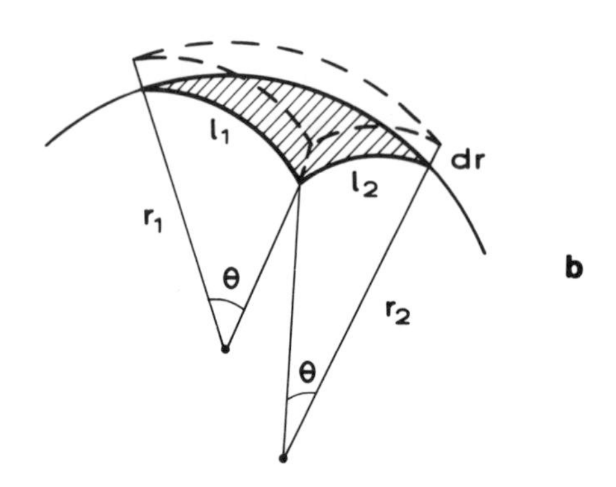

Figure A3.4

$$dA = l_1 dl_2 + l_2 dl_1 = r_1 \theta dr\theta + r_2 \theta dr\theta$$

$$dA = (r_1 + r_2) dr\theta^2 \qquad [A3.22]$$

$$dv = l_1 l_2 dr = (r_1 \theta)(r_2 \theta) dr$$

$$dv = r_1 r_2 dr\theta^2 \qquad [A3.23]$$

Therefore from equations A3.21, A3.22, and A3.23,

$$K = \left(\frac{1}{r_1} + \frac{1}{r_2}\right) \qquad [A3.24]$$

In the case of a sphere, $r_1 = r_2 = r$ so that $K = 2/r$. In the case of a cylinder, r_1 is infinite and $r_2 = r$ so that $K = 1/r$.

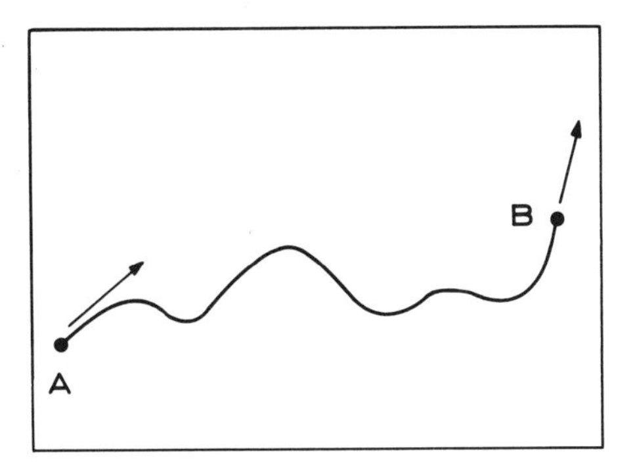

Figure A3.5

Thus, using equation A3.17, the decrease in melting point due to the curvature of a spherical crystal in a melt can be written:

$$\Delta T_r = \frac{2\Gamma}{r} \quad \text{[A3.25]}$$

It is supposed in the above calculations that the value of σ is isotropic. The effect of anisotropy is treated by Aaronson [10].

Mechanical Equilibrium at the Three-Phase Junction

A junction between two solid phases at the solid/liquid interface will form a groove. At this point, the surface forces will tend to impose an equilibrium (minimum energy) morphology in which:

$$\Sigma f = 0$$

From figure A3.6, it is evident that this condition will be satisfied when:

$$\sigma_{\alpha\beta} = \sigma_{\alpha l}\cos(\theta_1) + \sigma_{\beta l}\cos(\theta_2)$$

and [A3.26]

$$\sigma_{\alpha l}\sin(\theta_1) = \sigma_{\beta l}\sin(\theta_2)$$

In establishing mechanical equilibrium, it is important to consider the equilibrium of the moments acting on the junction (second equation above) since this affects the angle

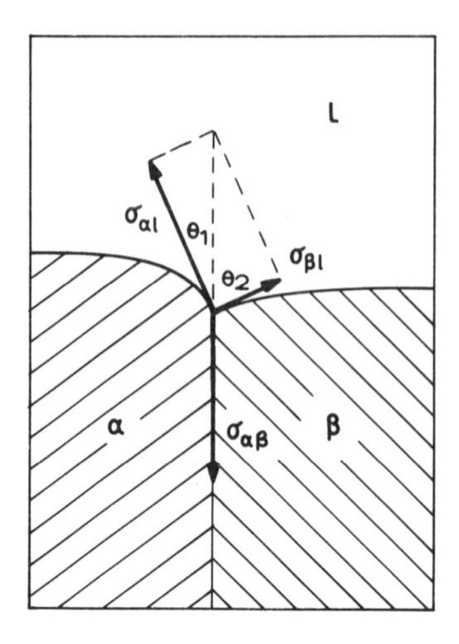

Figure A3.6

of eutectic interfaces or grain boundaries with respect to the solid/liquid interface. In most cases, the angle is expected to be close to 90°.

Calculation of $f(\theta)$ for Heterogeneous Nucleation [4]

The application of equations A3.26 to heterogeneous nucleation shows that true mechanical equilibrium (Fig. A3.7b) cannot be established, and that a surface stress will be set up (Fig. A3.7a). Due to the presence of foreign crystalline surfaces (crucible, surface oxide, inclusions) in a melt, nucleation may become much easier. The effect of these interfaces can be deduced from the energy balance:

ΔG_i = (interface energy creation due to nucleation) - (interface energy gained due to the substrate)

$$\Delta G_i = (A_{lc}\sigma_{lc} + A_{cs}\sigma_{cs}) - A_{cs}\sigma_{ls}$$

$$\Delta G_i = A_{lc}\sigma_{lc} + \pi R^2(\sigma_{cs} - \sigma_{ls}) \quad [A3.27]$$

$$\sigma_{ls} = \sigma_{cs} + \sigma_{lc}\cos\theta$$

$$\Delta G_i = A_{lc}\sigma_{lc} - \pi R^2\cos\theta\,\sigma_{lc} \quad [A3.28]$$

$$\Delta G = \Delta G_v + \Delta G_i = v_c\Delta g + [A_{lc} - \pi R^2\cos\theta\,]\sigma_{lc} \quad [A3.29]$$

$$v_c = \frac{\pi r^3[2 - 3\cos\theta + \cos^3\theta]}{3}$$

$$A_{lc} = 2\pi r^2[1 - \cos\theta]$$

$$R = r\sin\theta$$

$$\sin^2\theta = 1 - \cos^2\theta$$

$$\Delta G = \left(\frac{4\pi r^3\Delta g}{3} + 4\pi r^2\sigma_{lc}\right)\left[\frac{2 - 3\cos\theta + \cos^3\theta}{4}\right]$$

$$\Delta G_{het} = \Delta G_{hom}\cdot f(\theta) \qquad \text{[A3.30]}$$

where:

$$f(\theta) = \frac{2 - 3\cos\theta + \cos^3\theta}{4} = \frac{[2 + \cos\theta][1 - \cos\theta]^2}{4}$$

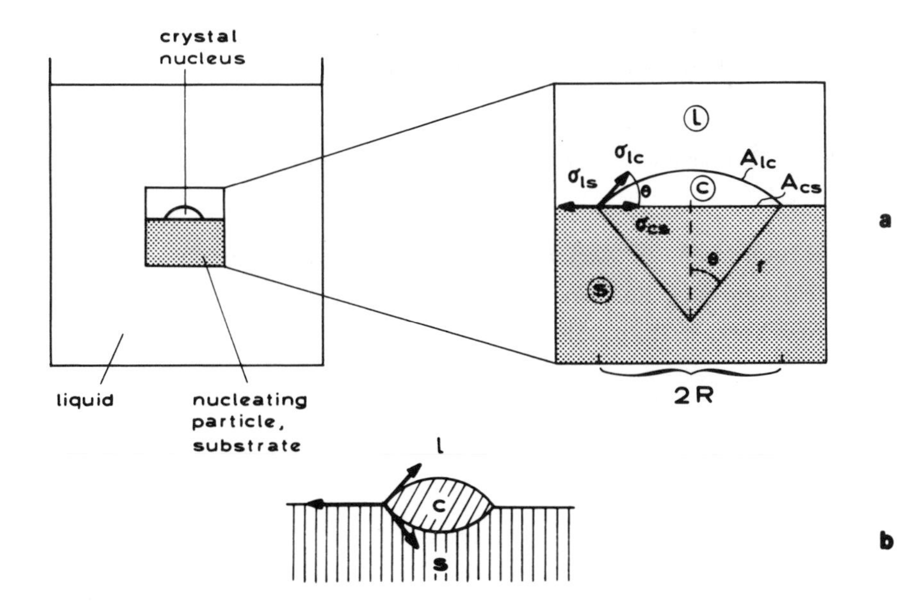

Figure A3.7

References

[1] M.J.Aziz, Journal of Applied Physics **53** (1982) 1158. K.Jackson, G.H.Gilmer, H.J.Leamy, in *Laser and Electron Beam Processing of Materials* (C.W.White, P.S.Peercy, Eds.), Academic Press, New York, 1980, p. 104. R.F.Wood, Physical Review **B25** (1982) 2786.

[2] D.R.Gaskell, *Introduction to Metallurgical Thermodynamics*, McGraw-Hill, New York, 1981.

[3] M.Hillert, in *Lectures on the Theory of Phase Transformations*, (H.I.Aaronson, Ed.) Transactions of the Metallurgical Society of AIME, New York, 1975, p. 1.

[4] J.S.Kirkaldy, in *Energetics in Metallurgical Phenomena*, Volume IV (W.M.Mueller, Ed.), Gordon & Breach, New York, 1968, p. 197.

[5] C.V.Thompson, F.Spaepen, Acta Metallurgica **27** (1979) 1855.

[6] K.S.Dubey, P. Ramachandrarao, Acta Metallurgica **32** (1984) 91.

[7] M.C.Flemings, *Solidification Processing*, McGraw-Hill, New York, 1974.

[8] W.W.Mullins, in *Metal Surfaces - Structure, Energetics, Kinetics*, American Society for Metals, Metals Park, 1963, p. 17.

[9] R.Trivedi, in *Lectures on the Theory of Phase Transformations*, (H.I.Aaronson, Ed.) Transactions of the Metallurgical Society of AIME, New York, 1975, p. 51.

[10] H.I.Aaronson, in *Lectures on the Theory of Phase Transformations*, (H.I.Aaronson, Ed.) Transactions of the Metallurgical Society of AIME, New York, 1975, p. 158.

APPENDIX 4

NUCLEATION KINETICS IN A PURE SUBSTANCE

In order to determine the nucleation rate, it is necessary to determine the number of critical nuclei and the rate of arrival of atoms necessary to make up these nuclei [1 - 3].

Equilibrium Distribution of Nuclei in an Undercooled Melt

The system shown in figure A4.1 represents a mixture of N_l atoms in the liquid state and N_n small crystal clusters, each of which contains n atoms. The free enthalpy change of such a system when compared with another one, at the same temperature, which contains only atoms and no crystal nuclei is:

$$\Delta G = N_n \Delta G_n - T\Delta S_n \qquad \text{[A4.1]}$$

Here, ΔG_n is the free enthalpy change due to the formation of one nucleus containing n atoms and ΔS_n is the entropy of mixing of N_n clusters with N_l atoms. (In this equation, the existence of an ideal mixture of crystal clusters and atoms of liquid is assumed. Therefore, the mixing enthalpy, ΔH_n, is equal to zero.)

In the case of sub-critical clusters (known as embryos), ΔG_n is positive due to the work of interface creation. This is demonstrated in figure A4.2 and corresponds to figure 2.2 where n and r, characterising the difference in size, are related by

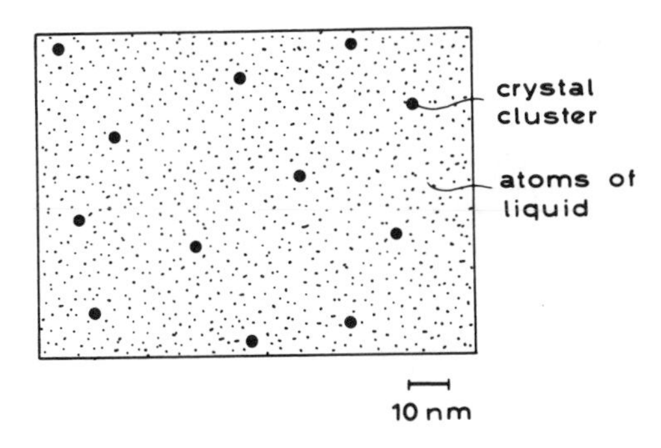

Figure A4.1

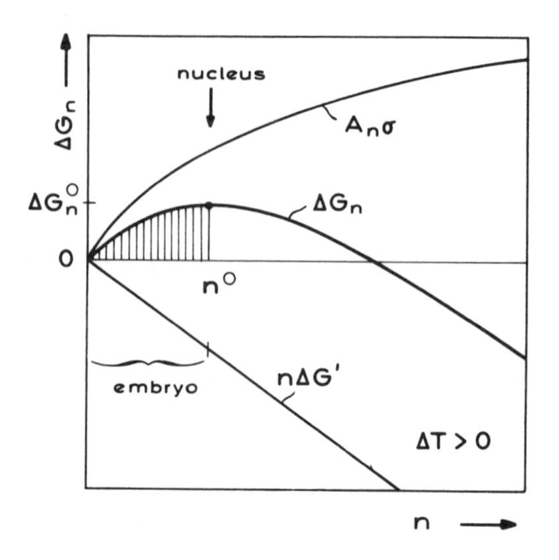

Figure A4.2

equation 2.2. The value of ΔG_n can be determined in an analogous manner by using equation 2.3, and leads to the relationship:

$$\Delta G_n = n\Delta G' + A_n \sigma$$
$$A_n = \eta n^{2/3} \qquad [A4.2]$$

where $\Delta G'$ is the atomic free energy difference, A_n is the interface area of the cluster, η is a form-factor which depends upon the shape of the cluster, and $n^{1/3}$ is proportional to the cluster diameter.

The mixing entropy of equation A4.1 can be derived by using the well-known relationship:

$$\Delta S_n = k_B \ln\left[\frac{(N_l + N_n)!}{N_n! N_l!}\right] \qquad [A4.3]$$

and therefore

$$\Delta G = N_n \Delta G_n - k_B T \ln[(N_l + N_n)!] + k_B T \ln[N_n!] + k_B T \ln[N_l!] \qquad [A4.4]$$

where ΔG_n is always positive for critical nuclei. An increase in N_n will first decrease the total free energy of the system, due to the mixing of clusters and atoms (Fig. A4.3), and then reach a minimum value which represents the equilibrium concentration of clusters for a given undercooling. Applying Stirling's approximation for large values of N, i.e. $\ln(N!) = N\ln(N) - N$, differentiating equation A4.4 with respect to N_n, and setting the result equal to zero gives:

$$\Delta G_n - k_B T[\ln(N_l + N_n) - \ln[N_n)] = 0$$

from which:

$$\frac{N_n}{N_l + N_n} = \exp[-\frac{\Delta G_n}{k_B T}] \qquad [A4.5]$$

Since $N_l >> N_n$

$$\frac{N_n}{N_l} = \exp[-\frac{\Delta G_n}{k_B T}]$$

The number of nuclei (critical clusters) in equilibrium is therefore:

$$N_n^{o} = N_l \exp[-\frac{\Delta G_n^{o}}{k_B T}] \qquad [A4.6]$$

where ΔG_n^{o} is the energy barrier for nucleation as defined in figure A4.2. These relationships are given in figure 2.3 of the main text.

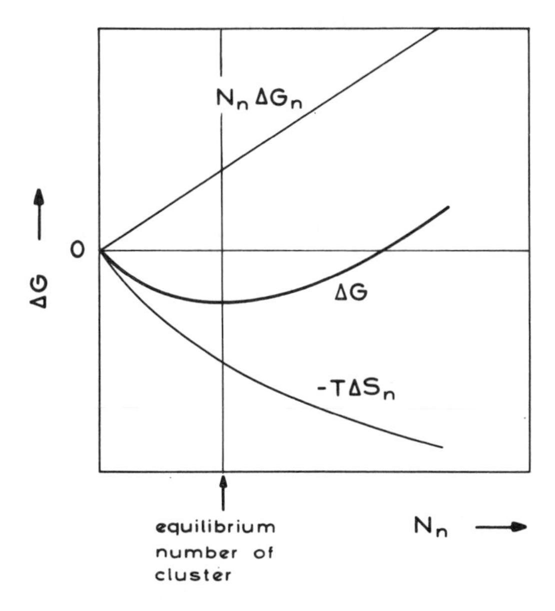

Figure A4.3

Rate of Formation of Stable Nuclei

The nucleation rate must be proportional to the number of crystals of critical size, N_n^o. However, in order for these crystals to grow, the addition of further atoms is required. The system, containing clusters of size, n^o, is in an unstable state and can produce smaller or larger clusters (Fig. A4.2) in order to decrease its energy. Therefore, it is necessary to determine the rate of incorporation, dn/dt, of new atoms into the nuclei. The nucleation rate is then:

$$I = N_n^o \frac{dn}{dt} \qquad [A4.7]$$

where the adsorption rate, dn/dt, is the product of an adsorption frequency, v, and the density of sites at which the atoms can be adsorbed by the critical nucleus, n_s^o:

$$\frac{dn}{dt} = v n_s^o \qquad [A4.8]$$

The adsorption frequency is:

$$v = v_0 \exp\left[-\frac{\Delta G_d}{k_B T}\right] p$$

where v_0 is the atomic vibration frequency, $\exp[-\Delta G_d/k_B T]$ is the fraction of atoms in the liquid which are sufficiently activated to surmount the interface addition activation energy, ΔG_d, and p is the adsorption probability.

The site density is given by:

$$n_s^o = A_n^o n_c \qquad [A4.9]$$

where A_n^o is the surface area of the critical nucleus and n_c is the capture site density per unit area.

The nucleation rate is therefore:

$$I = I_0 \exp\left[-\frac{\Delta G_n^o + \Delta G_d}{k_B T}\right] \qquad [A4.10]$$

and

$$I_0 = N_l \cdot v_0 p A_n^o n_c$$

Since the exponential term is extremely sensitive to small variations in the argument (Fig. 2.5), the exact value of I_0 is relatively unimportant and, for metals, is often approximated by:

$$I_0 = N_l \left(\frac{k_B T}{h}\right) = N_l \cdot \nu_0 \cong 10^{42}/\mathrm{m}^3\mathrm{s} \qquad [A4.11]$$

All of these relations assume the existence of a steady state and are not of general applicability. Nevertheless, they give a good guide to the principles which are involved. For a more complete treatment of nucleation theory, the reader should consult reference 3. In the case of alloys, the present approach has to be modified [4].

References

[1] D.Turnbull, J.C.Fisher, Journal of Chemical Physics **17** (1949) 71.

[2] D.R.Uhlmann, B.Chalmers, Industrial and Engineering Chemistry **57** (1965) 19.

[3] J.W.Christian, *The Theory of Transformations in Metals and Alloys* (2nd edition), Pergamon Press, Oxford, 1975.

[4] C.V.Thompson, F.Spaepen, Acta Metallurgica **31** (1983) 2021.

APPENDIX 5

ATOMIC STRUCTURE OF THE SOLID/LIQUID INTERFACE

In order to explain the principles involved, the simple case of a two-dimensional crystal consisting of 'square atoms' of only one type, which interact only with their nearest neighbours and form single-layer interfaces, will be treated here. It is assumed that there are three different structural elements making up the surface structure (three-site model - Fig. A5.1). This situation has been treated in detail by Jackson [1]. The model is an improved version of a previous one which was developed by the same author [2].

There is a continuous interchange of sites due to the thermally activated adsorption or desorption of atoms. For example, an $n = 1$ site will be transformed into an $n = 2$ site by adsorption of one neighbouring atom. It is assumed that, after a short time, the interface structure reaches a steady state, i.e., the overall density of the three types of site does not change with time. The rate of adsorption of atoms is independent of the rate of departure. Both are activated processes which are similar to diffusion in the liquid, but the desorption of atoms is more difficult from an undercooled crystal, due to

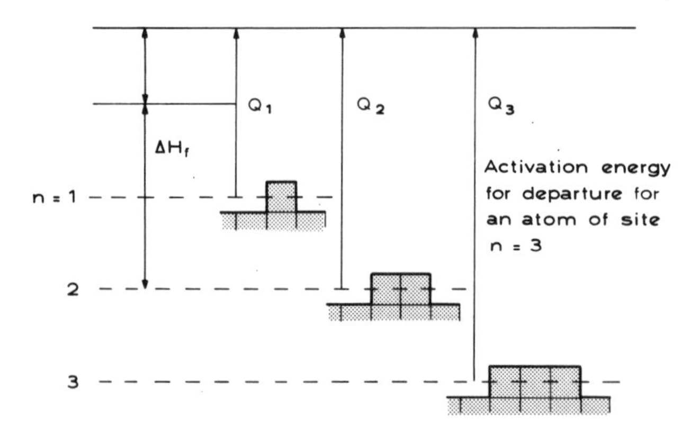

Figure A5.1

the gain in energy which occurs when an atom is added to the solid/liquid interface. The probability of adsorption will depend upon the roughness of the interface; the greater the density of steps (and therefore the greater the number of exposed atomic bonds presented to the liquid), the higher is the probability that an atom will be incorporated into the crystal; as a result of the stronger bonding involved.

The rate of atom arrival, J^+, is governed by:

$$J^+ = J_0^+ \exp[-\frac{Q}{RT}] \qquad [A5.1]$$

where Q is the activation energy for diffusion in the liquid. The flux of atoms leaving is determined by:

$$J^- = J_0^- \exp[-\frac{Q}{RT}]\exp[-\frac{2n\Delta H_f}{zRT}] \qquad [A5.2]$$

where ΔH_f is the latent heat of fusion, n is the number of bonds which have to be broken by an atom in leaving the crystal, and z is the coordination number (4 in the case of a two-dimensional crystal). An atom which makes two bonds with the crystal (i.e. half of the maximum possible number of bonds) can be viewed as being half in the solid and half in the liquid. At the melting point, T_f, the rates of arrival and departure from such sites ($n = 2$) should be equal. Thus,

$$J_0^+ \exp[-\frac{Q}{RT_f}] = J_0^- \exp[-\frac{Q}{RT_f}]\exp[-\frac{\Delta H_f}{RT_f}] \qquad [A5.3]$$

Therefore:

$$\frac{J_0^-}{J_0^+} = \exp[\frac{\Delta H_f}{RT_f}] \qquad [A5.4]$$

Combining equations A5.2 and A5.4, and rearranging the exponential terms:

$$J^- = J_0^+ \exp[-\frac{Q}{RT}]\exp[\frac{\Delta H_f}{RT_f} - \frac{n\Delta H_f}{2RT}] \qquad [A5.5]$$

The second exponential term can be regarded as being a measure of the probability that an atom which has n neighbours will leave the site, that is:

$$p_n = \exp[\frac{\Delta H_f}{RT_f} - \frac{n\Delta H_f}{2RT}]$$

The net flux is determined by the difference between equations A5.1 and A5.5:

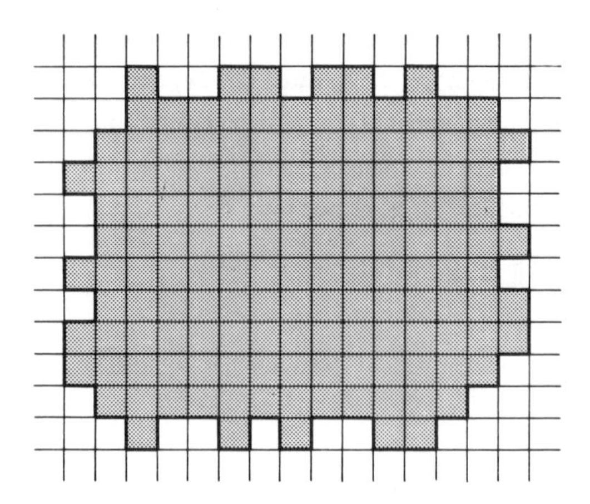

Figure A5.2

$$J = J_0^+ \exp[-\frac{Q}{RT}][1-p_n] \qquad [A5.6]$$

It is now possible to set up balance equations for the arrival and departure of atoms at various types of site. The detailed calculations have been given by Jackson [1]. Carrying these through and realising that, if $(\Delta H_f/RT_f)(\Delta T/T) << 1$,

$$\exp[-\frac{\Delta H_f \Delta T}{RT_f T}] \cong 1 - \frac{\Delta H_f \Delta T}{RT_f T}$$

the net growth rate, V, of the crystal becomes:

$$V = Jv' = v'J_0^+ \exp[-\frac{Q}{RT}]\alpha(\frac{\Delta T}{T})f[hk] \qquad [A5.7]$$

where $\alpha = \Delta H_f/RT_f = \Delta S_f/R$ (the dimensionless melting entropy) and v' is the atomic volume.

Therefore, to a first approximation, the growth rate is a linear function of the undercooling. It also depends upon the magnitude of the crystallographic factor, $f[hk]$, which is a function of the interface structure. It thus depends upon the value of α and upon the Miller indices, (hk), of the crystallographic plane. When α takes on high values (e.g. 10), the fraction of extra atoms or holes in the interface is proportional to $\exp[-\alpha]$.

For (11)-faces of the present square crystal (Fig. A5.2), f is independent of α but this is not so for the (10)-face (table A5.1). The reason for this is that a 45° edge

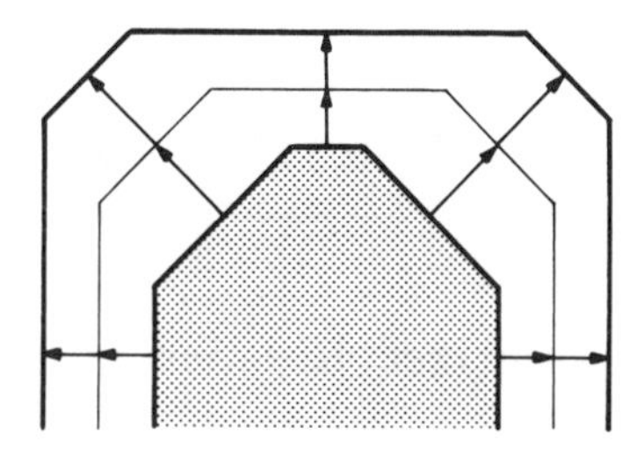

Figure A5.3

exhibits many growth steps, which cannot be removed by atom addition. When the value of α is large, the (11)-edges will grow much more quickly and leave the crystal bounded by slow-growing (10)-edges (Fig. A5.3 - see also Fig. 2.9). Note also in table A5.1 that, for small values of α, the growth is almost isotropic ($f[10] \sim f[11]$). This behaviour is typical of metals.

Table A5.1 Crystallographic Factor, $f[hk]$, of two Faces of a Square Crystal

α	$f[10]$	$f[11]$
1	0.56	0.60
5	0.30	0.60
10	0.039	0.60

References

[1] K.A.Jackson, Journal of Crystal Growth **3/4** (1968) 507.

[2] K.A.Jackson, in *Liquid Metals and Solidification*, American Society for Metals, Cleveland, 1958, p. 174.

APPENDIX 6

THERMODYNAMICS OF RAPID SOLIDIFICATION

Interface Kinetics

As explained in appendix 3, under normal solidification conditions the non-faceted solid/liquid interface itself is close to equilibrium even if concentration and temperature gradients exist in the adjacent phases. Therefore, the assumption of local equilibrium is useful when the short-range diffusional rate of atoms crossing the interface is much higher than the rate of arrangement of atoms into a configuration of low mobility (e.g. crystallisation).

Consider the case of an alloy, and assume now that the time required to freeze-in solute atoms is, due to rapid interface movement, much smaller than the time taken to diffuse through a characteristic distance, δ_i, which is of the order of the interatomic distance (Fig. A6.1). As a result, solute is incorporated into the solid at a concentration which is different to the equilibrium concentration. Such solute trapping will always occur, at least partially, if the following condition is satisfied:

$$\tau = \delta_i/V \leq \delta_i^2/D_i \qquad \text{[A6.1]}$$

where D_i is the interface diffusion coefficient. From this relationship, one can also see

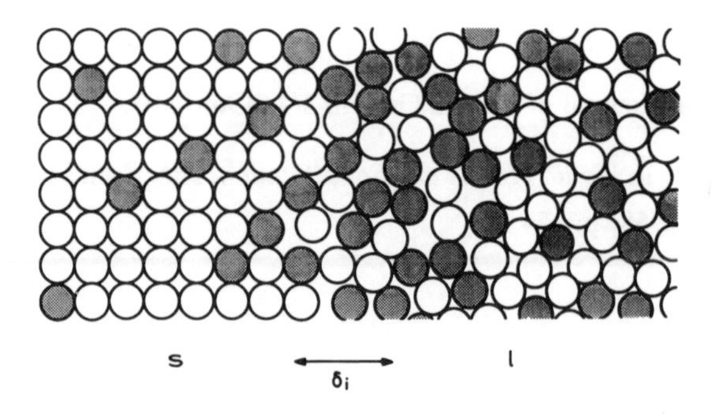

Figure A6.1

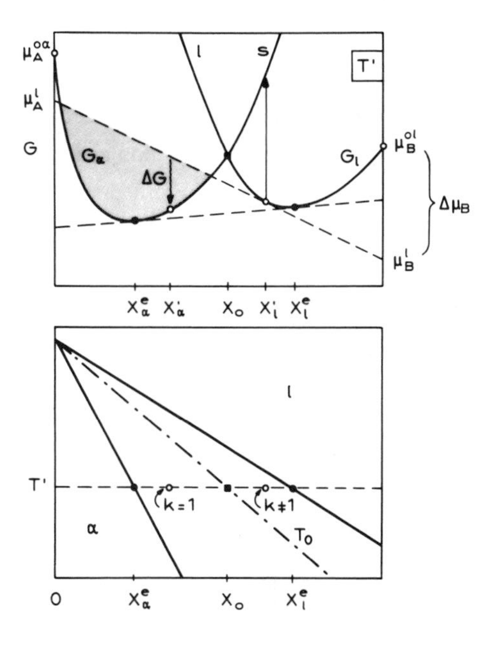

Figure A6.2

that local equilibrium will be lost when the interface Péclet number, P_i (= $\delta_i V/D_i$) satisfies:

$$P_i \geq 1 \qquad [A6.2]$$

At very high interface Péclet numbers, k_v is equal to unity (see below). In order that this should occur, another condition must be satisfied. This is that the change in free energy must still be negative (at least slightly) when the liquid transforms to a solid. This is best seen with the aid of free energy diagrams (Fig. A6.2).

T_0 - Condition

For a given interface temperature, $T_i = T'$, between solid (α) and liquid (l), the equilibrium compositions are X^e_α and X^e_l. Now, if the interface composition is X_l' and $k_v = 1$, solid α of the same composition as the liquid, X_l', cannot form since its free energy is higher than that of the liquid phase. However, solid of composition, X_α', can form from liquid of composition, X_l', at T' since its free energy will decrease. The range of compositions of α which can crystallise from liquid of composition, X_l', at T'

is given by the intersection of the tangent at $G(X_l^{'})$ with the curve, G_{α},.

In order to trap solute completely ($k_v = 1$) at $T_i = T'$, the composition must become equal to, or smaller than, X_0. That is, the maximum solid composition at T' corresponds to X_0 ($= X_l = X_\alpha$) and $T' = T_0$.

Following the approach of Baker and Cahn [1] and Boettinger and Coriell [2], one can now calculate the driving force for crystallisation as a function of k_v. This tends towards unity when V becomes very large. From the definition of the chemical potential, one can write for solute B:

$$\begin{aligned} \mu_B^\alpha &= \mu_B^{o\alpha} + RT\ln[a_\alpha] \\ \mu_B^l &= \mu_B^{ol} + RT\ln[a_l] \end{aligned} \qquad [A6.3]$$

where the activity, a, of the solute equals γX. Since, for simplicity, only dilute solutions will be considered here one can assume that the activity coefficient, γ, is constant.

From the condition of equilibrium, one has:

$$\mu_B^{o\alpha} + RT\ln[\gamma_\alpha X_\alpha^e] = \mu_B^{ol} + RT\ln[\gamma_l X_l^e] \qquad [A6.4]$$

and, under non-equilibrium conditions,

$$\mu_B^{o\alpha} + RT\ln[\gamma_\alpha X_\alpha] = \mu_B^{ol} + RT\ln[\gamma_l X_l] + \Delta\mu_B \qquad [A6.5]$$

From equations A6.4 and A6.5, one deduces that:

$$\Delta\mu_B = RT\ln[k_v/k] \qquad [A6.6]$$

where the non-equilibrium distribution coefficient, k_v, is equal to X_α/X_l, and the equilibrium distribution coefficient, k, is equal to X_α^e/X_l^e.

The same calculation can be carried out for the solute, A, in a binary system of concentration, $(1-X)$. From the relationship:

$$\Delta G = (\mu_A^\alpha - \mu_A^l)(1 - X_\alpha) + (\mu_B^\alpha - \mu_B^l)X_\alpha \qquad [A6.7]$$

it follows that:

$$\frac{\Delta G}{RT} = [1 - k_v X_l]\ln\left[\frac{1 - k_v X_l}{1 - X_l} \cdot \frac{1 - X_l^e}{1 - kX_l^e}\right] + k_v X_l \ln\left[\frac{k_v}{k}\right] \qquad [A6.8]$$

and one obtains for $X_l = X_l^e$, in the case of dilute solutions ($X_l << 1$) and after approximating $\ln[z]$ (when $z \rightarrow 1$) by $z - 1$,

$$\frac{\Delta G}{RT} = \frac{(1-k_v X_l)X_l(k-k_v)}{1-kX_l} + k_v X_l \ln\left[\frac{k_v}{k}\right] \qquad [A6.9]$$

The first term on the right-hand side of equation A6.9 is approximately equal to $X_l[k-k_v]$, and one finally obtains, for the thermodynamic relationship between k_v and ΔG (at the interface),

$$\Delta G^*/RT^* = X_l^*\{k - k_v\,[1-\ln(\frac{k_v}{k})]\} \qquad [A6.10]$$

Using the relationship, $\Delta G = \Delta S_f \Delta T$, one can relate the interface undercooling to the deviation from local equilibrium, as expressed by k_v.

A further undercooling is associated with the kinetics of rearrangement of the atoms at the interface (Fig. A6.1). This value can be obtained using:

$$\Delta G_k = \Delta S_f \Delta T_k$$

where $\Delta S_f \cong RT_f(1-k)/m$ (see equation A3.14). This gives, for the kinetic driving force:

$$\Delta G_k/RT_f = [(1-k)/m](T_f + mX_l^* - T_i) \qquad [A6.11]$$

Adding equations A6.10 and A6.11 one finds, for small undercoolings, $T^* \sim T_f$:

$$\Delta G/RT^* = [(1-k)/m](T_f + mX_l^* - T_i) + X_l^*\{k - k_v[1-\ln(\frac{k_v}{k})]\} \qquad [A6.12]$$

From the relationship between ΔG and the growth rate [3],

$$V = fV_0[1-\exp(\Delta G/RT^*)] \qquad [A6.13]$$

where f is the fraction of interface sites which are growth sites (close to unity for metals), and V_0 is the upper limit of interface advance (comparable to the velocity of sound). Since $V << V_0$, one can write:

$$V/V_0 = -\Delta G/RT^* \qquad [A6.14]$$

and, combining equations A6.12 and A6.14, one obtains for the temperature of a planar interface:

$$T_i = T_f + mX_l^* + [mX_l^*/(1-k)]\{k - k_v[1-\ln[\frac{k_v}{k}]]\} + (V/V_0)[m/(1-k)]$$

[A6.15]

and, for a dendrite tip of radius, R:

$$T_i = T_f - 2\Gamma/R + mX_l^*[1 + f(k)] + (V/V_0)[m/(1-k)] \qquad [A6.16]$$

where $f(k) = \{k - k_v[1 - \ln(k_v/k)]\}/(1 - k)$ and the kinetic liquidus slope is $m' = m[1 + f(k)]$.

Setting $R = \infty$ in equation A6.16, $V_0 = \infty$ (infinitely rapid interface kinetics), and $k_v = 1$, one obtains the equation for the T_0-line:

$$T_0 = T_f + \frac{X_l^* m \ln(k)}{k-1} \qquad [A6.17]$$

Because of the simplifications made, this equation holds only for dilute solutions.

Non-Equilibrium Distribution Coefficient

The last equation to be developed is that which describes the relationship between the non-equilibrium distribution coefficient and the rate of interface movement. This can be easily found by considering the two extreme cases shown in figure 7.1. At $V = 0$, the interface is at equilibrium: $\mu_\alpha = \mu_l$ and $k_v = k$. Here, the net flux of atoms crossing the interface is zero. The other extreme case, $V = \infty$, leads to complete solute trapping and $k_v = 1$ ($X_\alpha^* = X_l^*$). This sets up a gradient of chemical potential which tends to drive atoms from α to l. This flux is proportional to the chemical potential gradient:

$$J_d = -XM_i(d\mu/dz) \qquad [A6.18]$$

For dilute solutions, the interface diffusion coefficient is related to the interface mobility, M_i, via:

$$D_i = M_i RT \qquad [A6.19]$$

giving:

$$J_d = -(D_i X/RT)(d\mu/dz) \qquad [A6.20]$$

Note that the interface diffusion coefficient can be substantially lower than the bulk liquid diffusion coefficient. The chemical potential gradient at the interface, in the case of dilute solutions (see equation A6.6), is:

$$(d\mu/dz) \cong (\Delta\mu_B/\delta_i) = RT\ln(k_v/k)/\delta_i \qquad [A6.21]$$

Setting the composition of the interface, X, equal to its mean value, $(X_l^* + X_\alpha^*)/2 = X_l^*(1 + k_v)/2$ leads to:

$$J_d = -(D_i/2\delta_i)X_l^*(1+k_v)\ln(k_v/k) \qquad [A6.22]$$

This flux in the growth direction, which arises from the chemical potential gradient, must be equal to a source term if steady-state conditions are to prevail at the interface. This source term is simply the solute rejected at the interface:

$$J_r = VX_l^*(1-k_v) \qquad [A6.23]$$

Setting J_d equal to J_r, one obtains:

$$V\delta_i/D_i = (1/2)[(1+k_v)/(1-k_v)]\ln(k_v/k) \qquad [A6.24]$$

This is just one relationship between the interface Péclet number ($P_i = V\delta_i/D_i$) and the non-equilibrium distribution coefficient. Others have been derived by Aziz [4]. For example, again for dilute solutions:

$$k_v = (k+P_i)/(1+P_i) \qquad [A6.25]$$

The general form of all of these relationships is such that $k_v = k$ when P_i is much smaller than unity, and $k_v = 1$ when P_i is much greater than unity (Fig. 7.3). The critical growth rate at which k_v changes markedly depends mainly upon the interface diffusion rate, D_i/δ_i. This is of the order of 0.1 to 1 m/s.

References

[1] J.C.Baker, J.W.Cahn, in *Solidification*, ASM, Metals Park, Ohio, 1971, p. 23.

[2] W.J.Boettinger, S.R.Coriell, in *Science and Technology of the Undercooled Melt* (P.R.Sahm, H.Jones, C.M.Adam, Eds.), Martinus Nijhoff Publications, Dordrecht, 1986, p. 81.

[3] D.Turnbull, Journal of Physical Chemistry **66** (1962) 609.

[4] M.J.Aziz, Journal of Applied Physics **53** (1982) 1158.

APPENDIX 7

INTERFACE STABILITY ANALYSIS

The constitutional undercooling criterion provides a useful means for estimating whether a solid/liquid interface will be planar under directional solidification conditions. However, from a theoretical point of view it has several faults. Firstly, it does not take account of the effect of surface tension. It is expected that this will tend to inhibit the formation of perturbations. Secondly, it takes account only of the temperature gradient in the liquid. Thirdly, the constitutional undercooling theory does not give any indication of the *scale* of the perturbations which will develop if an interface becomes unstable. In order to deduce more information about the morphological instability of a plane interface than that provided by the constitutional undercooling theory, one must suppose that the interface has already been slightly disturbed and then ask whether the disturbance will grow or disappear (chapter 3). To this end, it is sufficient to consider a sinusoidal interface form having a very small, time-dependent amplitude (Fig. A7.1). (Provided that the amplitude is small, a sinusoidal form represents the most general disturbance possible; it can be supposed to be one term of the Fourier series describing any possible disturbance.)

The object of the present calculations is to determine the conditions which govern the growth or decay of a perturbation at the solid/liquid interface (Fig. 3.1). In order that the results should be applicable to high rates of solidification, it is necessary to consider both solute and heat diffusion. In the case of solute diffusion, the solid phase can be neglected because the rate of diffusion in the solid is so slow. In the case of heat diffusion, both the liquid and solid phases have to be considered. The reader is advised to consult the section on the 'slightly perturbed interface' in appendix 2. The latter

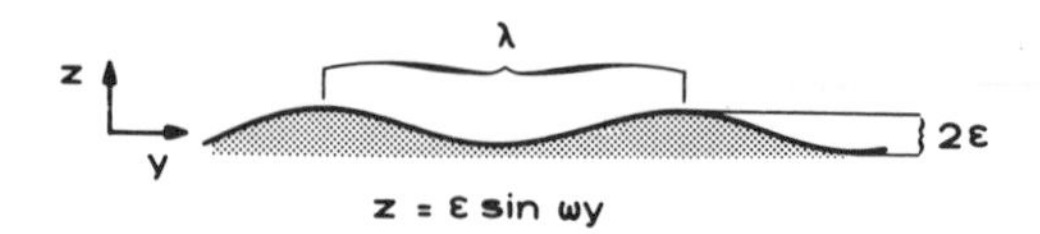

Figure A7.1

section introduces, in the context of a simpler problem, all of the techniques which will be used below.

Three equations have to be solved. These are, for two-dimensional diffusion:

$$\frac{\partial^2 C}{\partial y^2} + \frac{\partial^2 C}{\partial z^2} + \frac{V}{D}\frac{\partial C}{\partial z} = 0 \qquad \text{[A7.1a]}$$

which describes the solute distribution in the liquid,

$$\frac{\partial^2 T_l}{\partial y^2} + \frac{\partial^2 T_l}{\partial z^2} + \frac{V}{a_l}\frac{\partial T_l}{\partial z} = 0 \qquad \text{[A7.1b]}$$

which describes the temperature distribution in the liquid, and

$$\frac{\partial^2 T_s}{\partial y^2} + \frac{\partial^2 T_s}{\partial z^2} + \frac{V}{a_s}\frac{\partial T_s}{\partial z} = 0 \qquad \text{[A7.1c]}$$

which describes the temperature distribution in the solid. Here, V is the growth rate and D, a_l, and a_s are solute and thermal (liquid and solid) diffusivities respectively.

For a plane interface, suitable solutions to these equations are, respectively:

$$C = C_0 - \frac{DG_c}{V}\exp[-\frac{Vz}{D}] \qquad \text{[A7.2a]}$$

$$T_l = T_0 + \frac{G_l a_l}{V}[1 - \exp(-\frac{Vz}{a_l})] \qquad \text{[A7.2b]}$$

$$T_s = T_0 + \frac{G_s a_s}{V}[1 - \exp(-\frac{Vz}{a_s})] \qquad \text{[A7.2c]}$$

where G_c, G_l, and G_s are solute and temperature gradients (liquid and solid) respectively at the plane interface. Note that C_0 is the concentration at infinity, but T_0 is the temperature at the interface. As in appendix 2, the plane interface is assumed to be perturbed so that it has the form shown in figure A7.1. According to perturbation theory (appendix 2), the solute and temperature distributions can be assumed to be of the form:

$$C = C_0 - \frac{DG_c}{V}\exp\left(-\frac{Vz}{D}\right) + A\varepsilon\sin\omega y\exp(-b_c z) \qquad \text{[A7.3a]}$$

$$T_l = T_0 + \frac{G_l a_l}{V}\left[1 - \exp\left(-\frac{Vz}{a_l}\right)\right] + B\varepsilon\sin\omega y\exp(-b_l z) \qquad \text{[A7.3b]}$$

$$T_s = T_0 + \frac{G_s a_s}{V}[1 - \exp(-\frac{Vz}{a_s})] + R\varepsilon \sin\omega y \exp(-b_s z) \qquad \text{[A7.3c]}$$

where A, B, R, b_c, b_l, and b_s have to be determined by applying the various physical constraints which are imposed by the situation:

Each of the b-values can be found by inserting the derivatives of the relevant equation into the corresponding differential equation (A7.1). In this way (appendix 2), it is found that:

$$b_c = \frac{V}{2D} + [(\frac{V}{2D})^2 + \omega^2]^{1/2} \qquad \text{[A7.4a]}$$

$$b_l = \frac{V}{2a_l} + [(\frac{V}{2a_l})^2 + \omega^2]^{1/2} \qquad \text{[A7.4b]}$$

$$b_s = -\frac{V}{2a_s} + [(\frac{V}{2a_s})^2 + \omega^2]^{1/2} \qquad \text{[A7.4c]}$$

The value of R can be found by recalling that, at the perturbed interface, the temperatures in the solid and in the liquid must be the same. That is (putting $S = \varepsilon \sin\omega y = z$ as in appendix 2):

$$T_0 + \frac{G_l a_l}{V}[1 - \exp(-\frac{VS}{a_l})] + BS\exp(-b_c S) =$$

$$= T_0 + \frac{G_s a_s}{V}[1 - \exp(-\frac{VS}{a_s})] + RS\exp(-b_s S)$$

Since ε is very small, one can write $\exp(-\varepsilon \sin\omega y) \approx 1 - \varepsilon \sin\omega y$, and terms in ε^2 disappear. This leads to:

$$R = (G_l - G_s) + B \qquad \text{[A7.5]}$$

A second condition is imposed by the coupling of the absolute temperatures and concentrations at the interface. That is, the actual temperature at the perturbed interface must correspond to that which is given by the sum of the constitutional supercooling and capillarity effects if the attachment kinetics are very rapid. This can be written:

$$T^+ = T_f + mC^+ - \Gamma\omega^2 S \qquad \text{[A7.6]}$$

where the index, +, indicates a value determined at the perturbed interface, and T_f is the melting point of the pure substance. The last term on the RHS involves just the second differential of the perturbing function rather than the full curvature expression, $K = z''/(1+z'^2)^{1/2}$. This is permissible because of the smallness of ε. Thus:

$$T_0 + \frac{G_l a_l}{V}\left[1-\exp\left(-\frac{VS}{a_l}\right)\right] + BS\exp(-b_l S) =$$

$$= T_f + m\left[C_0 - \frac{DG_c}{V}\exp\left(-\frac{VS}{D}\right) + AS - \exp(-b_c S)\right] - \Gamma\omega^2 S$$

That is:

$$T_0 + G_l S + BS = T_f + mC_0 - \frac{mDG_c}{V} + mG_c S + mAS - \Gamma\omega^2 S$$

Since, for the perturbed interface, $T_0 = T_f + mC_0 - mDG_c/V$, one can write:

$$G_l S + BS = mG_c S + mAS - \Gamma\omega^2 S$$

Cancelling S throughout leads to:

$$B = mG_c + mA - \Gamma\omega^2 - G_l \qquad [A7.7]$$

Thus, one now has B in terms of A. In order to eliminate this last remaining unknown, one imposes the condition that the gradients of solute and temperature must satisfy solute and heat conservation at the perturbed interface *for the same growth rate*. This can be written:

$$\frac{\kappa_s G_s^+ - \kappa_l G_l^+}{\Delta h_f} = \frac{DG_c^+}{(k-1)C^+} \qquad [A7.8]$$

where κ_s and κ_l are the conductivities of solid and liquid, respectively, and Δh_f is the volumic heat of fusion. The quantities which are marked +, must be evaluated at the perturbed interface. That is:

$$G_s^+ = G_s - \frac{G_s}{a_s}VS - b_s RS \qquad [A7.9a]$$

$$G_l^+ = G_l - \frac{G_l}{a_l}VS - b_l BS \qquad [A7.9b]$$

$$G_c^+ = G_c - \frac{G_c}{D}VS - b_c AS \qquad [A7.9c]$$

$$C^{+} = C_0 - \frac{DG_c}{V} + G_c S + AS \qquad [A7.9d]$$

Inserting equations A7.9 into equation A7.8 gives:

$$n_1 + n_2 VS + (\kappa_l b_l B - \kappa_s b_s R)S = \frac{\Delta h_f}{(k-1)} \cdot \frac{DG_c - G_c VS - Db_c AS}{C^{*} + G_c S + AS}$$

where

$$n_1 = (\kappa_s G_s - \kappa_l G_l), \qquad n_2 = \left(\frac{\kappa_l G_l}{a_l} - \frac{\kappa_s G_s}{a_s}\right)$$

and C^{*} is the liquid composition at a planar solid/liquid interface. Multiplying throughout by the denominator on the RHS, and remembering that terms in S^2 can be neglected, gives:

$$n_1 C^{*} + n_1 G_c S + n_1 AS + n_2 VC^{*}S + (\kappa_l b_l B - \kappa_s b_s R)C^{*}S =$$

$$= \frac{\Delta h_f}{k-1}(DG_c - G_c VS - Db_c AS)$$

but, from equation A7.8, it follows that

$$n_1 C^{+} = \frac{\Delta h_f}{k-1} DG_c$$

Therefore, after dividing throughout by S, one obtains:

$$n_1 G_c + n_1 A + n_2 VC^{*} + (\kappa_l b_l B - \kappa_s b_s R)C^{*} = \frac{\Delta h_f}{p}(G_c V + Db_c A)$$

where $p = 1 - k$. By substituting for R from equation A7.5, and for B from equation A7.7, one can obtain the value of the last unknown:

$$A = \frac{-n_1\left(\frac{VC^{*}}{D} + G_c\right) - n_2 VC^{*} - n_3 C^{*}(\phi - \omega^2 \Gamma) + \kappa_s b_s (G_l - G_s)C^{*}}{n_1 + n_3 mC^{*} + n_1\left(\frac{C^{*}}{G_c}\right) b_c}$$

[A7.10]

where $n_3 = \kappa_l b_l - \kappa_s b_s$ and $\phi = mG_c - G_l$ (the degree of constitutional supercooling).

In order to study the time dependence, it is assumed that the local velocity of the perturbed interface is of the form:

$$V + \dot{\varepsilon}\sin\omega y \qquad [A7.11]$$

This must be equal to the heat conservation term on the LHS of equation A7.8. By inserting the values from equations A7.9a and A7.9b, one obtains:

$$V + \dot{\varepsilon}\sin\omega y =$$

$$= \frac{(\kappa_s G_s - \kappa_l G_l) + \left(\frac{\kappa_l G_l}{a_l} - \frac{\kappa_s G_s}{a_s}\right)V\varepsilon\sin\omega y + (\kappa_l b_l B - \kappa_s b_s R)\varepsilon\sin\omega y}{\Delta h_f}$$

or

$$V + \dot{\varepsilon}\sin\omega y = \frac{n_1 + n_2 V\varepsilon\sin\omega y + (\kappa_l b_l B - \kappa_s b_s R)\varepsilon\sin\omega y}{\Delta h_f} \qquad [A7.12]$$

The above equation expresses the identity of two polynomials (in $\varepsilon\sin\omega y$). Therefore, the equivalent coefficients on the two sides must be equal. Thus:

$$V = \frac{\kappa_s G_s - \kappa_l G_l}{\Delta h_f} \qquad [A7.13a]$$

and

$$\dot{\varepsilon}\sin\omega y = \frac{n_2 V\varepsilon\sin\omega y + (\kappa_l b_l B - \kappa_s b_s R)\varepsilon\sin\omega y}{\Delta h_f} \qquad [A7.13b]$$

That is,

$$\frac{\dot{\varepsilon}}{\varepsilon} = \frac{n_2 V + \kappa_l b_l B - \kappa_s b_s R}{\Delta h_f} \qquad [A7.14]$$

The sign of $\dot{\varepsilon}/\varepsilon$ determines the stability of the interface. If it is positive for any value of ω, then perturbations with that wave-number will be amplified. In order to study the stability behaviour in detail, it is necessary to use equations A7.5 and A7.7 in order to substitute for R and B, respectively, in equation A7.14. That is:

$$\frac{\dot{\varepsilon}}{\varepsilon} = \frac{n_2 V - \kappa_s b_s (G_l - G_s) + (\kappa_l b_l - \kappa_s b_s)(\phi - \omega^2\Gamma + mA)}{\Delta h_f} \qquad [A7.15]$$

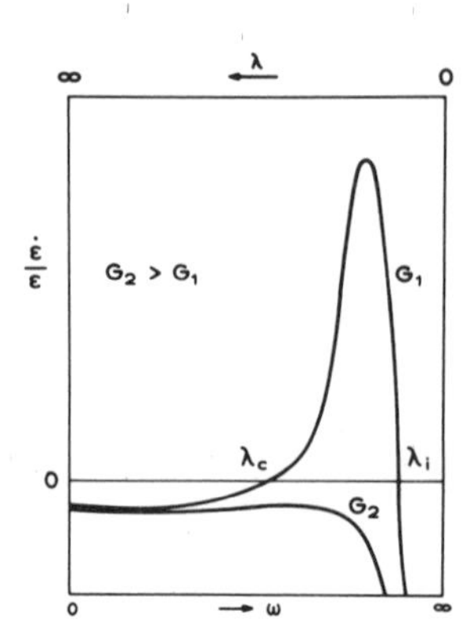

Figure A7.2

This equation is shown in figure A7.2 for two temperature gradients ($G = G_l = G_s$). At the lower gradient G_1, the function has two zeros; limiting the range of instability to a band of ω. The largest marginally stable ω-value defines the smallest λ-value which is used in dendrite growth theory. The denominator of equation A7.15, after substituting for A, will be the same as the denominator of equation A7.10. Examination of the denominator shows that it is always positive and therefore cannot affect the stability of the interface. Therefore, attention will now be restricted to the numerator. This can be written as:

$$\frac{\dot{\varepsilon}}{\varepsilon} = [n_1 n_2 V - n_1 n_4 + n_1 n_3(\phi - \omega^2 \Gamma)]\left(\frac{1}{C^*} + \frac{b_c}{G_c}\right) - n_1 n_3\left(\frac{V}{D} + \frac{G_c}{C^*}\right)m$$

where $n_4 = \kappa_s b_s (G_l - G_s)$.

Further rearrangement gives the final result for marginal stability, i.e. for $\dot{\varepsilon}/\varepsilon = 0$:

$$-\Gamma\omega^2 - [\bar{\kappa}_l G_l \xi_l + \bar{\kappa}_s G_s \xi_s] + mG_c \xi_c = 0 \qquad \text{[A7.16]}$$

where

$$\xi_l = [b_l - (V/a_l)] / [\bar{\kappa}_l b_l + \bar{\kappa}_s b_s]$$

$$\xi_s = [b_s - (V/a_s)] / [\bar{\kappa}_l b_l + \bar{\kappa}_s b_s]$$

$$\xi_c = [b_c - \frac{V}{D}] / [b_c - \frac{Vp}{D}]$$

$$\bar{\kappa}_l = \kappa_l/(\kappa_l + \kappa_s)$$

$$\bar{\kappa}_s = \kappa_s/(\kappa_l + \kappa_s)$$

In order to understand the implications of this very important result, it is helpful to assume, for the moment, that $\kappa_s = \kappa_l$, $a_s = a_l$, $G_s = G_l = G$, and $\xi_s = \xi_l$. The latter is true when $\omega >> V/a$. (It is also true for the case of constrained growth with $G > 0$ when $G_s = G_l$.). This corresponds to the small thermal Péclet number solution of Mullins and Sekerka [1] but allows for large solutal Péclet numbers. Under these conditions, equation A7.16 becomes:

$$-\omega^2\Gamma\left(b_c - \frac{Vp}{D}\right) - G\left(b_c - \frac{Vp}{D}\right) + mG_c\left(b_c - \frac{V}{D}\right) = 0 \qquad [A7.17]$$

Upper Limit of Stability (Absolute Stability)

Writting equation A7.17 in the form:

$$-\Gamma\omega^2 - G + mG_c\xi_c = 0$$

and assuming that the Péclet number is much larger than one, i.e. $\xi_c \rightarrow \pi^2/kP_c^2$, one obtains:

$$-\Gamma\omega^2 - G + mG_c\pi^2/kP_c^2 = 0$$

In most cases, it can be assumed that G can be neglected at very high growth rates, so that:

$$-\Gamma\omega^2 + mG_c\pi^2/kP_c^2 = 0$$

Substituting for ω and P_c gives:

$$-4\pi^2\Gamma/\lambda^2 + mG_c(\pi^2/k)(4D^2/V^2\lambda^2) = 0$$

Because it is assumed throughout that the long-range diffusion field is the same as that for the plane front, one can substitute $\Delta T_0V/D$ for mG_c. Thus:

$$kV\Gamma/\Delta T_0D = 1$$

When this condition is satisfied or exceeded, the interface will be stable, regardless of the value of G, if $G << G_a$ (Fig. 7.5). This is referred to as 'absolute stability'. One can define a critical velocity for absolute stability:

$$(V_a)_c \geq \frac{\Delta T_0 D}{k\Gamma} \qquad [A7.18]$$

Using typical values for metals: $\Delta T_0 = 10K$, $D = 5 \times 10^{-3}mm^2/s$, $k = 0.5$, and $\Gamma = 10^{-4}Kmm$, the limit of absolute stability should be attained when V is greater than 1m/s. Such rates are very high but can be obtained in rapid solidification processes such as laser surface melting. At these rates, account must be taken of the value of the distribution coefficient, which is a function of the growth rate and may approach unity (see appendix 6)# .The change in k will decrease $(v_a)_c$ via the decreased ΔT_0 and increased k values.

Although the gradient, G_l, is considered to have a negligible effect in the above case, it is still positive. One can also consider the case where the temperature gradient in the solid is equal to zero and the gradient, G_l, in the liquid is negative; as in the case of equiaxed grain growth. The growth rate is still assumed to be very high. Equation A7.16, in the case where $G_s = 0$ and $\kappa_s = \kappa_l$, becomes:

$$-\Gamma\omega^2 - 0.5G_l\xi_l + mG_c\xi_c = 0$$

Substituting the high Péclet number approximations, $\xi_l = 2\pi^2/P_l^2$ and $\xi_c = \pi^2/kP_c^2$ gives:

$$-\Gamma\omega^2 - G_l(\pi^2/P_l^2) + mG_c(\pi^2/kP_c^2) = 0$$

This expression determines the form (values of ω) which the perturbed interface must assume in order to satisfy all of the conditions of the problem.

From the signs of the terms in equation A7.16, one can immediately deduce the effects of the parameters involved. For example, increasing $\omega^2\Gamma$ (curvature effect) or G (imposed temperature gradient) tends to decrease the value of $\dot{\varepsilon}/\varepsilon$ and thus increases the stability, whereas mG_c (liquidus temperature gradient) is always positive (since m and G_c always have the same sign) and therefore decreases stability.

If equation A7.16 is further simplified by making the assumption that k is equal to zero (that is, $p = 1$), one can immediately obtain the result:

$$\omega^2 = \phi/\Gamma \qquad [A7.19]$$

which, as pointed out in section 3.4, is the same as the result derived there using very

Note that the limit in equation A7.18 can also be derived by assuming that, in equation A7.17, the value of G becomes irrelevant when the de-stabilizing term, mG_c, is balanced by the term, $V^2k\Gamma/D^2$. Equating the two terms again gives equation A7.18.

simple arguments.

Another relatively simple step which one can take is to replace the exact expression for b_c (which leads to an intractable cubic equation) by $b_c \sim V/D + (D/V)\omega^2$ (which implies $\omega << V/2D$). This leads to the equation:

$$\frac{D\Gamma}{V}\omega^4 - \left(\phi\frac{D}{V} - \Gamma\frac{Vk}{D}\right)\omega^2 + G\frac{Vk}{D} = 0 \qquad [A7.20]$$

Using Descartes' rule of signs, if the equation is to have no positive roots (implying stability) then the coefficient of ω^2 must be negative. This leads to the stability condition:

$$G > mG_c - \frac{V^2 k\Gamma}{D^2} \qquad [A7.21]$$

Returning now to consider the full equation (A7.15), one question of interest is how ξ_l, ξ_s and ξ_c vary as the parameters change. This is most easily seen by again assuming that $a_l = a_s$ and $\kappa_l = \kappa_s$, which is true, for example, in the case of organic materials such as succinonitrile. This then leads to the equations:

$$\xi_l = 1 - (V/2a_l) / [(V/2a_l)^2 + \omega^2]^{1/2} \qquad [A7.22a]$$

$$\xi_s = 1 + (V/2a_s) / [(V/2a_s)^2 + \omega^2]^{1/2} \qquad [A7.22b]$$

$$\xi_c = [1 - (1 + 4D^2\omega^2/V^2)^{1/2}] / [1 - 2k - (1 + 4D^2\omega^2/V^2)^{1/2}] \qquad [A7.22c]$$

By substituting the relevant Péclet numbers, these equations can be written in the corresponding forms:

$$\xi_l = 1 - 1/[1 + (2\pi/P_l)^2]^{1/2} \qquad [A7.23a]$$

$$\xi_s = 1 + 1/[1 + (2\pi/P_s)^2]^{1/2} \qquad [A7.23b]$$

$$\xi_c = 1 - 2k/\{[1 + (2\pi/P_c)^2]^{1/2} - 1 + 2k\} \qquad [A7.23c]$$

As can be seen in figure A7.3, all three are equal to unity at low Péclet numbers, and the first and last tend towards zero as the Péclet number tends towards infinity. The second of the three tends towards 2 as the Péclet number approaches infinity. When the Péclet number is high but not infinite, the expressions can be further simplified since $1/[1 + (2\pi/P)^2]^{1/2}$ is approximately equal to $1 - 2\pi^2/P^2$ under these conditions. This gives:

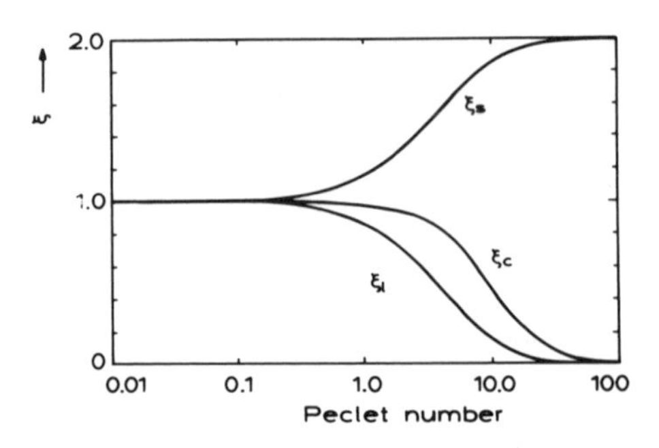

Figure A7.3

$$\xi_l \cong 2\pi^2/P_l^2$$

$$\xi_s \cong 2-2\pi^2/P_s^2 \qquad P >> 1$$

$$\xi_c \cong \pi^2/kP_c^2$$

Replacing mG_c, P_l, ω, and P_c by their respective definitions gives:

$$-G_l \frac{a_l^2}{\Gamma V^2} + \frac{\Delta T_0 D}{\Gamma k} = V$$

For growth in undercooled melts, with $G_s = 0$, the heat flux balance is $V\Delta h_f = -\kappa_l G_l$. Therefore:

$$\frac{\Delta h_f a_l^2}{\kappa_l \Gamma} + \frac{\Delta T_0 D}{\Gamma k} = V$$

or

$$V = \frac{D\Delta T_0}{\Gamma k} + \frac{a_l \theta_t}{\Gamma}$$

where θ_t is the unit thermal undercooling. This condition can be written [2]:

$$V_a = (V_a)_c + (V_a)_t \qquad [A7.24]$$

This equation shows that absolute stability is a general phenomenon and will always be observed when the growth rate of the interface is higher than that given by equation A7.24. The second term on the RHS of this equation is much larger than the first one (for metals where $a >> D$). Therefore, V_a in an undercooled melt (equiaxed growth) is larger than in the case of directional growth, where the temperature gradients are positive and $(V_a)_t = 0$.

References

[1] W.W.Mullins, R.F.Sekerka, Journal of Applied Physics **35** (1964) 444.

[2] R.Trivedi, W.Kurz, Acta Metallurgica **34** (1986) 1663.

APPENDIX 8

DIFFUSION AT A DENDRITE TIP

There are three different types of dendrite:

i equiaxed dendrites of pure substances - freely growing and governed by thermal diffusion (Fig. 4.7b);

ii equiaxed dendrites of alloys - freely growing and governed by solute and thermal diffusion (Fig. 4.7d);

iii columnar alloy dendrites - constrained in their growth by a positive temperature gradient and controlled by solute diffusion (Fig. 4.7c).

The first two types of growth form are similar and both lead to the creation of an equiaxed polycrystal in which each randomly arranged grain is made up of six orthogonal primary trunks (in the case of a cubic crystal). The space remaining between the trunks is filled with secondary and possibly higher-order branches (Fig. 4.16). These crystals grow in an undercooled melt, and this makes them inherently unstable. Therefore, no steady-state cellular morphologies can be observed and the spacing, λ_1, between the trunks corresponds approximately to the grain diameter. The growth of equiaxed dendrites of pure metals occurs under conditions where only heat flows from the interface to the surrounding liquid. That is, the temperature gradient is negative at the interface (left-hand side of figure A8.1) and a thermal undercooling, ΔT_t, exists. In the case of equiaxed alloy growth, there exists not only a negative temperature gradient but also a solute build-up (if k is less than unity) ahead of the dendrite tip. This changes the local liquidus temperature (right-hand side of figure A8.1). When an equiaxed grain of an alloy of composition, C_0, is growing it experiences an undercooling, ΔT, which is the sum of a solute undercooling, ΔT_c, a thermal undercooling, ΔT_t, and a curvature undercooling, ΔT_r. The temperature and solute fields of columnar dendrites, growing in an alloy melt have been depicted in figure 4.8.

In general it can be said that, due to the large dendrite tip radius which is predicted by the stability criterion (appendix 9), the curvature undercooling is small in comparison to the other contributions and can often be neglected to a first approximation under normal solidification conditions. At the tip, one can therefore write: $C_l^*(r) \cong C_l^*$. One

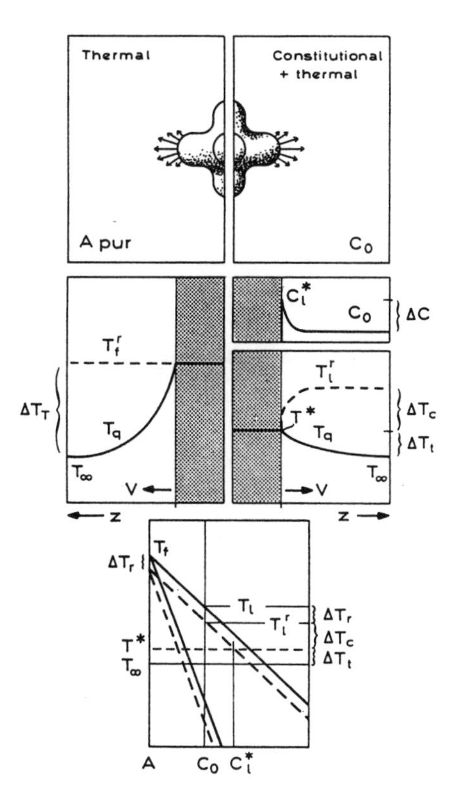

Figure A8.1

can now evaluate the solutal and thermal undercoolings, ΔT_c and ΔT_t. From the definition of the solutal supersaturation (Fig. A8.2):

$$\Omega = \frac{C_l^* - C_0}{C_l^* p} \qquad \text{[A8.1]}$$

$$C_l^* = \frac{C_0}{1 - \Omega p}$$

$$\Delta T_c = m(C_0 - C_l^*) = mC_0\left[1 - \frac{1}{1 - \Omega p}\right] \qquad \text{[A8.2]}$$

where $p = 1 - k$. At low supersaturations, where Ω is much smaller than unity:

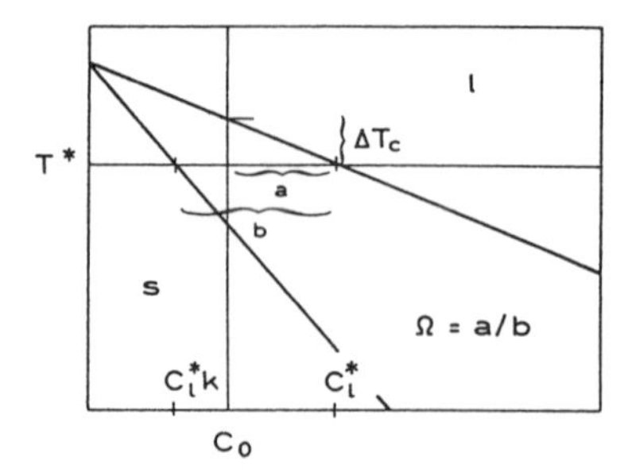

Figure A8.2

$$\Delta T_c \cong -mC_0 \Omega p$$

Substituting $-mC_0p = \Delta T_0 k$, one obtains:

$$\Delta T_c \cong \Omega \Delta T_0 k \qquad \Omega << 1 \qquad [A8.3]$$

At small supersaturations, the solutal supersaturation, Ω, can be defined as the ratio of two temperature differences

$$\Omega = \frac{\Delta T_c}{\Delta T_0 k} \qquad [A8.4]$$

(When Ω approaches unity, the unit undercooling $\Delta T_0 k$ approaches ΔT_0).

Similarly, the thermal supersaturation, Ω_t, can be defined as the ratio of the thermal undercooling to the unit undercooling ($\Delta h_f/c$)

$$\Omega_t = \frac{\Delta T_t}{\Delta h_f/c} \qquad [A8.5]$$

from which is obtained an equation that is analogous to equation A8.3:

$$\Delta T_t = \Omega_t \frac{\Delta h_f}{c} \qquad [A8.6]$$

Hemispherical Needle Approximation

A cylinder with a hemispherical tip, growing along its axis, is the simplest approximation which can be made to the problem of dendrite tip growth [1]. It permits a rapid assimilation of the physical factors which are important in dendrite growth. The cross-section of the cylinder, $A = \pi R^2$, determines the volume which grows in a time, dt, and which is responsible for the rejection of solute (Fig. A8.3a). The surface area of the hemispherical cap, $A_h = 2\pi R^2$, determines the amount of radial solute diffusion. Thus, a flux due to solute rejection, J_1, and one due to diffusion in the liquid ahead of the tip, J_2, can be identified:

$$J_1 = AV(C_l^* - C_s^*) \qquad [A8.7]$$

$$J_2 = -DA_h \left(\frac{dC}{dr}\right)_{r=R} \qquad [A8.8]$$

Under steady-state conditions, both of the fluxes must be equal; leading to the relationship:

$$VC_l^*(1-k) = -2D\left(\frac{dC}{dr}\right)_R \qquad [A8.9]$$

The concentration gradient at the tip can be approximated by the value which was found for a growing sphere (equation A2.26):

$$\left(\frac{dC}{dr}\right)_R = -\frac{C_l^* - C_0}{R}$$

Therefore, the diffusion equation reduces to:

$$\frac{VR}{2D} = \frac{C_l^* - C_0}{C_l^*(1-k)} \qquad [A8.10]$$

which is generally written in the abbreviated form:

$$P_c = \Omega \qquad [A8.11]$$

Here, P_c $(= VR/2D)$ is the solute Péclet number (the ratio of a characteristic

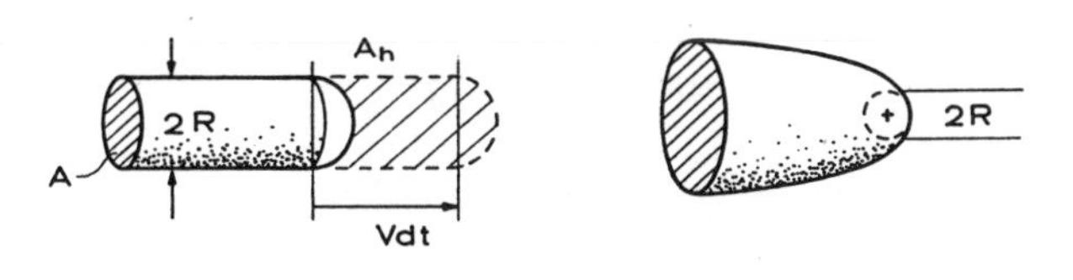

Figure A8.3

dimension, R, of the system to the solute diffusion distance, $2D/V$). In figure 4.9, this expression is represented by the straight line which runs from the upper left to the lower right.

In the case of thermal diffusion-limited dendrites, a similar flux balance to that above can be made and leads to the same relationship as that of equation A8.11; where the solute Péclet number is replaced by a thermal Péclet number:

$$P_t = \frac{VR}{2a} \qquad [A8.12]$$

and the solute supersaturation is replaced by the thermal supersaturation, as defined by equation A8.5. Therefore, the thermal case is defined by

$$P_t = \Omega_t \qquad [A8.13]$$

Paraboloid of Revolution

As proposed originally by Papapetrou [2], a much better model for the dendrite tip is a paraboloid of revolution (Fig. A8.3b). Ivantsov [3] was the first to develop a mathematical analysis for this shape; and his analysis has since been generalised by Horvay and Cahn [4]. They found that:

$$\Omega = \mathrm{I}(P) \qquad [A8.14]$$

where the Ivantsov function (a function of P) for a needle crystal corresponding to a paraboloid of revolution is given by:

$$\mathrm{I}(P) = P\exp(P)\,\mathrm{E}_1(P) \qquad [A8.15]$$

and for a parabolic cylinder (plate) by:

$$\mathrm{I}'(P) = (\pi P)^{1/2}\exp(P)\mathrm{erfc}(P^{1/2}) \qquad [A8.16]$$

Here, E_1 is the exponential integral function, defined by:

$$\mathrm{E}_1(P) = \int_P^\infty \frac{\exp(-z)}{z}dz = -\mathrm{Ei}(-P) \qquad [A8.17]$$

Its value (Fig. A1.3) can be determined from the series [5]:

$$E_1(P) = -0.5772157 - \ln(P) - \sum_{n=1}^{\infty} \frac{(-1)^n P^n}{n \cdot n!} \cong$$

$$\cong -0.577 - \ln(P) + \frac{4P}{P+4} \qquad \text{[A8.18]}$$

for intermediate values of P. For the purpose of numerical calculations, a good approximation to $E_1(P)$ is [5]; for $0 \leq P \leq 1$:

$$E_1(P) = a_0 + a_1 P + a_2 P^2 + a_3 P^3 + a_4 P^4 + a_5 P^5 - \ln(P) \qquad \text{[A8.19]}$$

where

$a_0 = -0.57721566$ $\quad a_1 = 0.99999193$

$a_2 = -0.24991055$ $\quad a_3 = 0.05519968$

$a_4 = -0.00976004$ $\quad a_5 = 0.00107857$

and, for $1 \leq P \leq \infty$

$$I(P) = P\exp(P)E_1(P) = \frac{P^4 + a_1 P^3 + a_2 P^2 + a_3 P + a_4}{P^4 + b_1 P^3 + b_2 P^2 + b_3 P + b_4} \qquad \text{[A8.20]}$$

where

$a_1 = 8.5733287401$ $\quad b_1 = 9.5733223454$

$a_2 = 18.0590169730$ $\quad b_2 = 25.6329561486$

$a_3 = 8.6347608925$ $\quad b_3 = 21.0996530827$

$a_4 = 0.2677737343$ $\quad b_4 = 3.9584969228$

(see also table A8.1).

The Ivantsov function, $I(P)$, can also be written as a continued fraction:

$$I(P) = \cfrac{P}{P + \cfrac{1}{1 + \cfrac{1}{P + \cfrac{2}{1 + \cfrac{2}{P + \dots}}}}} \qquad \text{[A8.21]}$$

Truncating the continued fraction at the zeroth, first, and higher terms will lead to various approximations:

$$I_0 = P$$

$$I_1 = \frac{P}{P+1}$$

$$I_2 = \frac{2P}{2P+1}$$

$$I_\infty = I(P) = P\exp(P)E_1(P)$$

These approximations are shown graphically in figure A8.4.

Substituting the zeroth approximation into equation A8.14 gives the solution which was obtained for the case of a hemispherical tip (equation A8.11). Furthermore, it is

Table A8.1

P	$\Omega = I(P)$	$P = -\Omega/\ln(\Omega)$#
0.001	0.006338	0.00125
0.002	0.01130	0.00252
0.004	0.01987	0.00507
0.006	0.02743	0.00763
0.008	0.03435	0.0102
0.010	0.04079	0.0127
0.020	0.06845	0.0255
0.040	0.1116	0.0509
0.060	0.1462	0.0761
0.080	0.1757	0.101
0.100	0.2015	0.126
0.200	0.2987	0.247
0.400	0.4191	0.482
0.600	0.4968	0.710
0.800	0.5530	0.933
1.000	0.5963	1.15
2.000	0.7227	2.22
4.000	0.8254	4.30
6.000	0.8716	6.34
8.000	0.8982	8.37
10.000	0.9156	10.4

In this column the Ω-values from column 2 have been used

interesting to note that the modification of Zener's analysis, made by Hillert [6] for high supersaturations, leads to:

$$\frac{VR}{D} = \frac{C_l^* - C_0}{C_0 - C_s^*} = \frac{\Omega}{1-\Omega} \qquad [A8.22]$$

which is equivalent to $2P = \Omega/(1 - \Omega)$ or $\Omega = 2P/(2P+1)$. This is the second approximation which arises from the continued fraction representation when it is substituted into equation A8.14. The paraboloid of revolution, for which the solution is given by equation A8.15, represents an isothermal/isoconcentrate dendrite, and closely approximates the experimentally observed form.

From figure A8.4, it can be seen that I_1 is a much better approximation than is I_0 because the latter cannot be used when P is greater than unity. As the series (equation A8.21) converges very slowly, it is usually preferable to use the polynomial expressions (equations A8.19 and A8.20).

From equation A8.22, one sees that another relationship can be obtained for the inverse Ivantsov solution [7]. This is a good approximation to equation A8.15 (see table A8.1 and figure A8.4):

$$P = \frac{\Omega}{-\ln(\Omega)} \qquad [A8.23]$$

More exact, non-isothermal, solutions have been developed by Trivedi [8].

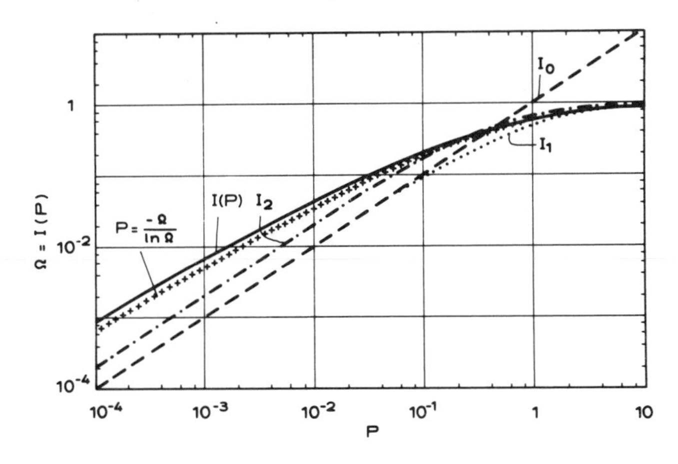

Figure A8.4

References

[1] J.C.Fisher, referred to by B.Chalmers in *Principles of Solidification*, Wiley, New York, 1966, p. 105.

[2] A.Papapetrou, Zeitschrift für Kristallographie **92** (1935) 89.

[3] G.P.Ivantsov, Doklady Akademii Nauk SSSR **58** (1947) 567.

[4] G.Horvay, J.W.Cahn, Acta Metallurgica **9** (1961) 695.

[5] M.Abramowitz, I.A.Stegun (Eds.), *Handbook of Mathematical Functions*, Dover, New York, 1965.

[6] M.Hillert, Jernkontorets Annaler **141** (1957) 757, and Metallurgical Transactions **6A**, (1975) 5.

[7] P.Pelcé, P.Clavin, Europhysics Letters **3** (1987) 907.

[8] R.Trivedi, Acta Metallurgica **18** (1970) 287, and Metallurgical Transactions **1** (1970) 921.

APPENDIX 9

DENDRITE TIP RADIUS AND SPACING

Growth at the Extremum

The relationships defining solutal (or thermal) diffusion at the hemispherical tip of a needle-like crystal (appendix 8):

$$\Omega = P \qquad [A9.1]$$

or a paraboloid of revolution

$$\Omega = \mathrm{I}(P) \qquad [A9.2]$$

do not specify an unique functional dependence of the tip radius upon the growth conditions (supersaturation, undercooling). They only relate the product, VR, to the supersaturation as shown in figure 4.9. Therefore, another equation which links the variables is required. For many years, this was done by adding a capillarity term to the diffusion equation and then determining the extremum of the corresponding function. This approach will be illustrated first by means of the simple solution (equation A9.1):

$$\Omega = P + \frac{2s}{R} \qquad [A9.3]$$

where s is the capillarity length (for the solutal case at low Ω-values, $s = s_c = \Gamma/\Delta T_0 k$, and for the thermal case, $s = s_t = \Gamma c/\Delta h_f$). In figure A9.1, relationships such as equation A9.3 for various models are presented, in terms of dimensionless variables, for free thermal dendrite growth and for a constant undercooling, $\Delta T_t = 0.05\Delta h_f/c$ [1]. These curves indicate that the addition of a capillarity term to equation A9.1 cuts off the diffusion solution at some tip radius (see also figure 4.9). The radius at which the cut-off occurs corresponds to the critical radius of nucleation. This can easily be verified by noting that this radius is the one which uses up the total supersaturation, indicated by equation A9.3, in satisfying the curvature. As a result, V (and therefore P) must be equal to zero at that point. Therefore:

$$\Omega = \frac{2s}{R^{\circ}} \qquad (V = 0)$$

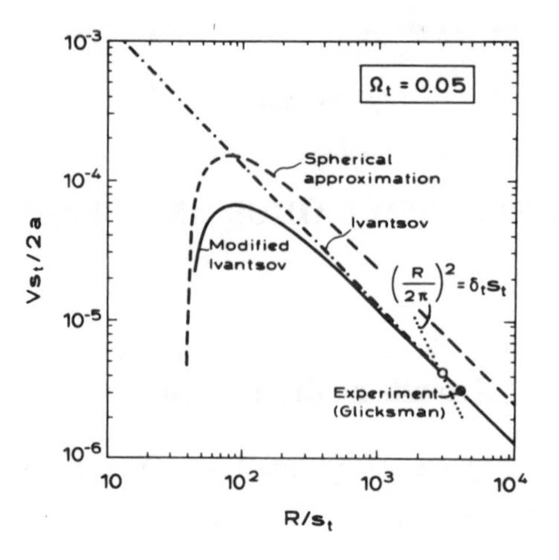

Figure A9.1

where the critical radius, R°, for growth is equal to the critical radius, r°, for nucleation. Substituting ΔT for Ω from equation A8.3 or A8.6 leads to:

$$R^{\circ} = \frac{2s}{\Omega} = \frac{2\Gamma}{\Delta T} = r^{\circ} \qquad [A9.4]$$

Examination of equation A9.3 for the thermal or solutal case reveals that:

$$\Delta T_t = \frac{VR}{2a}\left(\frac{\Delta h_f}{c}\right) + \frac{2\Gamma}{R} \qquad [A9.5]$$

$$\Delta T_c = \frac{VR}{2D}(\Delta T_0 k) + \frac{2\Gamma}{R} \qquad [A9.6]$$

The extremum (maximum) value of V was thought to define the tip radius (R_e, figure 4.9). It is shown in chapter 5 (Fig. 5.7) that the maximum value of V in an isothermal environment corresponds to a minimum in ΔT for constant-velocity growth. Therefore, minimising ΔT in equation A9.5 or A9.6 will give the extremum radius and undercooling. This leads to:

$$R_e^t = 2\left(\frac{\Gamma ac}{\Delta h_f}\right)^{1/2} \frac{1}{V^{1/2}} \qquad [A9.7]$$

$$\Delta T_t = 2\left(\frac{\Gamma \Delta h_f}{ac}\right)^{1/2} V^{1/2} \qquad [A9.8]$$

for thermal dendrites and to

$$R_e^c = 2\left(\frac{\Gamma D}{\Delta T_0 k}\right)^{1/2} \frac{1}{V^{1/2}} \quad \text{[A9.9]}$$

$$\Delta T_c = 2\left(\frac{\Gamma \Delta T_0 k}{D}\right)^{1/2} V^{1/2} \quad \text{[A9.10]}$$

for solutal dendrites when both are growing at the extremum. Equations A9.8 and A9.10 reflect the well-known square-root-of-V relationships obtained for free dendrite growth in a constant supersaturation environment. Others can be found in the review paper by Glicksman et al. [2].

Instead of using the extremum method, the recent trend has been to use stability criteria which are analogous to those used in treating the stability of a planar solid/liquid interface [1, 3-12]. Some simple cases will be treated here which demonstrate the essential points of the more complete treatments.

Growth at Marginal Stability#

Solution for Dendrite Growth in Undercooled Alloy Melts

The essentials of such an analysis can be rapidly demonstrated in the case of a thermal dendrite, for example. As the limit of stability of the dendrite tip lies at much larger tip radii than does the extremum values, the capillarity term in equation A9.5 can be neglected and one obtains instead (for a hemispherical needle at low Péclet numbers):

$$\Delta T_t = P_t \theta_t \quad \text{[A9.11]}$$

where θ_t is the unit thermal undercooling (= $\Delta h_f/c$). With $G_c = 0$ for a pure metal, the critical tip radius becomes (equation 4.8):

$$R = \left[\frac{\Gamma}{\sigma^*(-\bar{G})}\right]^{1/2} \quad \text{[A9.12]}$$

where $\sigma^* = 1/4\pi^2$ and the mean gradient at the tip, $\bar{G}$, for the thermal (undercooled) dendrite can be obtained from $G_s = 0$ (isothermal tip, figure A8.1) and $G_l = -2P_t\theta_t/R$ (see exercise 4.6) via the relationship:

See first footnote in section 4.4.

$$\overline{G} = \frac{\kappa_s G_s + \kappa_l G_l}{\kappa_s + \kappa_l} \qquad [A9.13]$$

When $\kappa_s = \kappa_l$:

$$\overline{G} = G_l/2 = -\frac{P_t \theta_t}{R} \qquad [A9.14]$$

and, from equation A9.12:

$$R_t = \left(\frac{2}{\sigma^*}\right)^{1/2} \left(\frac{a\Gamma}{\theta_t}\right)^{1/2} \frac{1}{V^{1/2}} \qquad [A9.15]$$

Substituting equation A9.15 into equation A9.11 leads to:

$$\Delta T_t = \left(\frac{1}{2\sigma^*}\right)^{1/2} \left(\frac{\theta_t \Gamma}{a}\right)^{1/2} V^{1/2} \qquad [A9.16]$$

A comparison of equations A9.15 and A9.16 with equations A9.7 and A9.8 shows that both the extremum and the stability arguments lead to qualitatively the same result (for low Péclet numbers) but with differing numerical constants. Thus, the tip radius for the marginally stable hemispherical dendrite is 4.4 times larger and the undercooling is 2.2 times larger than the corresponding extremum values.

After considering this very simple case, a more complete model will be developed which is useful for the interpretation of both low and high Péclet number cases; that is, dendrite growth under slow and rapid solidification conditions. In the case of *alloy growth from the undercooled melt*, the coupled transport problem (solute and heat diffusion) has to be solved. The total undercooling, ΔT (= $T_l - T_\infty$), is made up of three contributions when attachment kinetics are neglected (Fig. A8.1). These are: ΔT_t, the thermal undercooling ($T^* - T_\infty$); ΔT_c, the solutal undercooling ($T_l^r - T^*$); and ΔT_r, the curvature undercooling ($T_l - T_l^r$).

Two of these undercoolings can be found from the Ivantsov solution for the case of a paraboloidal tip,

$$\Delta T_t = \theta_t \mathrm{I}(P_t) \qquad [A9.17]$$

where θ_t is the unit thermal undercooling (= $\Delta h_f/c$), and:

$$\Delta T_c = mC_0 [1 - \mathrm{A}(P_c)] \qquad [A9.18]$$

where

$$A(P_c) = C_l^*/C_0 = [1 - pI(P_c)]^{-1} \qquad \text{[A9.18a]}$$

(see equation A8.2).

The third contribution to ΔT is the curvature undercooling, which is determined by the relationship:

$$\Delta T_r = 2\Gamma/R \qquad \text{[A9.19]}$$

The sum of the three contributions to ΔT gives the total undercooling [8]:

$$\Delta T = \Delta T_t + \Delta T_c + \Delta T_r \qquad \text{[A9.20]}$$

It has been seen previously that this expression is not, in itself, sufficient to solve the problem. In order to find an unique solution to equation A9.20, one can advantageously use the approach developed by Langer and Müller-Krumbhaar. These authors [1] proposed a criterion which was based upon extensive numerical calculations. Use of the criterion gives results which are close to those found experimentally. For instance, see the points on the curve in figure A9.1 where the open circle refers to Langer and Müller-Krumbhaar's theory, and the closed circle is due to Glicksman et al. [2]. This criterion supposes that the dendrite tip grows at a constant value of the stability constant, σ ($\equiv \delta s/R^2$: product of the diffusion length, δ, and the capillarity length, s; each made dimensionless by dividing by the tip radius, R) which is also given by the relationship, $\sigma = (\lambda_i/2\pi R)^2$. Taking the operating value of the stability constant, σ^*, to be about $1/4\pi^2$, as has been observed experimentally, it follows that:

$$R = \lambda_i \qquad \text{[A9.21]}$$

where λ_i is the lower limiting wavelength for a perturbation which can grow at the solid/liquid interface. Using this value for the wavelength at a planar interface (equation 3.22) as a zeroth approximation gives [9]:

$$\lambda_i = 2\pi \left(\frac{\Gamma}{\phi}\right)^{1/2} \qquad \text{[A9.22]}$$

Thus *at low Péclet numbers*:

$$R = \left[\frac{\Gamma}{\sigma^*(mG_c - \bar{G})}\right]^{1/2} \qquad (P_c, P_t << 1) \qquad \text{[A9.23]}$$

where $\sigma^* = 1/4\pi^2$. The concentration gradient ahead of an advancing dendrite can be found from a simple flux balance: $VC_l^*(1 - k) = -DG_c$.

$$G_c = -\frac{2P_c C_l^* p}{R} \qquad [A9.24]$$

Substituting for C_l^* from equation A8.1 and using Ivantsov's solution gives:

$$G_c = -2P_c p C_0 A(P_c)/R \qquad [A9.25]$$

In a similar manner, the temperature gradient, G_l, in the liquid ahead of a steadily growing dendrite is found to be:

$$G_l = -\frac{2P_t \Delta h_f}{cR} \qquad [A9.26]$$

In contrast to the solute diffusion field, which is evaluated only for the liquid phase, the temperature field has to be calculated for both phases because of the similarity in their thermal diffusivities. Therefore, one has to introduce a conductivity-weighted average temperature gradient. Assuming that $\kappa_s = \kappa_l$ and that the Ivantsov dendrite is isothermal, one finds that $\overline{G}$ at the tip is equal to $G_l/2$ (equation A9.14). It is this value which has to be substituted into equation A9.23. Substituting equation A9.23 into equation A9.19 then yields an expression for the tip radius:

$$R = \frac{\Gamma/\sigma^*}{\theta_t P_t - 2P_c m C_0 p \, A(P_c)} \qquad [A9.27]$$

where $A(P_c) = 1/[1-(1-k)I(P_c)]$.

The capillarity undercooling at the tip is therefore:

$$\Delta T_r = 2\sigma^*[\theta_t P_t - 2P_c m C_0 p \, A(P_c)] \qquad [A9.28]$$

The solutal Péclet number is simply related to the thermal Péclet number by $P_c = P_t(a/D)$. Therefore, the only unknown quantity in equations A9.17 to A9.20 is the product, VR (or the Péclet number). From the definition of P_t, for example, one can finally obtain the growth rate:

$$V = 2aP_t/R \qquad [A9.29]$$

At *high Péclet numbers*, this solution has to be generalised by using a modified stability criterion (appendix 7). Thus one can write, instead of equation A9.23 [10],

$$R = \left[\frac{\Gamma}{\sigma^*(mG_c^* - G^*)}\right]^{1/2} \qquad [A9.30]$$

where σ^* is again approximately equal to $1/4\pi^2$ and G_c^* and G^* are the effective concentration and temperature gradients as defined in equation 7.8. Note that, since the temperature gradient in the solid at the tip, G_S, is zero, the product, $\bar{\kappa}_S G_S \xi_S$, is equal to zero in equation A7.16. Therefore, one can set $\xi_l = \xi_t$ (for thermal diffusion in undercooled melts) from now on.

Equation A9.27 then becomes, in its general form:

$$R = \frac{\Gamma/\sigma^*}{\theta_t P_t \xi_t + 2P_c \theta_c \xi_c} \qquad \text{[A9.31]}$$

where $\theta_c = \Delta T_0 k A(P_c)$ and $\xi_t = \xi_l$ and ξ_c are defined by equation A7.23. Substituting equation A9.31 into the curvature undercooling term (equation A9.19) and using the other undercooling relationships (equations A9.17 and A9.18) in combination with equation A9.20 gives the *general solution* for slow or rapid dendrite growth in undercooled alloy melts.

This solution is depicted in figure A9.2, which shows the variation in the dimensionless growth rate, $\bar{V} = Vs_t/2a$ (a), and dimensionless tip radius, $\bar{R} = R/s_t$ (b), with dimensionless concentration $\bar{C}_0 = C_0|m|\theta_t$ for succinonitrile-acetone mixtures at an undercooling of 0.5K [8]. The solid line gives the predictions of the present model and the broken line gives those of a more complicated model due to Karma and Langer [7]. The points are experimental results from Chopra and Glicksman. For a given (small) ΔT, the solution of these equations reveals a maximum in V and a minimum in R with increasing solute content in a given alloy system [8]. For a detailed discussion of the behaviour of this model for small undercoolings, see reference [8] and for rapid solidification conditions (large undercoolings), see references [10] and [11].

Constrained (Alloy) Dendrite Growth

During the directional growth of dendrites, the heat generated at the growing interface does not depend on the tip radius. The latent heat flows into the solid, due to the imposed temperature gradient. The moving isotherms force the tips to grow at a given rate into the liquid. Therefore, the resultant growth behaviour (e.g. the tip undercooling) is determined by the solute flux at the tip. In order to treat this case, one starts again with equation A9.20, but neglects the thermal contribution to the tip diffusion fields ($\Delta T_t = 0$).

Therefore, for the transport solution one has:

$$\Delta T = \Delta T_c + \Delta T_r \qquad (G > 0) \qquad \text{[A9.32]}$$

The second equation is deduced from the general stability relationship (equation

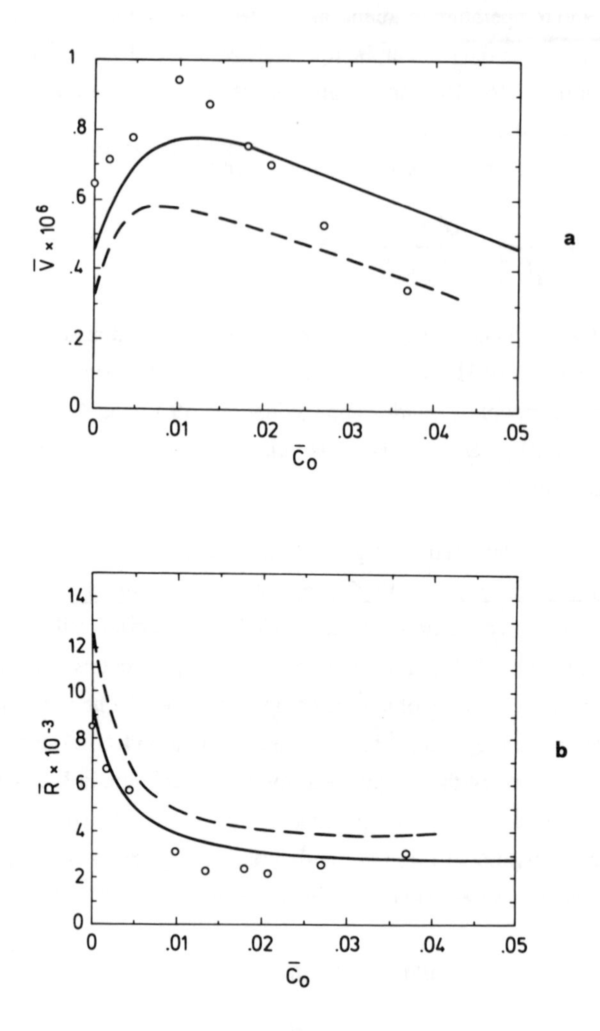

Figure A9.2

A9.30), using the appropriate temperature gradient. Assuming that, at the very tip of the dendrite, the imposed temperature gradient is the same in both liquid and solid ($G = G_l = G_s$) and that the thermal conductivities of the solid and the liquid are equal ($\kappa = \kappa_l = \kappa_s$), one obtains for the effective temperature gradient:

$$G^* = (G/2)(\xi_l + \xi_s) \quad [A9.33]$$

From equation A7.23, one sees that $\xi_l + \xi_s$ is equal to 2, and the effective gradient is thus equal to the temperature gradient imposed by the furnace, for example. That is:

$$G^* = G \qquad [A9.34]$$

Therefore, the case of columnar dendritic growth is simpler than that of growth from an undercooled melt. Finally, using the stability expression gives:

$$R = \left[\frac{\Gamma}{\sigma^*(mG_c\xi_c - G)}\right]^{1/2} \qquad [A9.35]$$

By substituting equations A9.18, A9.19, and A9.35 into equation A9.32, one finally obtains the result:

$$V^2A' + VB' + G = 0 \qquad [A9.36]$$

where $A' = \pi^2\Gamma/(P_c^2D^2)$ and $B' = \theta_c\xi_c/D$. This corresponds to equation 7.13.

Since, at high growth rates, the temperature gradient has little effect upon the V-R relationship of equation A9.36, it can be neglected. This further simplifies the relationship, to give:

$$R^2V = D\Gamma/\sigma^*\theta_c\xi_c \qquad [A9.37]$$

Note that, at small P_c values, ξ_c tends towards unity, $I(P_c)$ tends towards zero, and $\theta_c = \Delta T_0k/[1 - pI(P_c)]$ tends towards ΔT_0k. Therefore, setting $\sigma^* = 1/4\pi^2$,

$$R^2V = 4\pi^2D\Gamma/\Delta T_0k, \qquad (P_c \ll 1) \qquad [A9.38]$$

which is exactly the same as equation 4.16.

The undercooling can be obtained from equation 7.15 by using the rate-dependent distribution coefficient of equation 7.3 and the temperature-dependent liquid-state diffusion coefficient. Figures 7.6 to 7.8 show some results derived by using equation A9.36. For more details see reference [12].

Primary Spacing in Constrained Growth

It has been assumed [9] that the shape of a fully developed dendrite, including the mean volume of its branches, can be approximated by an ellipsoid of revolution (Fig. 4.14). The radius of an ellipse is given by its semi-axes, a and b:

$$R = \frac{b^2}{a} \qquad [A9.39]$$

In the case of an hexagonal array, $b = \lambda_1/\sqrt{3}$ and $a = \Delta T'/G$, where $\Delta T'$ is the non-equilibrium solidification range. Therefore:

$$\lambda_1 = \left(\frac{3\Delta T'R}{G}\right)^{1/2} \qquad \text{[A9.40]}$$

Substituting equation A9.38 into equation A9.40 and replacing, to a first approximation, $\Delta T'$ by ΔT_0 gives a relationship for the primary trunk spacing:

$$\lambda_1 = 4.3\left(\frac{D\Gamma\Delta T_0}{k}\right)^{1/4} V^{-1/4} G^{-1/2} \qquad \text{[A9.41]}$$

This relationship is very similar to those derived by Hunt [13] and by Trivedi [14]. The latter extended Hunt's treatment by using the marginally stable tip radius for low Péclet numbers. Hunt and Trivedi consider the flux balance for a small interdendritic volume element (Fig. 6.6), where the cell/dendrite axis is the z-axis, and r is the radial distance. Summing the radial solute concentration change,

$$f_l(\partial C_l/\partial t) = (\partial f_s/\partial t)C_l(1-k)$$

and longitudinal solute concentration change,

$$(\partial C_l/\partial t) = \partial[D(\partial C_l/\partial x)]/\partial x$$

gives:

$$f_l(\partial C_l/\partial t) = (\partial f_s/\partial t)C_l(1-k) + \partial[Df_l(\partial C_l/\partial x)]/\partial x \qquad \text{[A9.42]}$$

Setting $G = m(\partial C_l/\partial x)$, and assuming the existence of a constant interdendritic concentration gradient ($\partial^2 C_l/\partial x^2 \rightarrow 0$) leads, since $\partial f_l/\partial x = -(1/V)(\partial f_l/\partial t)$ under steady-state conditions, to:

$$f_l \partial C_l = [C_l(k-1) - (DG/mV)]\partial f_l \qquad \text{[A9.43]}$$

Integrating from $f_l = 1$ to f_l and from $C_l = C_l^*$ to C_l gives:

$$f_l(k-1) = [C_l(k-1) - (DG/mV)]/[C_l^*(k-1) - (DG/mV)] \qquad \text{[A9.44]}$$

For a cylindrical geometry, one has $f_l = [1 - r^2/(\lambda_1/2)^2]$. Substituting this expression for f_l and fitting a spherical tip to the 'Scheil-cell' finally leads to [13]:

$$R = \frac{-G\lambda_2^2}{5.66[mC_l^*(1-k) + (DG/mV)]} \qquad \text{[A9.45]}$$

Following Trivedi [14], and using the results derived for the dendrite itself, one can substitute for the tip radius and concentration (equations A9.37 and A9.18a, respectively). Using the low Péclet number approximations: $A(P_c) \rightarrow 1$ and $\xi_c \rightarrow 1$, gives:

$$\lambda_1 = 6(\Delta T_0 kD\Gamma)^{1/4} V^{-1/4} G^{-1/2}[1-(DG/V\Delta T_0 k]^{1/2} \quad [A9.46]$$

A comparison of equations A9.46 and A9.41 reveals a similar behaviour at the small G/V ratios which are typical of dendrite growth. If $DG/V\Delta T_0 k$ tends towards unity, cells will be the preferred morphology and the spacing increases with increasing V.

In summary, it should be pointed out that, with regard to λ_1, none of these models represents well the real behaviour, as described by Esaka and others [15]. They are useful only for making qualitative estimates of λ_1, rather than for making precise predictions.

Secondary Spacing in Constrained or Unconstrained Growth

In order to simplify the model, it is assumed that only two dendrite arms of different diameters need to be considered. In reality, a distribution of arms of various thickness will exist.

Following Kattamis and Flemings [16] and Feurer and Wunderlin [17], the situation illustrated in figure A9.3 will be analysed in a very approximate manner. Two arms of radius, R and r, are placed in a locally isothermal melt. Since, at the interface between the solid and the liquid, local equilibrium will be established very rapidly, the concentration along the surface of the cylindrical arms will differ; the thinner arms will be in liquid of lower solute concentration. That is:

$$T' = T_f + mC_l^R - \frac{\Gamma}{R}$$

$$T' = T_f + mC_l^r - \frac{\Gamma}{r}$$

For an isothermal system, one therefore obtains:

$$m(C_l^R - C_l^r) = \Gamma\left(\frac{1}{R} - \frac{1}{r}\right) \quad [A9.47]$$

Solute will diffuse along the concentration gradient from thick to thin arms while the solvent will diffuse from the thin to the thick arms. The thin arms therefore tend to dissolve while the thicker arms tend to thicken.

For simplicity, it is assumed that the concentration gradient between two arms is constant and that the diffusion is unidirectional. Now, if R is assumed to be much greater than r, dR/dt can be neglected with respect to dr/dt, and the two fluxes existing between the arms are:

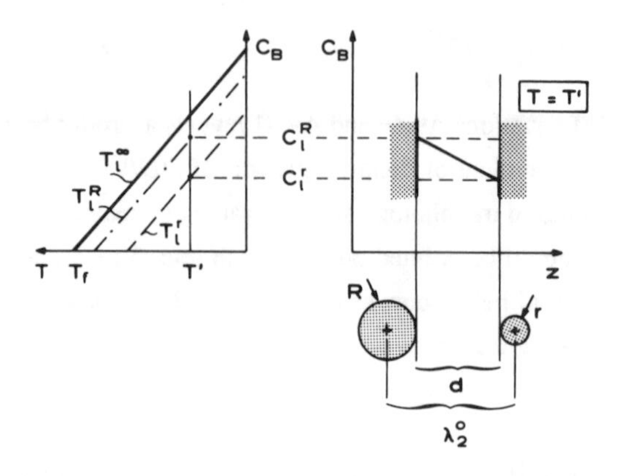

Figure A9.3

$$J = D\frac{(C_l^R - C_l^r)}{d} \qquad [A9.48]$$

$$J = -C_l^r(1-k)\frac{dr}{dt} \qquad [A9.49]$$

Combining equations A9.47 to A9.49:

$$\frac{dr}{dt} = \frac{\Gamma D}{mC_l^r(1-k)d}\left(\frac{1}{r} - \frac{1}{R}\right) \qquad [A9.50]$$

For small compositional differences in the liquid between the arms, C_l^r is approximately equal to C_l; the interdendritic concentration due to segregation. It is further assumed that d is approximately equal to the arm spacing, leading to:

$$\frac{dr}{dt} = \frac{\Gamma D}{mC_l(1-k)\lambda_2^o}\left(\frac{1}{r} - \frac{1}{R}\right) \qquad [A9.51]$$

where $\overset{o}{\lambda_2}$ is the arm spacing before ripening. All of the parameters, except C_l on the right-hand side of equation A9.51, are assumed to be constant. Furthermore, it is assumed that $R/\overset{o}{\lambda_2}$ and r_0/R are constant and that the interdendritic liquid concentration is a linear function of time; starting with the alloy concentration, C_0, and ending (due to segregation) at the composition, C_l^m:

$$C_l = C_0 + (C_l^m - C_0)\frac{t}{t_f} \qquad [A9.52]$$

where t is the time elapsed since the start of solidification, and t_f is the local solidification time. If $C_l^m = C_e$, t_f is approximately equal to $(T_l - T_e)/\dot{T}$.

Rearranging equation A9.51 and integrating from $t = 0$ to $t = t_f$ and from $r = r_0$ to $r = 0$ gives:

$$\lambda_2^{o}R^2 \left[\frac{r_0}{R} + \ln(1 - \frac{r_0}{R})\right] = Mt_f \quad [A9.53]$$

where:

$$M = \frac{-\Gamma D \ln(C_l^m / C_0)}{m(1-k)(C_l^m - C_0)} \quad [A9.54]$$

Feurer and Wunderlin [17] then assume that $R/\lambda_2 \cong 0.5$ and that $r_0/R \cong 0.5$, which gives $Mt_f = 0.1(\lambda_2^{o})^3$. Furthermore, they suppose that, when the arms have melted, $\lambda_2 = 2\lambda_2^{o}$ and therefore:

$$\lambda_2 = 5.5(Mt_f)^{1/3} \quad [A9.55]$$

Due to the extreme simplification of the ripening phenomena described, the constant factor of 5.5 in equation A9.55 should not be accorded too much significance. This applies to both its value and its constancy. Nevertheless, the coarsening parameter, M, has been shown to be of some use in estimating λ_2 values in Al alloys [17]. However, it should be kept in mind that the coarsening described here relates to the secondary branch spacing next to the primary trunk. This is so because, when one measures λ_2 at points far from the trunk, one misses the finer dissolving branches.

Because the real situation is here again very complicated, numerical studies of the behaviour of distributions of arms or particles as a function of time can be very useful in obtaining a better understanding of coarsening in general [18]. As a result, improved modelling of microstructural changes becomes possible [19].

References

[1] J.S.Langer, H.Müller-Krumbhaar, Journal of Crystal Growth **42** (1977) 11.

[2] M.E.Glicksman, R.J.Schaefer, J.D.Ayers, Metallurgical Transactions **7A** (1976) 1747.

[3] W.Oldfield, Materials Science and Engineering **11** (1973) 211.

[4] J.S.Langer, Reviews of Modern Physics **52** (1980) 1.

[5] J.S.Langer, H.Müller-Krumbhaar, Acta Metallurgica **26** (1978) 1681, 1689 and 1697.

[6] J.S.Langer, Physico-Chemical Hydrodynamics **1** (1980) 41.

[7] A.Karma, J.S.Langer, Physical Review **A30** (1984) 3147.

[8] J.Lipton, M.E.Glicksman, W.Kurz, Metallurgical Transactions **18A** (1987) 341.

[9] W.Kurz, D.J.Fisher, Acta Metallurgica **29** (1981) 11.

[10] J.Lipton, W.Kurz, R.Trivedi, Acta Metallurgica **35** (1987) 957.

[11] R.Trivedi, J.Lipton, W.Kurz, Acta Metallurgica **35** (1987) 965.

[12] W.Kurz, B.Giovanola, R.Trivedi, Acta Metallurgica **34** (1986) 823, and Journal of Crystal Growth **91** (1988) 123.

[13] J.D.Hunt, *Solidification and Casting of Metals*, The Metals Society, Book 192, London, 1979, p. 3.

[14] R.Trivedi, Metallurgical Transactions **15A** (1984) 977.

[15] H.Esaka, W.Kurz, R.Trivedi, in *Solidification Processing 1987*, The Institute of Metals, London, 1988, p. 198.

[16] T.Z.Kattamis, M.C.Flemings, Transactions of the Metallurgical Society of AIME **233** (1965) 992.

[17] U.Feurer, R.Wunderlin, *Fachbericht DGM*, 1977 (see detailed reference in chapter 4).

[18] P.W.Voorhees, M.E.Glicksman, Metallurgical Transactions **15A** (1984) 1081.

[19] W.Kurz, M.Rappaz, in *Solidification des Alliages*, Ecole d'été, Carry-le-Rouet, 1985, Les Éditions de Physique, France, 1988, p. 191.

APPENDIX 10

EUTECTIC GROWTH

Low Péclet Number Solution

This appendix follows, in its first part, the treatment of Jackson and Hunt [1], but is a simplified version of the latter paper. Figure 5.2 shows the corresponding phase diagram and interface geometry. Using symmetry arguments (appendix 2), the analysis of a solidifying lamellar eutectic interface can be reduced firstly to the consideration of a pair of lamellae. No net mass transport can occur between this pair and another pair under steady-state conditions because this would cause changes in the morphology and violate the assumption of steady-state behaviour. Thus, the concentration gradient in the y-direction is zero at the mid-point of each lamella (Fig. A10.1). Secondly, attention can be further restricted to half of each lamella because, again, there can be no net transport of solute in or out of this symmetry element.

The differential equation describing the solute distribution in the melt ahead of a steadily advancing interface is:

$$\frac{\partial^2 C}{\partial y^2} + \frac{\partial^2 C}{\partial z^2} + \frac{V}{D} \cdot \frac{\partial C}{\partial z} = 0 \qquad \text{[A2.6]}$$

and it has been shown (appendix 2) that the general solution satisfying this equation consists of products of circular and exponential functions. It is logical to associate the exponential function with the z-variation of the solution, since it is known that the boundary layer has this form for a single-phase interface. It is equally logical to associate the circular functions with the alternating pattern of eutectic phases. Thus, merely by inspection, the distribution of component B can be deduced to be of the form:

$$C = C_e + A \exp\left(-\frac{Vz}{D}\right) + B\exp(-\mathrm{bz})\cos(gy) \qquad \text{[A10.1]}$$

This approximate solution# can be seen to be made up of an exponential term which reflects the planarity of the interface as a whole, and another term which reflects the

The exact solute profile must be described by a sum of circular functions rather than by just one term. However, this choice simplifies the treatment without losing sight of the important physical phenomena.

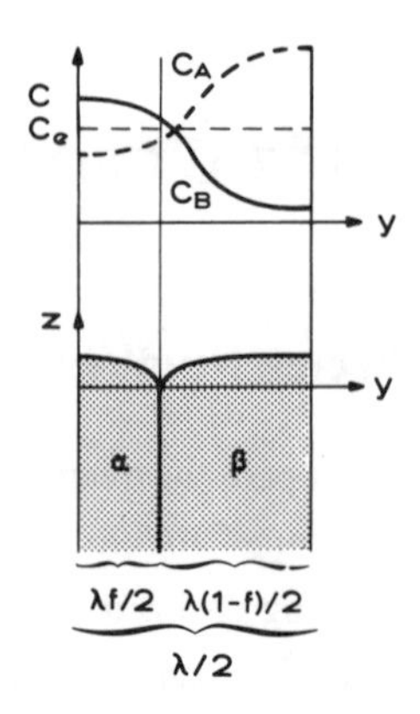

Figure A10.1

alternating pattern of the two phases. The cosine, rather than sine, function is chosen because the expression describing the gradient of the interface concentration in the y-direction must be able to take zero values at the origin of the region considered and at $\lambda/2$ (Fig. A10.1). Another way of regarding the choice of the function in equation A10.1 is to imagine that it reflects the interface solute distributions depicted in figure 5.3. Thus, equation A10.1 without the third term on the right-hand side, and with a suitable choice of value for A, could describe the solute distribution ahead of either of the single phases (Fig. 5.3a). The re-introduction of the third term can then be looked upon as being the expected solution when the originally single phase interface is 'perturbed' by the addition of the second phase (Fig. 5.3b).

It is now necessary to determine the constants, A, B, b, and g. The value of g can be immediately determined because, again due to symmetry considerations, the derivative of the solution with respect to y, $\partial C/\partial y$, must be equal to zero when $y = \lambda/2$. Thus:

$$C = C_e + A\exp\left(-\frac{Vz}{D}\right) + B\exp(-bz)\cos\left(\frac{2\pi y}{\lambda}\right) \qquad [A10.2]$$

The derivation of b has already been discussed (equation A2.10). It is equal to:

$$b = \frac{V}{2D} + \left[\left(\frac{V}{2D}\right)^2 + \left(\frac{2\pi}{\lambda}\right)^2\right]^{1/2} \qquad [A10.3]$$

This relatively complicated expression can be simplified on the basis of experimental observation. Firstly, it is noted that the term, $V/2D$, is the inverse of the equivalent solute boundary layer of a planar interface. At low growth rates its value will be relatively small, whereas $2\pi/\lambda$ is the wave number of the eutectic and will be large, due to the small values of λ which are usually encountered. Therefore:

$$\frac{4\pi^2}{\lambda^2} \gg \frac{V^2}{4D^2}$$

and equation A10.3 can be simplified to:

$$b \cong \frac{2\pi}{\lambda} \qquad (P \ll 1)$$

by neglecting terms in $V/2D$ ($P = V\lambda/2D$). Thus, equation A10.2 can now be written:

$$C = C_e + A\exp\left(-\frac{Vz}{D}\right) + B\exp\left(-\frac{2\pi z}{\lambda}\right)\cos\left(\frac{2\pi y}{\lambda}\right) \qquad [A10.4]$$

The values of the constants, A, B, can be found by forcing the above solution to satisfy the flux condition at the interface. The fluxes can be deduced from the phase diagram. The general flux boundary condition is:

$$V(k-1)C^* = D\left(\frac{\partial C}{\partial z}\right)_{z=0} \qquad [A10.5]$$

but, because the variation of the interface concentration with respect to C_e is relatively small in most cases except rapid solidification, it can be supposed that $C^* \cong C_e$.

Thus, the flux conditions for the two eutectic phases can be written in terms of the weight fraction of element B as:

$$\alpha\text{-phase}: \; V(k_\alpha - 1)C_e = D\left(\frac{\partial C}{\partial z}\right)_{z=0} \qquad [A10.6]$$

$$\beta\text{-phase}: \; -V(k_\beta - 1)(1 - C_e) = D\left(\frac{\partial C}{\partial z}\right)_{z=0} \qquad [A10.7]$$

The negative sign in the case of the β-phase appears because the interface gradient of C_B in the growth direction is expected to be positive. The concentration gradient at the solid/liquid interface can be found from equation A10.4:

$$\left(\frac{\partial C}{\partial z}\right)_{z=0} = -\frac{V}{D}A - \frac{2\pi}{\lambda}B\cos\left(\frac{2\pi y}{\lambda}\right) \qquad [A10.8]$$

The α-phase extends from the origin to the point, $f\lambda/2$, where f is the volume fraction of the α-phase (as has been defined in reference [1]).

Using the method of 'weighted residuals' (appendix 2), the values of A and B (the 'weights') can be found by satisfying the flux boundary conditions (equations A10.6 and A10.7) in an average fashion over the α and β interfaces, i.e.:

$$\int_0^{\frac{f\lambda}{2}} V(k_\alpha - 1)C_e dy = \int_0^{\frac{f\lambda}{2}} D\left(\frac{\partial C}{\partial z}\right)_{z=0} dy \qquad [A10.9]$$

$$\int_{\frac{f\lambda}{2}}^{\frac{\lambda}{2}} V(k_\beta - 1)(1 - C_e)dy = \int_{\frac{f\lambda}{2}}^{\frac{\lambda}{2}} D\left(\frac{\partial C}{\partial z}\right)_{z=0} dy \qquad [A10.10]$$

Inserting the value of $(\partial C/\partial z)_{z=0}$ from equation A10.8, and carrying out the integrations yields the simultaneous algebraic equations:

$$fV\lambda A + 2D\sin(\pi f)B = (1 - k_\alpha)C_e fV\lambda \qquad [A10.11]$$

$$(1 - f)V\lambda A - 2\sin(\pi f)B = (k_\beta - 1)(1 - C_e)(1 - f)V\lambda \qquad [A10.12]$$

and the values of A and B are easily found to be:

$$A = fC' - C^\beta \qquad [A10.13]$$

$$B = \frac{f(1-f)V\lambda C'}{2D\sin(\pi f)} \qquad [A10.14]$$

Note that A will be very small in general, and will be equal to zero if the phases have the same density. Here, C' is the difference in composition between the ends of the eutectic tie-line and C^β is the difference in composition between the eutectic and the maximum solid solubility in the β-phase (Fig. A10.2).

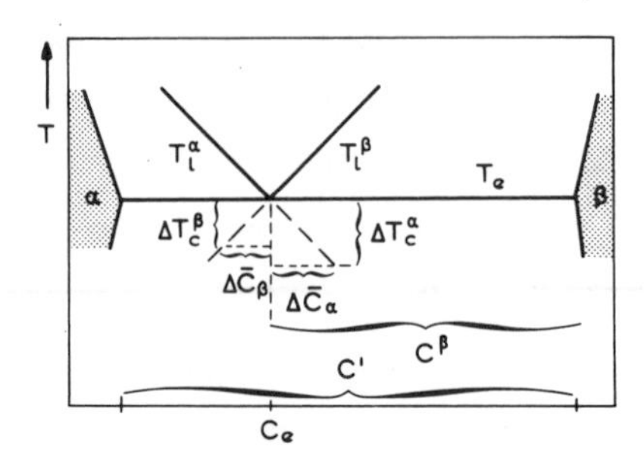

Figure A10.2

Thus, a solution has been obtained which satisfies most of the boundary conditions. The final condition to be satified is the coupling condition, which relates the local melting point to the imposed temperature distribution. This has already been depicted in a qualitative fashion in figure 5.5. If account were to be taken of the detailed shape of the eutectic interface, the problem would rapidly become intractable because of the need to satisfy the coupling condition at each point#. Instead, the eutectic interface is assumed to be perfectly planar, in order to make a solution possible, and the quantities involved are treated in an average fashion. The satisfaction of the coupling condition can then be achieved in three steps. These involve calculating the average solute undercooling and the average curvature undercooling for each phase, and then equating the two total undercoolings.

The average concentration differences of the liquid, $\overline{\Delta C} = \bar{C} - C_e$, at the α- and β-phase interfaces, relative to the eutectic composition will first be found. From equation A10.4, they can be shown to be, for $z = 0$:

$$\overline{\Delta C}_\alpha = \frac{2}{f\lambda}\int_0^{\frac{f\lambda}{2}}[A + B\cos(\frac{2\pi y}{\lambda})]dy$$

$$\overline{\Delta C}_\beta = \frac{2}{(1-f)\lambda}\int_{\frac{f\lambda}{2}}^{\frac{\lambda}{2}}[A + B\cos(\frac{2\pi y}{\lambda})]dy$$

giving:

$$\overline{\Delta C}_\alpha = A + B\frac{\sin(\pi f)}{\pi f} \qquad [A10.15]$$

$$\overline{\Delta C}_\beta = A - B\frac{\sin(\pi f)}{\pi(1-f)} \qquad [A10.16]$$

Since A is usually negligible and B is positive, it can be seen that the average concentration difference in component B at the α-phase interface is positive, while that at the β-phase is negative. The mean solute undercoolings, with respect to the eutectic temperature (Fig. 5.5), can be found by multiplying the above concentration differences by the absolute liquidus slopes, $|m_\alpha|$ and $|m_\beta|$. The solute undercoolings are

Series et al. [2] have solved this mathematically difficult problem by using a hybrid method in which an electrical analogue was used to provide data for further calculations.

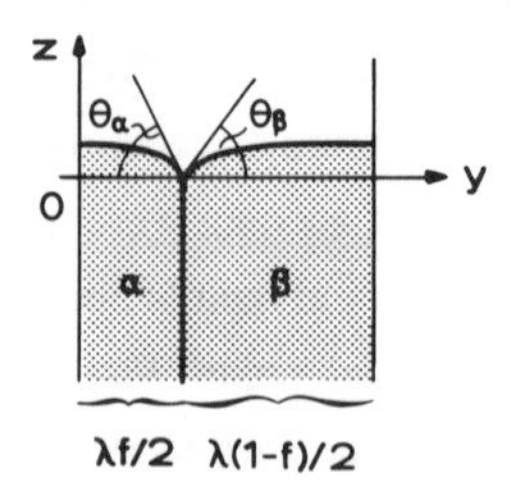

Figure A10.3

therefore:

$$\Delta T_c^{\alpha} = |m_{\alpha}| \left[A + B\frac{\sin(\pi f)}{\pi f}\right] \qquad \text{[A10.17]}$$

$$\Delta T_c^{\beta} = -|m_{\beta}| \left[A - B\frac{\sin(\pi f)}{\pi(1-f)}\right] \qquad \text{[A10.18]}$$

since both of them are positive.

The second source of undercooling to be considered is the average undercooling due to the curvature of the interface. Firstly, note that the average curvature of a line between two points which are a distance, L, apart can be found by integrating the general expression for the curvature (appendix 3),

$$K = \frac{z''}{(1 + z'^2)^{3/2}} \qquad \text{[A10.19]}$$

between the two points. Thus:

$$\overline{K} = \frac{1}{L}\int_0^L \frac{z''}{(1 + z'^2)^{3/2}}\, dy$$

Making the successive substitutions, $Z = z'$ and $\tan(\theta) = Z$ (Fig. A10.3) leads to:

$$\overline{K} = \frac{1}{L}\, [\sin(\theta)]_0^L$$

Again using the above identities leads to:

$$\overline{K} = \frac{1}{L}\, \sin[\arctan(z')]_0^L \qquad \text{[A10.20]}$$

The form of the eutectic interface is assumed to be as shown in figure A10.3. Thus, the average curvature of the β-phase can be found by substituting the slopes at $y = f\lambda/2$

and $\lambda/2$ into equation A10.20:

$$\overline{K}_{\alpha} = \frac{2\sin(\theta_{\alpha})}{f\lambda}$$

$$\overline{K}_{\beta} = \frac{2\sin(\theta_{\beta})}{(1-f)\lambda}$$

The curvature has been defined to be positive if the solid projects into the melt. Mathematically speaking, it is negative. The curvature is defined differently here so as to give a positive undercooling (for a solid projection) when combined with the (positive) Gibbs-Thomson coefficient. Therefore, using the associated Gibbs-Thomson coefficients, the curvature undercoolings become:

$$\Delta T_r^{\alpha} = \frac{2\Gamma_{\alpha}\sin(\theta_{\alpha})}{f\lambda} \qquad [A10.21]$$

$$\Delta T_r^{\beta} = \frac{2\Gamma_{\beta}\sin(\theta_{\beta})}{(1-f)\lambda} \qquad [A10.22]$$

Combining equations A10.17 or A10.18 with equations A10.21 or A10.22 gives the total undercooling of the two phases:

$$\Delta T_{\alpha} = |m_{\alpha}|\left[A + B\frac{\sin(\pi f)}{\pi f}\right] + \frac{2\Gamma_{\alpha}\sin(\theta_{\alpha})}{f\lambda} \qquad [A10.23]$$

$$\Delta T_{\beta} = -|m_{\beta}|\left[A - B\frac{\sin(\pi f)}{\pi(1-f)}\right] + \frac{2\Gamma_{\alpha}\sin(\theta_{\beta})}{(1-f)\lambda} \qquad [A10.24]$$

Multiplying equation A10.23 by $|m_{\beta}|$ and equation A10.24 by $|m_{\alpha}|$, and adding gives:

$$|m_{\beta}|\,\Delta T_{\alpha} + |m_{\alpha}|\,\Delta T_{\beta} = |m_{\alpha}|\,|m_{\beta}|\,B\frac{\sin(\pi f)}{\pi f(1-f)} + \\ + \frac{2|m_{\beta}|\Gamma_{\alpha}\sin(\theta_{\alpha})}{f\lambda} + \\ + \frac{2|m_{\alpha}|\Gamma_{\beta}\sin(\theta_{\beta})}{(1-f)\lambda} \qquad [A10.24a]$$

However, the coupling condition (appendix 2) requires that the α- and β-interfaces should lie on the same isotherm. Due to the very small distances between the phases (typically of the order of a few microns) and their high thermal conductivities, no substantial temperature difference can exist between the α and β phases. Therefore, $\Delta T_\alpha = \Delta T_\beta = \Delta T$ (Fig. 5.5):

$$\Delta T = \frac{|m_\alpha||m_\beta|}{|m_\alpha| + |m_\beta|} B \frac{\sin(\pi f)}{\pi f(1-f)} + \frac{2|m_\beta|\Gamma_\alpha \sin(\theta_\alpha)}{f\lambda(|m_\alpha| + |m_\beta|)} + \frac{2|m_\alpha|\Gamma_\beta \sin(\theta_\beta)}{(1-f)\lambda(|m_\alpha| + |m_\beta|)} \qquad [A10.25]$$

Substituting for B from equation A10.14 gives:

$$\Delta T = \frac{|m_\alpha||m_\beta|}{|m_\alpha| + |m_\beta|} V\lambda \frac{C'}{2\pi D} + \frac{2(1-f)|m_\beta|\Gamma_\alpha \sin(\theta_\alpha) + 2f|m_\alpha|\Gamma_\beta \sin(\theta_\beta)}{f(1-f)\lambda(|m_\alpha| + |m_\beta|)} \qquad [A10.26]$$

This can be written (see also equation 5.8):

$$\Delta T = K_c V\lambda + \frac{K_r}{\lambda} \qquad [A10.27]$$

where the physical constants of the alloy are:

$$K_c = \frac{|m_\alpha||m_\beta|}{|m_\alpha| + |m_\beta|} \cdot \frac{C'}{2\pi D} \qquad [A10.28]$$

and

$$K_r = \frac{2(1-f)|m_\beta|\Gamma_\alpha \sin(\theta_\alpha) + 2f|m_\alpha|\Gamma_\beta \sin(\theta_\beta)}{f(1-f)(|m_\alpha| + |m_\beta|)} \qquad [A10.29]$$

Assuming that the λ-value chosen by the eutectic is the one which makes ΔT a minimum, i.e.:

$$\frac{d(\Delta T)}{d\lambda} = K_c V - \frac{K_r}{\lambda^2} = 0$$

it is found that:

$$\lambda^2 V = \frac{K_r}{K_c} = \frac{4\pi D}{f(1-f)} \frac{f|m_\alpha|\Gamma_\beta \sin(\theta_\beta) + (1-f)|m_\beta|\Gamma_\alpha \sin(\theta_\alpha)}{|m_\alpha||m_\beta|C'} \qquad [A10.30]$$

Finally, it is interesting to compare the approximate solution above with the more exact derivation performed by Jackson and Hunt [1]. Writing their expression for $\lambda^2 V$ in terms of the nomenclature used in this book gives:

$$\lambda^2 V = \frac{2D}{P'} \frac{f|m_\alpha|\Gamma_\beta \sin(\theta_\beta) + (1-f)|m_\beta|\Gamma_\alpha \sin(\theta_\alpha)}{|m_\alpha||m_\beta|C'} \qquad [A10.31]$$

where $P' = \Sigma(1/n^3\pi^3)\sin^2(n\pi f)$.

Equations A10.30 and A10.31 differ only with respect to the first multiplying term on the right-hand side. The relationship is seen even more clearly if one writes their equivalent expressions for K_c and K_r for the J.H. model:

$$K_c = \frac{\bar{m}C'P'}{f(1-f)D}$$

$$K_r = 2\bar{m}\left[\frac{\Gamma_\alpha \sin(\theta_\alpha)}{f|m_\alpha|} + \frac{\Gamma_\beta \sin(\theta_\beta)}{(1-f)|m_\beta|}\right]$$

with $\bar{m} = |m_\alpha||m_\beta|/(|m_\alpha| + |m_\beta|)$.

If the complicated expression, P' (equation A10.31), is compared with the equivalent expression, $f(1-f)/2\pi$, the agreement is found to be quite reasonable (table A10.1) in spite of the comparative simplicity of the above derivation. As one would expect from the symmetrical form of the approximate solution used, the accuracy is greatest for f-values of about 0.5.

Table A10.1 Comparison of Terms Appearing in the Jackson-Hunt (JH) and Present Analyses#

f	$P' = \sum_{1}^{100}(1/n^3\pi^3)\sin^2(n\pi f)$ (JH)	$(f-f^2)/2\pi$	Error(%)
0.1	0.00626	0.01432	129
0.2	0.01633	0.02546	56
0.3	0.02553	0.03342	31
0.4	0.03174	0.03820	20
0.5	0.03392	0.03820	17

(the results for higher f-values are symmetrical about $f = 0.5$)

Another useful approximation is $P' = 0.335[f(1-f)]^{1.65}$. A similar expression can also be derived for rod eutectics [5].

Table A10.2 Values of the P' Function (J.-H. Model)

f	P'	f	P'	f	P'
0.00	0.00000	0.17	0.01327	0.34	0.02845
0.01	0.00014	0.18	0.01429	0.35	0.02909
0.02	0.00046	0.19	0.01532	0.36	0.02970
0.03	0.00091	0.20	0.01633	0.37	0.03027
0.04	0.00147	0.21	0.01734	0.38	0.03080
0.05	0.00212	0.22	0.01833	0.39	0.03129
0.06	0.00284	0.23	0.01930	0.40	0.03174
0.07	0.00362	0.24	0.02026	0.41	0.03215
0.08	0.00446	0.25	0.02120	0.42	0.03252
0.09	0.00534	0.26	0.02212	0.43	0.03285
0.10	0.00626	0.27	0.02301	0.44	0.03313
0.11	0.00721	0.28	0.02388	0.45	0.03337
0.12	0.00819	0.29	0.02472	0.46	0.03357
0.13	0.00918	0.30	0.02553	0.47	0.03372
0.14	0.01019	0.31	0.02631	0.48	0.03383
0.15	0.01121	0.32	0.02705	0.49	0.03390
0.16	0.01224	0.33	0.02777	0.50	0.03392

For the purposes of making more exact calculations using equation A10.31, an expanded table [3] of P' values is given in table A10.2.

Eutectic Growth at High Rates

The main difference between the analysis of eutectic growth at high and low rates is that, in the former case, one cannot make some of the simplifying assumptions which were made earlier in this appendix.

The first difference is that one cannot assume that $4\pi^2/\lambda^2 >> V^2/4D^2$, because V is now also very large. The second difference is that one can no longer assume that the interface composition in the liquid is close to the original eutectic one. Therefore, the interface gradients are now functions of the interface concentrations, and the exact solution of the problem becomes much more difficult. In fact, a solution has been developed [4] by making the assumption that $k_\alpha = k_\beta$.

In the approximate derivation which follows, it is not necessary to make this simplification, but it will be retained anyway in order to facilitate comparison with the more exact result [4].

Starting with equation A10.2, one again finds an expression for the concentration gradient in the liquid at the interface:

$$\partial C/\partial z = -(V/D)A - bB\cos(2\pi y/\lambda) \quad [A10.32]$$

Note that one cannot here assume that $b = 2\pi/\lambda$. One must later insert this expression for the gradient into equation A10.5 giving, since one has to assume that the liquid composition at the interface, C_l^*, is not equal to C_e:

$$V(k-1)C_l^* = D(\partial C/\partial z)_{z=0} \quad [A10.33]$$

$$-V(k-1)(1-C_l^*) = D(\partial C/\partial z)_{z=0} \quad [A10.34]$$

Recall that it is assumed that $k_\alpha = k_\beta = k$.

The composition at the interface is:

$$C_l^* = C_e + A + B\cos(2\pi y/\lambda) \quad [A10.35]$$

It is now necessary to set up equations which are equivalent to equations A10.9 and A10.10. This gives, upon substituting from equations A10.32 and A10.35:

$$V(1-k)\int_0^{\frac{f\lambda}{2}}(C_e + A + B\cos\frac{2\pi y}{\lambda})dy =$$

$$= D\int_0^{\frac{f\lambda}{2}}(\frac{V}{D}A + bB\cos\frac{2\pi y}{\lambda})dy \quad [A10.36]$$

$$V(1-k)\int_{\frac{f\lambda}{2}}^{\frac{\lambda}{2}}(1 - C_e - A - B\cos\frac{2\pi y}{\lambda})dy =$$

$$= D\int_{\frac{f\lambda}{2}}^{\frac{\lambda}{2}}(-\frac{V}{D}A - bB\cos\frac{2\pi y}{\lambda})dy \quad [A10.37]$$

This leads to two simultaneous algebraic equations which are equivalent to equations A10.11 and A10.12:

$$P(1-k)[(C_e+A)f+\frac{B}{\pi}\sin\pi f] = PAf + bB\frac{\lambda}{2\pi}\sin\pi f \qquad [A10.38]$$

$$P(1-k)[(C_e+A-1)(1-f)-\frac{B}{\pi}\sin\pi f] = PA(1-f)-bB\frac{\lambda}{2\pi}\sin\pi f \qquad [A10.39]$$

where P is the Péclet number (= $V\lambda/2D$). Adding the two equations and simplifying gives:

$$A = [(1-k)/k][fC_e - (1-f)(1-C_e)] \qquad [A10.40]$$

This is the equivalent equation to equation A10.13, and A is identical to B_0 in equation 15a of reference 4.

Multiplying equation A10.38 by $(1-f)$, multiplying equation A10.39 by f, and subtracting one from the other leads, after substituting for b from equation A10.3, to:

$$B = P\frac{(1-k)(f-f^2)(2\pi/P)}{\sin(\pi f)((1+(2\pi/P)^2)^{1/2}-1+2k)} \qquad [A10.41]$$

The calculations can now be carried forward in exactly the same way as for the low-velocity case, but using the new values of A and B. For instance, the average concentration difference ahead of the α phase is now:

$$\overline{\Delta C}_\alpha = \frac{1-k}{k}[fC_e-(1-f)(1-C_e)] +$$

$$+\frac{\lambda V}{fD}\left[\frac{\frac{(f-f^2)}{2\pi}(1-k)\frac{2\pi}{P}}{\left(1+(\frac{2\pi}{P})^2\right)^{1/2}-1+2k}\right] \qquad [A10.42]$$

The expression in the large bracket on the RHS of equation A10.42 is equivalent to the exact function, $P(f,p,k)$, to be found in reference [4]. It has been written in the present form so as to reveal the similarity. Note that the factor, $(f-f^2)/2\pi$, is the simple solution in table A10.1.

The overall effect of high growth rates can be seen by replacing B, in equation A10.24a, by the expression in equation A10.41. This gives:

$$\Delta T = \frac{|m_\alpha||m_\beta|}{|m_\alpha| + |m_\beta|} \cdot \frac{\frac{V\lambda}{2\pi D}(1-k)\, 2\pi/P}{[1+(2\pi/P)^2]^{1/2} - 2k + 1} +$$

$$+ \frac{2(1-f)|m_\beta|\Gamma_\alpha \sin(\theta_\alpha) + 2f|m_\alpha|\Gamma_\beta \sin(\theta_\beta)}{f(1-f)\lambda(|m_\alpha| + |m_\beta|)} \qquad [A10.43]$$

That is:

$$\Delta T = K_c V\lambda + K_r/\lambda$$

where

$$K_c = \frac{|m_\alpha||m_\beta|}{|m_\alpha| + |m_\beta|} \cdot \frac{1-k}{2\pi D} \cdot \frac{(2\pi/P)}{[1+(2\pi/P)^2]^{1/2} - 1 + 2k} \qquad [A10.44]$$

Compare this with equation A10.28 and note that $(1 - k)$ corresponds to C'. Also, K_r is given by equation A10.29. Hence, after optimisation:

$$\lambda^2 V = \frac{4\pi D}{f(1-f)} \cdot \frac{[1 + (2\pi/P)^2]^{1/2} - 1 + 2k}{(2\pi/P)} \cdot$$

$$\cdot \frac{f|m_\alpha|\Gamma_\beta \sin(\theta_\beta) + (1-f)|m_\beta|\Gamma_\alpha \sin(\theta_\alpha)}{|m_\alpha||m_\beta|(1-k)} \qquad [A10.45]$$

It is of interest is to see how this high growth rate (high P) solution differs from the low Péclet number solution. This is quickly done by separating out the P-dependent terms. Thus:

$$\lambda^2 V = D\Theta \frac{(P^2 + 4\pi^2)^{1/2} - P + 2kP}{2\pi(1-k)} \qquad [A10.46]$$

where Θ is a constant which groups together all of the fixed parameters which appear in equation A10.45.

Note firstly that the RHS of equation A10.46 tends to

$$\lambda^2 V = D\Theta/(1-k)$$

when P tends to zero; which corresponds to equation A10.30 and indicates that $\lambda^2 V$ is independent of the growth rate. By differentiating equation A10.46 with respect to P and equating the result to zero, one finds that the $\lambda^2 V$ versus P curve has a minimum when:

$$P = \frac{\pi(1-2k)}{[k(1-k)]^{1/2}} \qquad \text{[A10.47]}$$

In fact, there is no minimum if k is greater than 0.5 (Fig. 7.12).

As P tends to infinity, equation A10.46 tends to the value:

$$\lambda^2 V = D\Theta \frac{kP}{\pi(1-k)}$$

Thus, $\lambda^2 V$ is no longer constant, as in the low-P Jackson-Hunt solution, but increases with P at a rate which depends upon k. In equation A10.47, note that the minimum moves rapidly towards infinity as k tends to zero. As a result, the curve can be expected to decrease monotonically towards zero.

As might be expected, the predictions of the simple expression are very inaccurate for small values of f (as in table A10.1). They also diverge badly from the exact values when k is large. This is also to be expected in view of the difficulties of convergence which are encountered when using numerical methods [4]. However, even when the absolute values predicted by the simple equation are inaccurate, the *relative* values over a range of k- or f-values are generally adequate. Thus, the simple expression can be used with confidence in exploring the qualitative behaviour of the solutions, and aids physical understanding.

References

[1] K.A.Jackson, J.D.Hunt, Transactions of the Metallurgical Society of the AIME **236** (1966) 1129.

[2] R.W.Series, J.D.Hunt, K.A.Jackson, Journal of Crystal Growth **40** (1977) 221.

[3] J.N.Clark, J.T.Edwards, R.Elliott, Metallurgical Transactions **6A** (1975) 232.

[4] R.Trivedi, P.Magnin, W.Kurz, Acta Metallurgica **35** (1987) 971.

[5] R.Trivedi, W.Kurz, in *Solidification Processing of Eutectic Alloys* (Eds. D.M.Stefanescu, G.J.Abbaschian, R.J.Bayuzick), The Metallurgical Society, 1988, p. 3.

APPENDIX 11

TRANSIENTS IN SOLUTE DIFFUSION

Initial Transient

As shown previously (chapter 3), the build-up of solute ahead of a planar solid/liquid interface takes some time - the time required for the saturation of the diffusion boundary layer. It is assumed, to a first approximation, that diffusion is rapid enough to ensure, during this transient, the existence of a quasi-steady-state solute distribution. In this case, the concentration decreases exponentially from the interface into the liquid, and the equivalent boundary layer is always of the same thickness (appendix 2):

$$\delta_c = \frac{2D}{V}$$

Assuming that the growth rate, V, increases instantaneously to a constant value at the start of solidification the flux at an interface of unit area, due to the differing solubilities of the solute in the two phases, will be:

$$J_1 = VC_l^*(1-k) \qquad \text{[A11.1]}$$

Initially, this flux is greater than the flux created by the concentration gradient in the liquid:

$$J_2 = -DG_c \qquad \text{[A11.2]}$$

The difference between the two fluxes will be used to 'fill' the diffusion boundary layer, of thickness δ_c, to a mean concentration, $\overline{C}_l$:

$$J_1 - J_2 = \frac{d\overline{C}_l}{dt}\delta_c \qquad \text{[A11.3]}$$

Recalling that a quasi steady-state profile is assumed to exist, the concentration gradient at the solid/liquid interface is given directly by the first derivative (at $z = 0$) of equation 3.2, with ΔC_0 in this transient state being replaced by $\Delta C = (C_l^* - C_0)$ (Fig. A11.1):

$$G_c = (C_0 - C_l^*)\frac{V}{D} \qquad [A11.4]$$

Therefore, for unit area:

$$J_2 = (C_l^* - C_0)V$$

and from equation A11.3

$$\frac{2D}{V} \cdot \frac{d\overline{C}_l}{dt} = V(C_0 - kC_l^*) \qquad [A11.5]$$

Noting that, for the triangular equivalent boundary layer (fig. A11.1b), the variation in the mean concentration is related to the variation in interface concentration, e.g.:

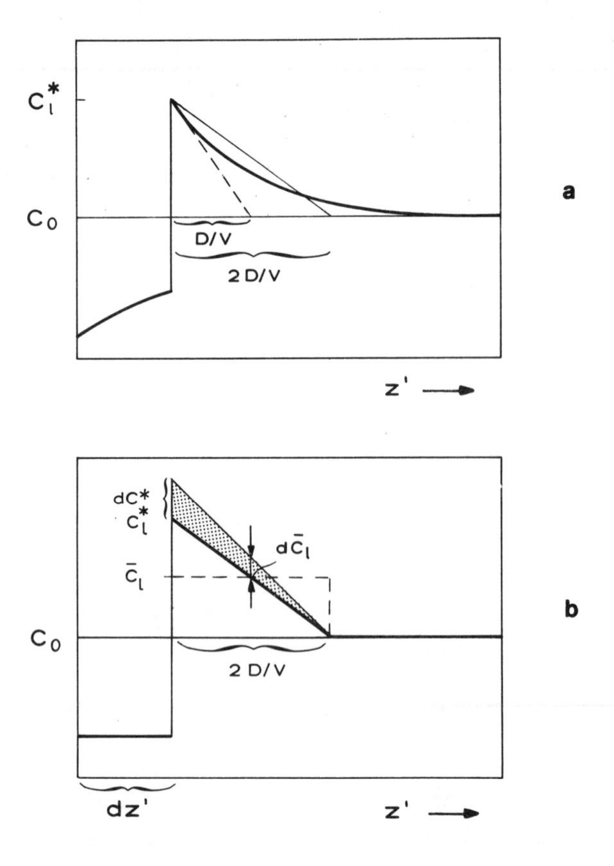

Figure A11.1

$$d\overline{C}_l = \frac{dC_l^*}{2}$$

with $V = dz'/dt$:

$$\frac{dC_l^*}{C_0 - kC_l^*} = \frac{V}{D}dz \qquad \text{[A11.6]}$$

Integrating:

$$\int_{C_0}^{C_l^*} \frac{dC_l^*}{C_0 - kC_l^*} = \frac{V}{D}\int_0^{z'} dz$$

gives:

$$\ln\left[\frac{C_0(1-k)}{C_0 - kC_l^*}\right]^{1/k} = \frac{Vz'}{D}$$

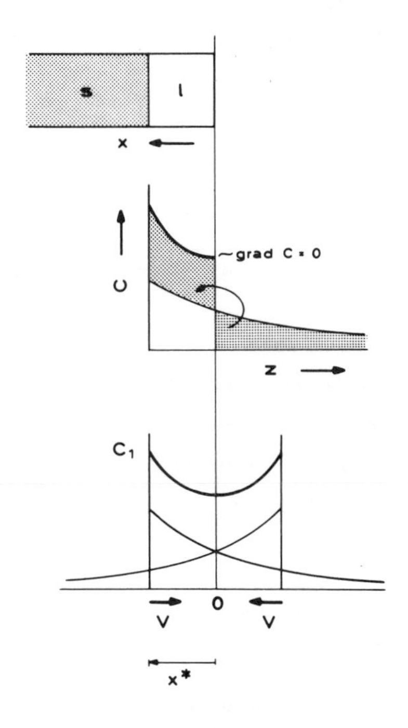

Figure A11.2

or

$$C_l^* = \frac{C_0}{k}[1-(1-k)\exp(-\frac{kz'V}{D})] \quad [A11.7]$$

Figure 6.2 illustrates the behaviour of this expression. It is a good approximation to the exact solution derived by Smith et al. [1].

Final Transient

While the initial transient is required to build up a boundary layer, the final transient is the result of the 'collision' of the boundary layer with the end of the specimen (Fig. A11.2). This problem has been solved by Smith et al. [1] in the following way for the case where the system has reached a steady state. The origin of the z-axis is placed at the end of the specimen and the diffusion equation becomes, in terms of the new x-coordinate (which does not move with the solid/liquid interface);

$$\frac{d^2C_l}{dx^2} = \frac{1}{D}\cdot\frac{dC_l}{dt} \quad [A11.8]$$

The boundary conditions apply for all values of t. At $x = 0$ (the end of the specimen) the diffusion flux must be equal to zero and:

$$\frac{dC_l}{dx} = 0 \quad [A11.9]$$

and at the solid/liquid interface, $x = x^*$:

$$\frac{dC_l}{dx} = \frac{V}{D}C_l^* p \quad [A11.10]$$

where x^* is the length of the liquid zone.

In order to meet the first boundary condition, an imaginary source placed in a symmetrical position with respect to the real solid/liquid interface is introduced. This is equivalent to a barrier to mass flux being placed at the end of the specimen (Fig. A11.2). The steady-state distributions at the two interfaces in an infinite specimen are:

$$C_l = C_0(1+\frac{p}{k})\exp[-\frac{V(x^* \pm x)}{D}] \quad [A11.11]$$

where $z = x^* \pm x$. In order to cope with the second boundary condition (equation A11.10) when the sources are close enough to interact, further sources (interfaces) are introduced at distances, nx^*, where n is an integer (Fig. A11.3). These sources travel

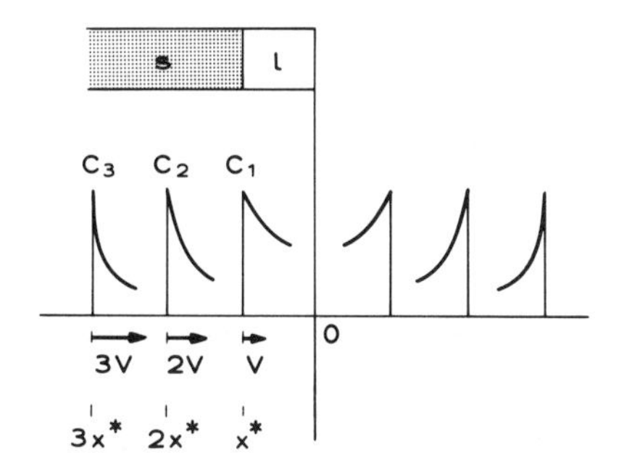

Figure A11.3

at speeds which are n times the speed of the real interface. This superposition permits the calculation of the necessary coefficients by constraining the interface to obey the flux balance (equation A11.10):

$$\frac{C_l}{C_0} = 1 + \sum_{n=1}^{\infty} C_n \left\{ \exp\left[-n\frac{V}{D}(nx^* - x)\right] + \exp\left[-n\frac{V}{D}(nx^* + x)\right] \right\} \quad [A11.12]$$

This procedure is given in more detail in the original reference [1] and leads to:

$$C_s(x) = 1 + 3\frac{1-k}{1+k}\exp\left(-\frac{2Vx}{D}\right) + 5\frac{(1-k)(2-k)}{(1+k)(2+k)}\exp\left(-\frac{6Vx}{D}\right) + \ldots$$

$$\ldots + (2n+1)\frac{(1-k)(2-k)\ldots(n-k)}{(1+k)(2+k)\ldots(n+k)}\exp\left[-\frac{n(n+1)Vx}{D}\right] \quad [A11.13]$$

The above equation describes the final transient in the solid for the case where a steady-state solute pile-up has been established at the solid/liquid interface.

References

[1] V.G.Smith, W.A.Tiller, J.W.Rutter, Canadian Journal of Physics **33** (1955) 723.

APPENDIX 12

MASS BALANCE EQUATIONS

The different solubilities of a solute in the liquid and solid phases, together with differences in mobility, lead to the spatial concentration variations known as segregation. Provided that the mass transport in the liquid is infinitely rapid (no concentration gradient in the liquid), the corresponding relationships can easily be derived from a mass balance (where C is expressed as a percentage):

$$f_s C_s + f_l C_l = 100$$

$$d(f_s C_s) + d(f_l C_l) = 0 \qquad [A12.1]$$

These equations state that all of the solute which cannot be incorporated into the solid must enter the liquid. No account is taken here of all of the other possible phenomena which might occur in a practical situation. For instance, solute vaporization or crucible reaction.

Lever Rule

The simplest case to which a mass balance can be applied is that of equilibrium solidification (no concentration gradient in the solid or liquid). This can be expressed by the relations:

$$D_l \gg D_s \gg LV \qquad [A12.2]$$

Since L is the length of the solidifying system (Fig. A12.1), equation A12.2 states that the diffusion boundary layer, $\delta_c = 2D/V$, is much larger than the maximum distance, L, over which either solid or liquid state diffusion can occur. Similarly, it can be supposed that if the growth rate, V, of the solid is constant and equal to L/t_f, the length of the specimen must be less than the characteristic diffusion length:

$$L \ll (D_s t_f)^{1/2} \qquad [A12.3]$$

Under these conditions:

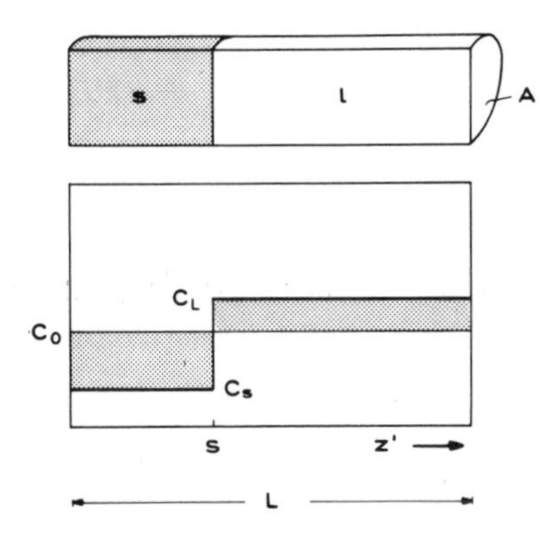

Figure A12.1

$$\frac{\partial C_l}{\partial z'} = \frac{\partial C_s}{\partial z'} \cong 0$$

Taking the differential form of equation A12.1, $C_s = kC_l$, $dC_s = kdC_l$, $f_s = 1 - f_l$, and $df_s = -df_l$ can be substituted and integrations performed:

$$\int_{C_0}^{C_l} \frac{dC_l}{pC_l} = \int_{0}^{f_s} \frac{df_s}{1 - f_s p}$$

giving, for the lever rule,

$$\frac{C_l}{C_0} = \frac{1}{1 - pf_s} \qquad [A12.4]$$

Solute Distribution with Back-Diffusion and Complete Mixing in the Liquid

The case of a bar having a constant cross-section, A, and a zero concentration gradient, due to rapid mass transport in the liquid ($C_l = C_l^*$) and limited diffusion in the solid, is shown in figure 6.3. For this case, Brody and Flemings [1] have developed a flux balance which, in a slightly modified form (sum of redistributed mass represented by the surfaces, $A_1 + A_2 + A_3 = 0$, in figure 6.3) is:

$$(C_l - C_s^*)Ads = (L - s)AdC_l + dC_s^* A \frac{\delta_s}{2} \qquad [A12.5]$$

Here, A_3 represents the surface of the equivalent boundary layer in the solid (appendix 2). Recognising that $f_s = s/L$, $df_s = ds/L$, $\delta_s = 2D_s/V = 2D_s dt/ds$, $C_s^* = kC_l$, and $dC_s^* = kdC_l$, then:

$$C_l(1-k)df_sL = L(1-f_s)dC_l + dC_lkD_s\frac{dt}{ds} \qquad [A12.6]$$

Dividing by L, and using a parabolic growth rate relationship (e.g. equation A1.22):

$$\frac{s}{L} = f_s = \left(\frac{t}{t_f}\right)^{1/2} \qquad [A12.7]$$

Evaluating ds/dt, substituting the results into equation A12.6, and rearranging gives:

$$\frac{dC_l}{pC_l} = \frac{df_s}{(1-f_s) + 2\alpha kf_s} \qquad [A12.8]$$

where α is a dimensionless solid-state back-diffusion parameter (dimensionless time = Fourier number):

$$\alpha = \frac{D_st_f}{L^2} \qquad [A12.9]$$

(compare with equation A12.3) from which

$$\frac{1}{p}\int_{C_0}^{C_l}\frac{dC_l}{C_l} = \int_0^{f_s}\frac{df_s}{1-f_s(1-2\alpha k)}$$

and integration leads to:

$$\frac{C_l}{C_0} = [1-f_s(1-2\alpha k)]^{\frac{k-1}{1-2\alpha k}} \qquad [A12.10]$$

Due to the simplifying assumptions made, this solution is limited to k-values which are smaller than 1. Apart from this limitation, equation A12.10 is an important one because it includes the two limiting cases (for $\partial C/\partial z' = 0$):

Lever rule: (equation A12.4) when $\alpha = 0.5$

Scheil's equation: when $D_s = 0$ (no solid-state diffusion) $\alpha = 0$. That is:

$$\frac{C_l}{C_0} = f_l^{(k-1)} \qquad [A12.11]$$

It can be seen that, according to equation A12.9, the case $\alpha = 0.5$ (lever rule) does

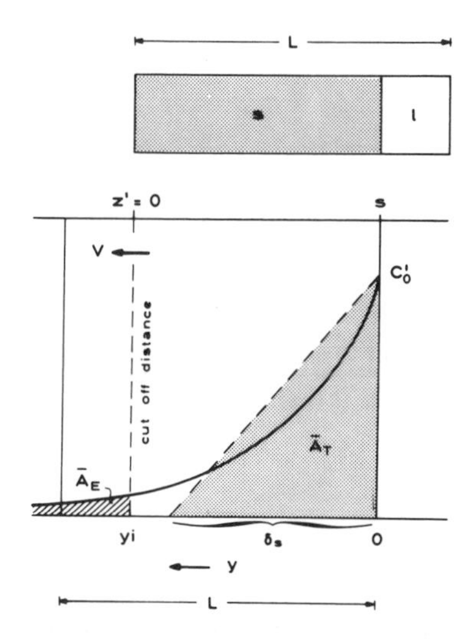

Figure A12.2

not correspond to the physical characteristics of equilibrium solidification. There, α should approach infinity (equation A12.3). Therefore, a modified back-diffusion parameter, α', has been proposed by Clyne and Kurz [2]. The basis of this calculation is that, in the original Brody-Flemings treatment for high α-values, solute was not conserved in the system. This can be easily understood with the aid of figure 6.3, where the solid diffusion boundary layer, δ_S, still has a small value. If this δ_S-value (proportional to α) becomes comparable in size to, or greater than, the solidified length, s, there is no longer any mass conservation according to equation A12.5 because the end effects are not considered. In reality, the initial specimen end ($z' = 0$) is an isolated system boundary and solute cannot leave the system at that point. In order to find a simple solution to this problem it will be assumed that diffusion is semi-infinite and that the tail of the diffusion boundary layer at z'-values less than zero (outside of the specimen) has to be taken into account.

For the purpose of the analysis, the coordinates are changed: $y = 0$ is fixed at the solid/liquid interface and only the solid concentration profile shown in figure A12.2 is considered. The end of the specimen ($z' = 0$) then corresponds to the cut-off distance on the y-axis (negative z'-axis). The total amount of solute in the boundary layer in the solid is $\bar{A}_T$. The neglected portion of the solute, due to the small cut-off distance, is $\bar{A}_E$. These can be obtained from [2]:

$$\bar{A}_T = \int_0^{\infty} C'dy = C_0' \frac{\delta_s}{2} \qquad [A12.12]$$

$$\bar{A}_E = \int_{y_i}^{\infty} C'dy = C_0' \frac{\delta_s}{2} \exp[-\frac{2y_i}{\delta_s}] \qquad [A12.13]$$

Here, it is assumed to a first approximation that the back-diffusion can be described by an exponential function. In order to estimate the correction factor necessary to cope with the cut-off effect, a parameter, Σ, is defined:

$$\Sigma(\alpha) = \frac{1}{t_f} \int_0^{t_f} \frac{\bar{A}_E}{\bar{A}_T} dt \qquad [A12.14]$$

It will be noted that, even if C_0' changes during solidification, the ratio contained in the integral will be independent of such changes. For parabolic growth laws (equation A12.7):

$$y_i = L \left(\frac{t}{t_f}\right)^{1/2} \qquad [A12.15]$$

Combining equations A12.12 to A12.15 gives:

$$\Sigma(\alpha) = \frac{1}{t_f} \int_0^{t_f} \exp[-\frac{2L}{\delta_s} \left(\frac{t}{t_f}\right)^{1/2}] dt \qquad [A12.16]$$

Recognising that the exponential term in the integral of equation A12.16 is equal to $(-1/2\alpha)$ leads to:

$$\Sigma(\alpha) = \exp[-\frac{1}{2\alpha}] \qquad [A12.17]$$

Equation A12.17 approaches zero asymptotically at low α-values, and unity at high α-values. Therefore, in the low α-range it represents the deviation from α and in the high α-range it represents the deviation from the limiting value, $\alpha' = 0.5$. A spline-like function which connects α' and α must therefore be found which satisfies the two boundary conditions: $\alpha' \rightarrow \alpha$ as $\Sigma(\alpha) \rightarrow 0$ and $\alpha' \rightarrow 0.5$ as $\Sigma(\alpha) \rightarrow 1$. An expression which has the required form (Fig. A12.3) is:

$$\alpha' = \alpha[1 - \exp(-\frac{1}{\alpha})] - \frac{1}{2}\exp(-\frac{1}{2\alpha}) \quad [A12.18]$$

Renaming α in equation A12.10 as α' (equation A12.18) permits the calculation of any solute distribution when diffusion in the liquid is very rapid. That is, it is not necessary to decide whether the lever rule or Scheil's equation has to be used in a given situation. This is of great importance, particularly for alloys which contain interstitial and substitutional solutes. In figure A12.3, typical values of α for the elements P and C in δ-Fe are given. It can be seen that carbon obeys the lever rule and phosphorus is intermediate in behaviour.

The most important application of equation A12.10 is related to the estimation of microsegregation, i.e. the modelling of the behaviour of the mushy zone. Replacing L by the characteristic diffusion distance, $\lambda/2$ (= $\lambda_1/2$ for cells and $\lambda_2/2$ for dendrites; figures 6.6 and 6.7) leads to:

$$\alpha = \frac{4D_s t_f}{\lambda^2} \quad [A12.19]$$

Rearranging A12.10, with α renamed α', permits the fraction solidified to be calculated:

$$f_s = \frac{1}{1 - 2\alpha' k}\{1 - [\frac{T_f - T}{T_f - T_l}]^{\frac{1 - 2\alpha' k}{k - 1}}\} \quad [A12.20]$$

Here, T_f is the melting point of the pure element, and C has been replaced by T, via the liquidus slope. The first derivative with respect to temperature is:

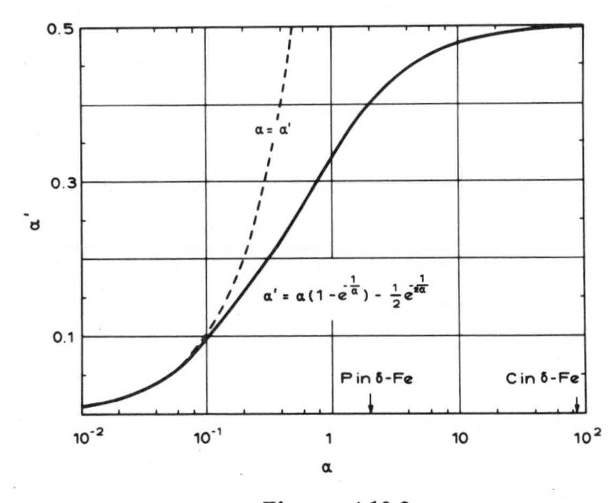

Figure A12.3

$$\frac{df_s}{dT} = \frac{1}{k-1}(T_f - T_l)^{\frac{2\alpha'k-1}{k-1}}(T_f - T)^{\frac{2-2\alpha'k-k}{k-1}} \qquad [A12.21]$$

This relationship can be used in order to make numerical calculations of the solidification microstructure of cast alloys (appendix 1).

Matsumiya et al. [3] have compared previously developed models with numerical results. They showed that, in some cases, equation A12.10 (with α' from equation A12.18) did not agree with the more exact calculations. Ohnaka [4] has proposed a more elegant approximation to microsegregation problem with back diffusion. Finally, Kobayashi [5] has developed an exact analytical solution to the microsegregation problem and has also provided some higher order approximations. The following one comes very close to the rather complicated exact solution:

$$C_s = kC_0\xi^{\eta}\left\{1 + U\left[\frac{1}{2}\left(\frac{1}{\xi^2} - 1\right) - 2\left(\frac{1}{\xi} - 1\right) - \ln\xi\right]\right\} \qquad [A12.22]$$

with

$$\xi = 1 - (1 - \beta k)f_s$$

$$\beta = 2\gamma/(1 + 2\gamma)$$

$$\gamma = 8Dt_f/\lambda^2$$

$$\eta = (k - 1)/(1 - \beta k)$$

$$U = \frac{\beta^3 k(k-1)[(1+\beta)k - 2]}{4\gamma(1 - \beta k)^3}$$

Solute Distribution under Rapid Solidification Conditions

In chapter 7 it has been shown that, under conditions of rapid solidification, the assumption of an homogeneous interdendritic liquid does not represent the observed microsegregation behaviour (Fig. 7.16). This is mainly due to the large solute pile-up around the rapidly growing dendrite tips (Fig. 7.17). Once these boundary layers interact at higher volume fractions ($f_S > f_X$) homogeneous interdendritic liquid can again be assumed. In reference [6], these effects have been treated in the following way:

for small solid fractions ($f_S < f_X$) the concentration is assumed to follow a second-order polynomial

$$f_s = a_1C_s^{*2} + a_2C_s^* + a_3 \qquad [A12.23]$$

and, for large solid fractions ($f_s > f_x$) Scheil's equation holds, i.e.

$$f_s = 1 + (f_x - 1)(C_s^*/C_x)^{\frac{1}{k-1}} \qquad [A12.24]$$

There are five unknowns to be determined a_1, a_2, a_3, f_x, C_x. The corresponding five equations which are necessary for the solution can be obtained as demonstrated in reference [6].

Precipitation in a Ternary System, A-B-C, in the Absence of Solid-State Diffusion

When the effect of back-diffusion is negligible, as in the case of a large volume of precipitating phase at the end of solidification (i.e. large concentration gradients close to $f_s = 1$ are avoided), Scheil's equation (A12.11) can be used. For two solute elements, B and C, the corresponding solute profiles (trace of interface concentration as a function of f_s) are:

$$\left(\frac{C_l}{C_0}\right)_B = f_l^{(k_B - 1)}$$

$$\left(\frac{C_l}{C_0}\right)_C = f_l^{(k_C - 1)} \qquad [A12.25]$$

When the solubility product of the phase, B_xC_y, is reached, precipitation will begin (if there is no difficulty of nucleation):

$$(C_l)_B^x \, (C_l)_C^y = K_{B_xC_y} \qquad [A12.26]$$

From equations A12.25 and A12.26:

$$K_{B_xC_y} = (C_0)_B^x \, (C_0)_C^y \, f_l^{(xk_B + yk_C - x - y)}$$

When precipitation begins, $f_l = f_p$ and the precipitate volume is:

$$f_p = \left[K_{B_xC_y} (C_0)_B^{-x} \, (C_0)_C^{-y}\right]^{\frac{1}{xk_B + yk_C - x - y}} \qquad [A12.27]$$

Solidification Path in Ternary Systems

In general, the path of solidification (trace of liquid or solid composition as a function of f_s) can be obtained in the case of solidification by relating the composition to f_l, which must be the same for all of the elements). Using equation A12.10 and A12.18:

$$\left(\frac{C_l}{C_0}\right)_B = (1 - u_B f_s)^{-\frac{p_B}{u_B}}$$

$$\left(\frac{C_l}{C_0}\right)_C = (1 - u_C f_s)^{-\frac{p_C}{u_C}} \qquad [A12.28]$$

where $u_i = 1-2\alpha_i' k_i$ and $p_i = 1-k_i$. Eliminating f_s:

$$\left(\frac{C_l}{C_0}\right)_B = \left\{1 - \frac{u_B}{u_C}\left[1 - \left(\frac{C_l}{C_0}\right)_C\right]^{-\frac{u_C}{p_C}}\right\}^{-\frac{p_B}{u_B}} \qquad [A12.29]$$

The latter equation relates the composition of solute B to that of solute C in the ternary system, A-B-C, with constant k and α.

References

[1] H.D.Brody, M.C.Flemings, Transactions of the Metallurgical Society of AIME **236** (1966) 615.

[2] T.W.Clyne, W.Kurz, Metallurgical Transactions **12A** (1981) 965.

[3] T.Matsumiya, H.Kajioka, S.Mizoguchi, Y.Ueshima, H.Esaka, Transactions of the Iron and Steel Institute of Japan **24** (1984) 873.

[4] I.Ohnaka, Transactions of the Iron and Steel Institute of Japan **26** (1986) 1045.

[5] S.Kobayashi, Transactions of the Iron and Steel Institute of Japan **28** (1988) 728.

[6] B.Giovanola, W.Kurz, Metallurgical Transactions **20A** (1989), in press.

APPENDIX 13

HOMOGENISATION OF INTERDENDRITIC SEGREGATION IN THE SOLID STATE

In order to determine the changes which occur during homogenisation of the cooling solid after solidification, only one dendrite arm need be considered (due to symmetry - appendix 2). In principle, the form of the solute segregation can be described approximately by the Scheil equation (Fig. A13.1). Here, it will be shown that all of the possible original distributions can be related to each other by using dimensionless constants.

The changes can be treated approximately by using the one-dimensional time-dependent diffusion equation:

$$D_s \frac{\partial^2 C}{\partial x^2} = \frac{\partial C}{\partial t} \quad \text{[A13.1]}$$

It is known (appendix 2) that a likely solution to this equation involves circular and exponential functions. The exponential function is more likely to be associated with the time dependence, and a cosine or sine function is more likely to reflect the characteristics of distributions such as those in figure A13.1. In practice, of course, the solute distribution may take any form at all. However, it will be shown below that analogous changes occur in the distribution, regardless of the initial state. Thus, one can suppose that:

$$C = C_0 + \delta C \exp(at)\cos(bx) \quad \text{[A13.2]}$$

where δC is the initial amplitude of the concentration variation (Fig. A13.1). Substitution of the derivatives of equation A13.2 into equation A13.1 shows that:

$$a = -D_s b^2$$

therefore:

$$C = C_0 + \delta C \exp(-D_s b^2 t)\cos(bx) \quad \text{[A13.3]}$$

The value of b can be evaluated by using the boundary conditions. Thus, the gradient

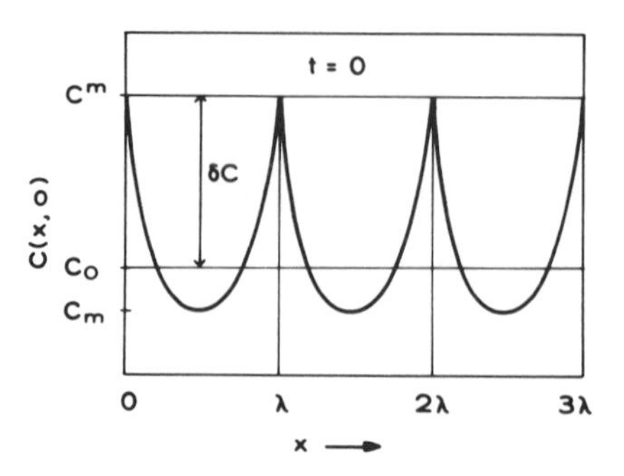

Figure A13.1

of the concentration at the origin must be zero at all times, t, since the origin is a point of symmetry. This can be assumed to be true even though the gradient of distributions such as those illustrated in figure A13.1 is not defined mathematically. The zero gradient condition is already satisfied by equation A13.2 at $\lambda/2$. The gradient of concentration must also be zero at λ(another point of symmetry) for all values of t. This is only true if:

$$\sin(b\lambda) = 0$$

so that:

$$b\lambda = n\pi$$

where n is an integer. The final general solution is:

$$C = C_0 + \delta C \exp\left(-D_s \frac{n^2\pi^2}{\lambda^2} t\right) \cos\left(\frac{n\pi x}{\lambda}\right) \qquad \text{[A13.4]}$$

The last condition to be satisfied is that the concentration distribution at the start of homogenisation (Fig. A13.1) should be described by equation A13.4 (with t equal to zero). Obviously, such a distribution cannot be described by the latter expression unless the initial distribution happens to be sinusoidal. However, because the diffusion equation (A13.1) is linear, any number of similar equations (with different values of n) can be added together. This is simply the technique of Fourier analysis (appendix 2) and leads to a solution, for any initial solute distribution described by $f(x)$, of the form:

$$C = C_0 + \frac{2}{\lambda} \sum \exp\left(-D_s \frac{n^2\pi^2}{\lambda^2} t\right) \cos\left(\frac{n\pi x}{\lambda}\right) \int_0^\lambda f(x) \cos\left(\frac{n\pi x}{\lambda}\right) dx \qquad \text{[A13.5]}$$

This solution can be used to determine the concentration distribution at any time. Recall that, using the method described in appendix 2, the result of performing the integration above can be written down immediately when the 'jumps' in the function and its derivatives are known.

Suppose that the measured distribution is parabolic (Fig. A13.1) with maximum, C^m, minimum, C_m, and average, C_0. There is no jump in the function at 0, λ, 2λ,... n, but there is a jump of $-(8/\lambda)(C^m - C_m)$ in the first derivative at 0 and at λ. There is no jump in the second and higher derivatives. Thus, using equation A2.29, one can find immediately that the Fourier coefficients are given by:

$$A_n = \frac{16}{n^2\lambda^2}(C^m - C_m)$$

Use of this technique is considerably easier than integrating expressions of the type, $x^2\cos nx$. Note that when one assumes an asymmetrical distribution such as the parabolic one, the maximum and minimum values of the distribution are not independent. Instead, they are related by the requirement that the average concentration should be equal to the original concentration, C_0.

One can introduce the dimensionless value, I, which is defined to be the instantaneous value of the amplitude of the distribution compared with its original value; say $\delta C/(C^m - C_m)$. Another dimensionless constant, $n^2\pi^2 D_s t/\lambda^2$ (compare with equation A12.9), arises naturally in the above calculation. Using this constant, the relaxation time, τ_n, of that component can be defined:

$$\tau_n = \frac{\lambda^2}{n^2\pi^2 D_s} \quad \text{[A13.6]}$$

For the present purpose, the simple sinusoidal concentration variation is quite useful because, as can be seen from equation A13.4, the higher-order terms (short wavelengths) decay much more rapidly than the longer ones, and the homogenisation process will therefore ultimately be determined by the relaxation time of the lowest-order term, i.e. by $\tau_1 = \lambda^2/\pi^2 D_s$. If the initial concentration variation is given approximately by:

$$C(x,0) = C_0 + \delta C\cos\left(\frac{\pi x}{\lambda}\right)$$

the solution for the lowest-order term is:

$$C(x,t) = C_0 + \delta C\cos\left(\frac{\pi x}{\lambda}\right)\exp\left(-\frac{t}{\tau}\right) \quad \text{[A13.7]}$$

where $\tau = \tau_1$. The maximum concentration at $x = 0$, C^m, changes with time according to

$$C^m(t) = C_0 + \delta C \exp\left(-\frac{t}{\tau}\right) \qquad [A13.8]$$

giving, after a time, $t = \tau$ (equation A13.6)

$$C^m(\tau) = C_0 + \frac{\delta C}{e} = C_0 + 0.37\,\delta C \qquad [A13.9]$$

or after a time, $t = 3\tau$

$$C^m(3\tau) = C_0 + \frac{\delta C}{e^3} = C_0 + 0.05\,\delta C \qquad [A13.10]$$

From equation A13.6 it can be seen that the secondary dendrite arm spacing will have a significant effect upon the annealing time since the relaxation time is proportional to λ^2. High solidification rates, which reduce λ, will have a marked effect upon the reduction of the annealing time. For example, in order to reduce the amplitude of the concentration variation to 5% of its initial value, the necessary annealing time can be obtained by using equations A13.10 and A13.7:

$$t_{0.05} \cong 0.3\frac{\lambda^2}{D_s} \qquad [A13.11]$$

Using equation A13.11, the annealing temperature which is required to homogenise (to less than a 5% variation) an alloy with a given dendrite arm spacing within a given time can be calculated:

$$T_{0.05} \cong \frac{Q}{R\ln\left(\frac{tD_0}{0.3\lambda^2}\right)} \qquad [A13.12]$$

where Q and D_0 are the activation energy and pre-exponential term respectively in the Arrhenius expression for the temperature dependence of the diffusivity.

APPENDIX 14

RELEVANT PHYSICAL PROPERTIES FOR SOLIDIFICATION

The properties given here are intended for use in the exercises. They are therefore consistent with each other but should not be assumed to be the most accurate ones available. Most of them are taken, or deduced, from data in the 5th edition of Metals Reference Book (C.J.Smithells, Butterworths, London, 1976). (Volumetric properties such as Δh_f have been calculated on the basis of the density at the melting point).

Properties of Pure Materials at the Melting Point

Property	Units	Al	Cu	δ-Fe (γ-Fe#)	SCN##
T_f	°C	660.4	1084.9	1538 (1526)	58.08
T_f	K	933.6	1358	1811 (1799)	331.23
Δh_f	J/m^3	9.5×10^8	1.62×10^9	1.93×10^9	4.6×10^7
Δs_f	J/m^3K	1.02×10^6	1.2×10^6	1.07×10^6	1.4×10^5
κ_l	W/mK	95	166	35	0.223
κ_s	W/mK	210	244	33	0.225
c_l	J/m^3K	2.58×10^6	3.96×10^6	5.74×10^6	2.0×10^6
c_s	J/m^3K	3.0×10^6	3.63×10^6	5.73×10^6	2.0×10^6
$\rho_l(T_f)$	kg/m^3	2.39×10^3	8.0×10^3	7.0×10^3	0.988×10^3
$\rho_s(T_f)$	kg/m^3	2.55×10^3	7.67×10^3	7.25×10^3	1.05×10^3
a_l	m^2/s	37×10^{-6}	42×10^{-6}	6.1×10^{-6}	0.116×10^{-6}
a_s	m^2/s	70×10^{-6}	67×10^{-6}	5.8×10^{-6}	0.112×10^{-6}
M	kg/mol	27×10^{-3}	63.5×10^{-3}	55.8×10^{-3}	80×10^{-3}
v_m^s	m^3/mol	11×10^{-6}	8.3×10^{-6}	7.7×10^{-6}	76×10^{-6}
$\Delta h_f/c$	K	368	409	336	23
σ	J/m^2	93×10^{-3}	177×10^{-3}	204×10^{-3}	9×10^{-3}
Γ	mK	2.4×10^{-7}	1.5×10^{-7}	1.9×10^{-7}	0.64×10^{-7}
s_T	m	0.24×10^{-9}	0.37×10^{-9}	0.57×10^{-9}	2.78×10^{-9}
$\Delta v/v$	–	6.5×10^{-2}	4.2×10^{-2}	$3.6(4.1) \times 10^{-2}$	6×10^{-2}

metastable, ## succinonitrile - $CH_2(CN)_2CH_2$

Properties of Aluminium Alloys and Succinonitrile-Acetone

Property	Units	Al-Cu	Al-Cu#	Al-Si	Al-Si#	SCN-ACE
C_0	wt%	2	33.1	6	12.6	1.3
T_l	°C(K)	656 (929)	–	624 (897)	–	54-8 (327.4)
ΔT_0	K	32	–	240##	–	32.8
T_l	°C(K)	–	548 (821)	–	577 (850)	–
C_e	wt%	–	33.1	–	12.6	–
C'	wt%	–	46.5	–	98.2	–
m_α	K/wt%	–2.6	– 4.9	– 6	–7.5	–2.8
m_β	K/wt%	–	3.3	–	17.5	–
k_α	–	0.14	0.18	0.13	0.13	0.1
k_β	–	–	0.05	–	2×10^{-4}	–
f_β	–	–	0.46	–	0.127	–
D_l	m^2/s	3×10^{-9}	3.4×10^{-9}	3×10^{-9}	5.5×10^{-9}	1.3×10^{-9}
D_s	m^2/s	3×10^{-13}	–	1×10^{-12}	–	–
Γ_α	mK	2.4×10^{-7}	2.4×10^{-7}	2×10^{-7}	1.96×10^{-7}	0.64×10^{-7}
Γ_β	mK	–	0.55×10^{-7}	–	1.7×10^{-7}	–
s_c###	m	54×10^{-9}	–	6.4×10^{-9}	–	19.5×10^{-9}
P'	–	–	3.36×10^{-2}	–	8.9×10^{-3}	–

#eutectic, ## metastable, ### $s_c \cong \Gamma/\Delta T_0 k$ when $\Omega << 1$

Properties of Iron Alloys

Property	Units	δFe-C	γFe-C	γFe-C#	γFe-Ni
C_0	wt%	0.09	0.6	4.26	10
T_l	°C(K)	1531 (1804)	1490 (1763)	–	1503 (1776)
ΔT_0	K	36	72	–	6
T_e or T_p	°C(K)	1493 (1766)[p]	1155 (1428)[e]	1155 (1428)[e]	–
C_e or C_p	wt%	0.53 [p]	4.26 [e]	4.26 [e]	–
C'	wt%	–	–	97.9	–
m_α	K/wt%	– 81	– 65	– 140	– 2.4
m_β	K/wt%	–	–	400	–
k_a	–	0.17	0.35	0.49	0.8
k_β	–	–	–	0.001	–
f_β	–	–	–	0.071	–
D_l	m^2/s	2×10^{-8}	2×10^{-8}	2×10^{-8}	7.5×10^{-9}
D_s	m^2/s	6×10^{-9}	1×10^{-9}	–	3×10^{-13}
Γ_α	mK	1.9×10^{-7}	1.9×10^{-7}	2×10^{-7}	2×10^{-7}
Γ_β	mK	–	–	2×10^{-7}	–
s_c	m	30×10^{-9}	7.5×10^{-9}	–	42×10^{-9}
P'	–	–	–	3.7×10^{-3}	–

eutectic

SYMBOLS

Symbol	Meaning	Definition	Units
A	surface or cross-sectional area	-	m^2
A	gradient term	$kVG_c/[Vp - Db]$	%/m
A'	surface area of casting	-	m^2
$A(P_c)$	normalised dendrite tip composition (C_l/C_0)	$[1-(1-k)I(P_c)]^{-1}$	-
B	constant	-	-
C	concentration	-	at%, wt%
C_e	eutectic composition	-	at%, wt%
C'	length of eutectic tie-line	-	at%, wt%
C^*	concentration at the solid/liquid interface	-	at%, wt%
C_0	initial alloy concentration	-	at%, wt%
D	diffusion coefficient in liquid	$-J/G_c$	m^2/s
D_0	pre-exponential term (diffusion)	-	m^2/s
D_i	interface diffusion coefficient	-	m^2/s
D_s	diffusion coefficient in solid	-	m^2/s
E	energy	-	J
E	internal energy	-	J/mol
E_1	exponential integral function	appendix 8	-
F	stability parameter	$(\dot{\varepsilon}/\varepsilon)(mG_c/V)$	K/m^2
G	Gibbs free energy	-	J/mol
G	interface temperature gradient	dT/dz	K/m
$\bar{G}$	mean temperature gradient	$\kappa_s G_s + \kappa_l G_l$	-
G_c	interface concentration gradient in liquid	$(dC_l/dz)_{z=0}$	at%/m, wt%/m
G_l	temperature gradient in liquid	-	K/m
G_s	temperature gradient in solid	-	K/m
G^*	effective temperature gradient	$(\kappa_l G_l \xi_l + \kappa_s G_s \xi_s)/(\kappa_s + \kappa_l)$	J/mol
H	enthalpy	-	J/mol
I	nucleation rate	appendix 2	$/m^3s$
I	Ivantsov function	appendix 8	-
J	mass flux	-	$/m^2s$
K	curvature	$1/r_1 + 1/r_2$	/m
K	constant	-	-
L	length	-	m

M	atomic (molecular)weight	-	g/mol
N	number	-	-
N_A	Avogadro's number	6.022×10^{23}	/mol
P	pressure	-	Pa
P	Péclet number	L/δ	-
P'	series in Jackson-Hunt eutectic growth model	appendix 10	-
P_c	solutal Péclet number	$VR/2D$	-
P_t	thermal Péclet number	$VR/2a$	-
Q	activation energy for diffusion	-	J/mol
Q	quantity of heat	-	J
R	gas constant	-	J/mol K
R	radius	-	m
S	entropy	-	J/mol K
S	perturbation term	$\varepsilon \sin(\omega y)$	m
T	temperature	-	K
$\dot{T}$	cooling rate	dT/dt	K/s
T_f	melting point of pure substance	-	K
T_l	liquidus temperature	-	K
T_m	mould temperature	-	K
T_0	temperature of equal free energy of two phases	-	K
T_q	measurable temperature	-	K
T_s	solidus temperature	-	K
T_s'	non-equilibrium solidus	-	K
V	rate of interface movement	-	m/s
V_a	critical growth rate for absolute stability	-	m/s
V_c	critical growth rate for constitutional undercooling	-	m/s
V'	rate of crucible movement	-	m/s
V_0	limiting crystallisation velocity (~velocity of sound)	-	m/s
X	mole fraction	-	-
Y	partial solution to differential equation	appendix 2	-
Z	partial solution to differential equation	appendix 2	-
a	thermal diffusivity	κ/c	m^2/s
a	separation constant	appendix 2	m
a	half-axis of ellipsoid	-	m
b	half-axis of ellipsoid	-	m
b	exponent in stability analysis	$(V/2D) + [(V/2D)^2 + \omega^2]^{1/2}$	/m

c	volumetric specific heat	-	J/m^3K
c_p	specific heat	-	J/kgK
c^*	effective specific heat	appendix 1	J/m^3K
d	exponent	-	-
d	distance	-	m
e	exponent	-	-
f	force	-	N
f	volume fraction of α-phase in eutectic	-	-
$f[hkl]$	crystallographic factor	appendix 5	-
f_i	fraction of solid within grains	-	-
f_l	volume fraction of liquid	$v_l/(v_l+v_s)$	-
f_s	volume fraction of solid	$1-f_l$	-
h	Planck's constant	6.63×10^{-34}	Js
h	heat transfer coefficient	$q/\Delta T_g$	W/m^2K
k	equilibrium distribution coefficient	C_s/C_l	-
k_B	Boltzmann's constant	1.38×10^{-23}	J/K
k_v	non-equilibrium distribution coefficient	appendix 6	-
l	length	-	m
m	liquidus slope	dT_l/dC	K/at%, K/wt%
m	mass	-	g
m	normal to isoconcentrates	-	-
n	number		
n	exponent	-	-
n	interface normal	-	-
n	grain number density	-	$/m^3$
n_s	adsorption site density	appendix 4	-
o	exponent	-	-
p	probability	-	-
p	complementary distribution coeff.	$1-k$	-
q	heat flux	-	W/m^2
r	radius	-	m
r^o	critical nucleation radius	-	m
s	position of s/l interface	-	m
s_c	solute capillarity length	$\Gamma/mC^*(k-1)$	m
s_t	thermal capillarity length	$-\Gamma c/\Delta h_f$	m
t	time	-	s
t_f	local solidification time	-	s
u	back-diffusion parameter	$(1-2\alpha' k)$	-
v	volume		m^3

v'	atomic volume	-	m^3
v_m	molar volume	-	m^3/mol
w	work	-	J
x	coordinate in s/l interface	-	m
y	coordinate in s/l interface	-	m
z	coordinate perpendicular to a planar solid/liquid interface	-	m
z'	system coordinate	-	m
α	dimensionless entropy of fusion	$\Delta S_f/R$	-
α	dimensionless coefficient for back-diffusion	$D_s t_f/L^2$	-
α'	dimensionless coefficient for interdendritic back-diffusion	appendix 12	-
Γ	Gibbs-Thomson coefficient	$\sigma/\Delta s_f$	Km
δ_c	solute boundary layer thickness	$2D/V$	m
δ_c	characteristic diffusion length	D/V	m
δ_i	interatomic jump distance	-	m
δ_s	solute boundary layer thickness in solid	$2D_s/V$	m
δ_t	thermal boundary layer thickness	$2a/V$	m
ΔC_0	concentration difference between liquidus and solidus at solidus temperature of alloy	$C_0(1-k)/k$	at%, wt%
ΔG	total Gibbs free energy	-	J/mol
ΔG_d	activation free energy for diffusion across solid/liquid interface	-	J/mol
$\Delta G^\bullet$	standard free energy	-	J/mol
ΔG°	activation energy for the nucleation of the cluster of critical radius	-	J
ΔG_n°	activation energy for the nucleation of a critical number of clustered atoms	-	J
Δg	Gibbs free energy per unit volume	$\Delta G_v/v_m$	J/m^3
$\Delta H^\bullet$	standard enthalpy	-	J/mol
ΔH_f	latent heat of fusion per mole (positive for solidification)	-	J/mol
Δh_f	latent heat of fusion per unit volume	$\Delta H_f/v_m$	J/m^3
ΔH_v	latent heat of vaporization	-	J/mol
$\Delta S^\bullet$	standard entropy	-	J/mol
ΔS_f	entropy of fusion per mole	$\Delta H_f/T_f$	J/mol K
Δs_f	entropy of fusion per unit volume	$\Delta h_f/T_f$	J/m^3K
ΔT	undercooling	$T_f - T$	K
$\Delta T'$	dendrite tip-to-root temperature difference	-	K

ΔT^+	critical undercooling	chapter 7	K
ΔT_0	liquidus-solidus range at C_0	$T_l - T_s$	K
ΔT_r	undercooling due to curvature	appendix 3	K
ΔT_c	temperature difference due to solute diffusion	-	K
ΔT_t	temperature difference due to heat flow	-	K
γ	surface tension	-	N/m
ε	amplitude of perturbation	-	m
$\dot{\varepsilon}$	growth rate of amplitude	$d\varepsilon/dt$	m/s
η	dynamic viscosity	-	Pas
η	shape factor	appendix 4	-
θ	angle	-	°
θ_c	unit solutal undercooling	$\Delta T_0 k A(P_c)$	K
θ_t	unit thermal undercooling	$\Delta h_f/c$	K
κ	thermal conductivity	-	W/Km
$\bar{\kappa}$	relative thermal conductivity	appendix 7	W/Km
λ	wavelength	-	m
λ	spacing	-	m
μ	chemical potential	-	J/mol
ν	adsorption frequency	appendix 4	/s
ν	kinematic viscosity	-	m^2/s
ν_0	atomic frequency	-	/s
ξ	stability parameter	appendix 7	-
ρ	density	-	kg/m^3
σ	solid/liquid interface energy	-	J/m^2
σ^*	stability constant	appendix 9	-
Σ	correction term for back-diffusion	appendix 12	-
τ	reduced temperature	$RT/\Delta H_v$	-
τ	relaxation time	-	s
ϕ	degree of constitutional supercooling	$mG_c - G$	K/m
Φ	reduced coordinate	appendix 1	-
ψ	azimuthal angle	-	°
ψ'	series in high Péclet number eutectic model	-	-
Ω	dimensionless solutal supersaturation	- appendix 8	- -
Ω	interaction parameter	appendix 3	-
Ω_t	dimensionless thermal supersaturation	appendix 8	-
ω	wave number	$2\pi/\lambda$	/m

INDEX